AF764191

Centre de Recherches Mathématiques

PROCEEDINGS OF THE CRM WORKSHOP
ON

HAMILTONIAN SYSTEMS, TRANSFORMATION GROUPS AND SPECTRAL TRANSFORM METHODS

Edited by:

J. HARNAD and J.E. MARSDEN

Les publications **CRM**
Directeur: Francis Clarke

ISBN 2-921120-05-4

Dépôt légal, 2e trimestre 1990

Tous droits de reproduction, d'adaptation ou de traduction réservés

© Les publications **CRM**

Université de Montréal
C.P. 6128-A
Montréal, QC H3C 3J7
Canada

Proceedings of the CRM WORKSHOP

on

HAMILTONIAN SYSTEMS, TRANSFORMATION GROUPS AND SPECTRAL TRANSFORM METHODS

Montréal, October 20 – 26, 1989

Edited by **J. Harnad* and J.E. Marsden****

*Department of Mathematics and Statistics, Concordia University
Montréal, QC H3G 1M8, CANADA
and
Centre de Recherches Mathématiques, Université de Montréal
Montréal, QC H3C 3J7, CANADA

** Department of Mathematics, University of California at Berkeley
Berkeley, CA 94720, USA

TABLE OF CONTENTS

ORGANIZING COMMITTEE

F.H. Clarke — Director, CRM
CRM
Université de Montréal

J. Harnad — Scientific Director
Department of Mathematics and Statistics
Concordia University
and
CRM
Université de Montréal

J.E. Marsden — Scientific Director
Department of Mathematics,
University of California, Berkeley

PREFACE

This volume represents the proceedings of the Workshop on Hamiltonian Systems, Transformation Groups and Spectral Transform Methods held at the Centre de Recherches Mathématiques, Université de Montréal, October 20 – 26, 1989. Approximately fifty participants were involved, including some of the most active researchers in this area.

As the table of contents indicates, a rather varied range of topics was addressed, the common element being the role of transformation groups in Hamiltonian systems. Over half the papers presented related to integrable systems, including some unusual interconnections with control theory and sorting algorithms. The remaining topics included: symmetry considerations and relative equilibria, quantization, R-matrices and classical gauge theory.

Between the time of a scientific conference and the appearance of the published proceedings a number of transformations may take place. Authors will often wish, for example, to update or modify their contributions or group together collaborative work. In some cases, they are unable to furnish a written account in time for inclusion in the published volume, or choose to publish it elsewhere. To preserve as much of the original flavour of this event as possible, we have included, as a supplement to the detailed papers, a list of titles and abstracts as they were presented at the meeting.

The detailed accounts included in this volume represent the contributions of 24 of the originally scheduled 31 invited speakers, with the inevitable transformations present. Thus, some talks that originally appeared as distinct contributions have been recombined into a single joint paper, perhaps including further authors, while others have changed somewhat in content or title relative to the original presentation.

In view of the range of topics covered, it is difficult to give a homogeneous overview or a natural partition of the material included. The groupings used in the table of contents are a bit arbitrary, since some papers overlap categories or represent very distinct approaches, though grouped under the same headings. Still, we hope such groupings may be of assistance in guiding the attention of potential readers to topics of interest.

The atmosphere of the original meeting was found, by all accounts, to have been both congenial and stimulating. We hope this volume may adequately convey the scientific content to the reader and also serve as a guide to the more detailed work of the various authors represented.

We thank all of the lecturers and participants for their hard work and enthusiasm, the coordinators S. Chênevert and J. Roy for their considerable help in all organizational matters, and Louise Letendre who, with the secretarial team at the CRM managed efficiently and good-naturedly to convert the diverse collection of inputs to these proceedings into a single coherent volume. Thanks go as well to the Natural Sciences and Engineering Research Council for its generous support of the event.

J. Harnad and J.E. Marsden

Montreal, May 1990

I. SYMPLECTIC GEOMETRY AND INTEGRABLE SYSTEMS

M.R. ADAMS, R.L. ANDERSON, and Robert VARLEY

M.R. ADAMS, J. HARNAD, and J. HURTUBISE

Larry M. BATES

L.A. DICKEY

Pierre MATHIEU

Emma PREVIATO

A.G. REYMAN

George WILSON

Pavel WINTERNITZ

REMARKS ON INTEGRABLE HIERARCHIES IN FINITE DIMENSIONS*

M.R. ADAMS, R.L. ANDERSON and R. VARLEY

Department of Mathematics, University of Georgia
Athens, GA 30602, USA

ABSTRACT. In this paper we consider the hierarchy of vector fields generated by recursively applying an operator to a given Hamiltonian vector field. We give a simple set of conditions which guarantee that the hierarchy is integrable. The result, formulated as Theorem 1 in §1, is similar to the theorem of Magri [M] on biHamiltonian systems. In §2 we illustrate the use of a recursion operator for certain oscillator systems and in §3 we treat the Korteweg-de Vries 2-soliton sector. One of the examples of 2 harmonic oscillators is presented as an analogue for a system of 2 interacting KdV 1-solitons. Our results on the 2-soliton sector are obtained by direct calculation with Hirota's soliton formulas and are similar to the n=2 case of work on the KdV n-solitons in the book [N] and in the preprint [F-O].

§1. Integrable hierarchies

An Hamiltonian system described by a vector field K on a 2n–dimensional phase space M is (Liouville) completely integrable if there are n functionally independent, mutually Poisson commuting constants of motion $F_1,\ldots,F_n$ (first integrals). In this situation, the flow of K on M preserves the common (n–dimensional) level surfaces of $F_1,\ldots,F_n$ and is linear with respect to the $\mathbb{R}^n$–action on these level surfaces given by the Hamiltonian flows for $F_1,\ldots,F_n$. By solving an associated Hamilton–Jacobi problem, one can often find explicit Darboux coordinates in a tubular neighborhood of a level surface so that the Hamiltonian flow is linearized in these coordinates, hence the flow is completely integrated.

* Supported by U.S. Army Grant DAAL03-87-K-0110

In recent years Magri [M,M–M] (see also [F], [G–D], [O], [tE], [K–S–M]) has studied complete integrability as a reflection of the existence of more than one symplectic structure with respect to which K is Hamiltonian. This is called a multi–Hamiltonian structure. The main component of this theory is a recursion operator $\phi \in \mathrm{End}(TM)$ which relates two symplectic structures ω_1, ω_2 by $\omega_2(X,Y) = \omega_1(X,\phi Y)$ for $X,Y \in \chi(M)$, the smooth vector fields on M. If ϕ satisfies certain compatibility conditions it follows that the vector fields K, ϕ K, ϕ^2 K,... are Hamiltonian with respect to ω_1 and that the Hamiltonians functions all Poisson commute. This puts the focus on the " hierarchy " of commuting vector fields K, ϕ K, ϕ^2 K,... generated by the recursion operator ϕ. Many examples of completely integrable systems (in both finite and infinite dimensions) have been shown to have natural multi–Hamiltonian structures and so their complete integration is given by an hierarchy of vector fields.

In this paper we note that the multi–Hamiltonian structure is not required to produce an integrable hierarchy of vector fields by recursion (see Theorem 1 below). We then describe some examples of such recursion operators for certain systems of harmonic oscillators and the 2–soliton sector of the Korteweg– de Vries equation. In these examples we will see that when the recursion operator produces " twin hierarchies ", linearizing Darboux coordinates can be easily found.

Here is the result on generation of an integrable hierarchy by recursion.

Theorem 1: Let (M,ω) be a symplectic manifold. Consider $K \in \chi(M)$ and $\phi \in \mathrm{End}(T(M))$ and define the "hierarchy" $K_j = \phi^{j-1}K$, $j = 1,2,\dots$, of vector fields on M. Assume

1. a) ϕ is symplectically self–adjoint, i.e., $\omega(\phi X,Y) = \omega(X,\phi Y)$ for all $X,Y \in \chi(M)$.

   b) ϕ is Nijenhuis, i.e., $\phi^2[X,Y] = \phi[X,\phi Y] + \phi[\phi X,Y] - [\phi X,\phi Y]$ for all $X,Y \in \chi(M)$.

2. K and ϕK are symplectic.

Then, there exist locally on M (and globally if the 1st deRham cohomology group of M is 0), functions $\{h_j\}$ in involution such that $K_j = X_{h_j}$ (the symplectic vector field corresponding to the 1–form dh_j). In particular, $\{K_j\}$ is an isotropic set of commuting vector fields on M, i.e., for any pair K_i, K_j (i,j = 1,2,...) of the vector fields, $\omega(K_i,K_j) = 0$ and $[K_i, K_j] = 0$.

Proof: We claim that

(1) $\omega(K_i, K_j) = 0$ for all i, j.

(2) K_j is a symplectic vector field for all j, i.e., $\mathscr{L}_{K_j}(\omega) = 0$

(For standard notation and properties used in this proof, we refer to [A–M], however we have adopted Arnold's [A] convention for ω^b. We set $\mathscr{Q} = \omega^b : TM \to T^*M$ and $\mathscr{P} = \omega^{\#} : T^*M \to TM$. Thus, for a vector field X, $\mathscr{Q}(X) = \omega^b(X) = \omega(.,X) = -\iota_X\omega$, and for a function h, $\mathscr{P}(dh) = \omega^{\#}(dh) = X_h$.) Then, from (2), $\mathscr{Q}(K_j)$ is closed, hence locally exact on M, i.e., locally $\mathscr{Q}(K_j) = dh_j$ for some function h_j on M. Then $\{h_i, h_j\} = -\omega(K_i, K_j) = 0$ by (1) so the functions $\{h_j\}$ are in involution. (In particular, $[K_i, K_j] = \mathscr{P}(d\{h_i, h_j\}) = 0$.)

To prove (1), first consider the case i + j = 2k. Then

$$\omega(K_i, K_j) = \omega(\phi^{i-1}K, \phi^{j-1}K) = \omega(\phi^{k-1}K, \phi^{k-1}K) = 0.$$

On the other hand, if i + j = 2k + 1,

$$\omega(K_i, K_j) = \omega(\phi^{k-1}K, \phi^k K) = \omega(X, \phi X),$$

where $X = \phi^{k-1}K$. But $\omega(X, \phi X) = \omega(\phi X, X) = -\omega(X, \phi X)$, so $\omega(X, \phi X) = 0$.

To prove (2), we use induction on j. For j = 1 and 2, we have $\mathscr{L}_{K_1}(\omega) = 0$ and $\mathscr{L}_{K_2}(\omega) = 0$ by hypothesis. Now fix j > 2, assume that $\mathscr{L}_{K_k}(\omega) = 0$ for $k \le j$, and consider $\mathscr{L}_{K_{j+1}}(\omega)$:

$\mathscr{L}_{K_{j+1}}(\omega) = \mathscr{L}_{\phi^j K}(\omega) = d(\iota_{\phi^j K}(\omega))$ since $\iota_{\phi^j K}(d\omega) = 0$, so we have to show that $\iota_{\phi^j K}(\omega) = -\mathscr{Q}(\phi^j K) = \omega(\phi^j K, \cdot)$ is closed. To do this, take arbitrary vector fields X, Y and compute

$$\begin{aligned} -d(\mathscr{Q}(\phi^j K))(X,Y) &= X(\omega(\phi^j K,Y)) - Y(\omega(\phi^j K,X)) - \omega(\phi^j K,[X,Y]) \\ &= X(\omega(\phi^{j-1}K,\phi Y)) - Y(\omega(\phi^{j-1}K,\phi X)) - \omega(\phi^{j-2}K,\phi^2[X,Y]). \end{aligned}$$

Notice we have the following equalities:

(A) $X(\omega(\phi^{j-1}K,\phi Y)) = \phi Y(\omega(\phi^{j-1}K,X)) + \omega(\phi^{j-1}K,[X,\phi Y])$ since

$\mathcal{L}_{\phi^{j-1}K}(\omega) = 0$ implies $d(\iota_{\phi^{j-1}K}(\omega)) = 0$ so $d(\mathcal{Q}(\phi^{j-1}K))(X,\phi Y) = 0$;

(B) $Y(\omega(\phi^{j-1}K, \phi X)) = \phi X(\omega(\phi^{j-1}K,Y)) + \omega(\phi^{j-1}K,[Y,\phi X])$ likewise;

(C) $\omega(\phi^{j-2}K, \phi^2[X,Y]) =$

$\omega(\phi^{j-2}K, \phi[X,\phi Y]) + \omega(\phi^{j-2}K, \phi[\phi X,Y]) - \omega(\phi^{j-2}K, [\phi X,\phi Y]) =$

$\omega(\phi^{j-2}K, \phi[X,\phi Y]) + \omega(\phi^{j-2}K, \phi[\phi X,Y]) - \phi X(\omega(\phi^{j-2}K,\phi Y) + \phi Y(\omega(\phi^{j-2}K,\phi X))$.

Then we can see, (A) – (B) – (C) = 0. Q.E.D.

Remarks:

1.1: Our hypotheses are weaker, hence easier to check, than those of Magri and our conclusion is also weaker than that of Magri, the conclusion of his theorem being that the system is multiHamiltonian. For other discussions of the theory of recursion, see [F], [G–D], [M], [O], [tE].

1.2: It is known that the eigenvalues of a symplectically self–adjoint operator occur with even multiplicity [M–M, p.27].

1.3: Let (M,ω) be a 2n–dimensional symplectic manifold. Assume that $\phi \in \mathrm{End}(TM)$ satisfies $\phi^n = 1$ and is symplectically self–adjoint with respect to ω. For $K \in \chi(M)$, let $K_i = \phi^{i-1} K$, for $i = 1,\ldots,n$, and suppose that $Y \in \chi(M)$ satisfies $\omega(K_i,Y) = \delta_{i1}$, for $i = 1,\ldots,n$. Now, it follows immediately that if $Y_j = \phi^{1-j} Y$, for $j = 1,\ldots,n$, then $\omega(K_i,Y_j) = \delta_{ij}$ (for all $i,j = 1,\ldots,n$).

1.4: An endomorphism of the tangent bundle which is constant in some coordinate system, is automatically Nijenhuis.

Before turning to examples and applications, we note the following two lemmas. They provide simple tests for conditions 1a) and 2) of Theorem 1, respectively, when $M = \mathbb{R}^{2n}$ and $\omega = dp \wedge dq$.

Lemma 1: The map $\phi \in \text{End}(T\mathbb{R}^{2n})$ is symplectically self–adjoint with respect to $dp \wedge dq$, if and only if its matrix representative at each base point $u = (p,q) \in \mathbb{R}^{2n}$ is of the form

$$\begin{bmatrix} \alpha(u) & \beta(u) \\ \gamma(u) & \delta(u) \end{bmatrix}$$

where α, β, γ, δ are n×n matrices and $\beta(u) = -\beta^t(u)$, $\gamma(u) = -\gamma^t(u)$, $\delta(u) = \alpha^t(u)$.

Lemma 2: The vector field $X \in \chi(\mathbb{R}^{2n})$ is symplectic with respect to $dp \wedge dq$, if and only if the derivative ∇X has a matrix representative at each base point $u = (p,q) \in \mathbb{R}^{2n}$ of the form

$$\begin{bmatrix} \alpha(u) & \beta(u) \\ \gamma(u) & \delta(u) \end{bmatrix},$$

where α, β, γ, δ are n×n matrices and $\beta(u) = \beta^t(u)$, $\gamma(u) = \gamma^t(u)$, $\delta(u) = -\alpha^t(u)$.

§2. Examples

Example 1: Consider a completely integrable 2n–dimensional Hamiltonian system $(M = \mathbb{R}^{2n}, \omega, H)$. Let K_F denote the Hamiltonian vector field $\omega^{\#}dF$ and $\{F,G\} = \omega(K_G,K_F) =$ the usual Poisson bracket. Then there exist local coordinates $I_1,\ldots,I_n$ and $\theta_1,\ldots,\theta_n$, i. e., Darboux coordinates such that

$$\omega = dI_1 \wedge d\theta_1 + \ldots + dI_n \wedge d\theta_n, \tag{2.1a}$$

and

$$H = H(I_1,\ldots,I_n). \tag{2.1b}$$

Throughout the rest of this paper, we will call coordinates satisfying eq.(2.1) linearizing Darboux coordinates. Note that action–angle variables have the additional property that the θ's are required to be periodic with the same peroid 2π. In the literature, coordinates satisfying eq.(2.1) are sometimes referred to as action–angle type variables.

Assume the coordinates $I_1,\ldots,I_n$ and $\theta_1,\ldots,\theta_n$ are global, then define $\phi \in \text{End}(T\mathbb{R}^{2n})$, so that in the basis $(\partial/\partial I_1,\ldots,\partial/\partial I_n,\partial/\partial\theta_1,\ldots,\partial/\partial\theta_n)$, it has the matrix representative

$$\begin{bmatrix} A^T & 0 \\ 0 & A \end{bmatrix},$$

where A is an n×n matrix whose only nonzero matrix elements are

$A_{1n} = A_{21} = A_{32} = \dots = A_{n\,n-1} = 1$, A^T is the transpose of A, and 0 is the n×n zero matrix. Direct computation shows that the triples $(\phi, K_{I_1}, \phi K_{I_1})$ and $(\phi, K_{\theta_1}, \phi K_{\theta_1})$ satisfy Theorem 1. Hence each triple generates a hierarchy of mutually commuting Hamiltonian vector fields such that the corresponding Hamiltonians Poisson commute.

Remark 2.1: Theorem 1 treats the twin hierarchies, the action hierarchy (K_I's) and the "angle" hierarchy (K_θ's), on equal footing.

Example 2: Consider a dynamical system of two mass points interacting via a linear spring force, but otherwise free to move along a line. An Hamiltonian formulation (M,ω,H) is given in global coordinates $(p_1,p_2,q_1,q_2) \in \mathbb{R}^4$ by

$$M = \mathbb{R}^4, \tag{2.2a}$$

$$\omega = dp_1 \wedge dq_1 + dp_2 \wedge dq_2, \tag{2.2b}$$

$$H = \beta_1\, p_1^2/2 + \beta_2\, (p_2^2 + q_2^2)/2\,, \tag{2.2c}$$

where (p_1,q_1) describes the center–of–mass motion and (p_2,q_2) the relative motion. Let K_F denote the Hamiltonian vector field $\omega^{\#}dF$ and $\{F,G\} = \omega(K_G,K_F)$ = the usual Poisson bracket. Consider trajectories such that $p_1 \neq 0 \neq p_2^2 + q_2^2$ and let

$$I_1 = p_1^2/2\,, \qquad \theta_1 = q_1/p_1\,,$$
$$I_2 = (p_2^2 + q_2^2)/2, \qquad \theta_2 = \arctan q_2/p_2, \tag{C1}$$

and

$$J_1 = p_1\,, \qquad \varphi_1 = q_1 - p_1 \arctan q_2/p_2,$$
$$J_2 = p_1^2/2 + (p_2^2 + q_2^2)/2\,, \qquad \varphi_2 = \arctan q_2/p_2. \tag{C2}$$

By direct computation

$$dI_1 \wedge d\theta_1 + dI_2 \wedge d\theta_2 = dp_1 \wedge dq_1 + dp_2 \wedge dq_2 = dJ_1 \wedge d\varphi_1 + dJ_2 \wedge d\varphi_2\,,$$

i.e. (C1) and (C2) are canonical transformations. Thus, we have the usual Poisson commutation relations for the equivalent sets $\left[I_1, \theta_1, I_2, \theta_2\right], \left[J_1, \varphi_1, J_2, \varphi_2\right]$ of canonical variables. Thus, $\left[K_{I_1}, K_{I_2}, K_{\theta_1}, K_{\theta_2}\right]$ and $\left[K_{J_1}, K_{J_2}, K_{\varphi_1}, K_{\varphi_2}\right]$ are sets of mutually commuting vector fields with respect to the usual Lie bracket for vector fields. Because, $\left\{H, I_i\right\} = \left\{H, J_i\right\} = 0$, i=1,2, (C1) and (C2) are, in fact, transformations to linearizing Darboux coordinates. Now we show how the structures associated with (C1) and (C2) fit into the framework developed in the preceding paragraphs. Define $\phi_1, \phi_2 \in \mathrm{End}(T\mathbb{R}^4)$, so that in the basis $(\partial/\partial p_1, \partial/\partial p_2, \partial/\partial q_1, \partial/\partial q_2)$, they have the matrix representatives,

$$\phi_1 = \begin{bmatrix} 0 & \alpha_2 & 0 & \alpha_1 \\ \alpha_3 & 0 & -\alpha_1 & 0 \\ 0 & -\alpha_4 & 0 & \alpha_3 \\ \alpha_4 & 0 & \alpha_2 & 0 \end{bmatrix},$$

where

$$\alpha_1 = q_2/p_1,\ \alpha_2 = p_2/p_1,\ \alpha_3 = p_1p_2/(p_2^2 + q_2^2) + q_1q_2/p_1^2,\ \alpha_4 = p_1q_2/(p_2^2 + q_2^2) - q_1p_2/p_1^2,$$

and

$$\phi_2 = \begin{bmatrix} p_1 & p_2 & 0 & q_2 \\ \varphi_2 q_2 + a p_2 & -p_1 & -q_2 & 0 \\ 0 & \varphi_2 p_2 - a q_2 & p_1 & \varphi_2 q_2 + a p_2 \\ -\varphi_2 p_2 + a q_2 & 0 & p_2 & -p_1 \end{bmatrix}$$

where $a = (1-p_1^2)/(p_2^2+q_2^2)$. Clearly ϕ_1 and ϕ_2 satisfy lemma 1. Direct computation shows that the triple $(\phi_1, K_{I_1}, \phi_1 K_{I_1})$ satisfies the other conditions of Theorem 1, hence it generates a commuting hierarchy $\left[K_{I_1}, K_{I_2}\right]$ with $\left\{I_1,I_2\right\}=0$. Similarly, $(\phi_1, K_{\theta_1}, \phi_1 K_{\theta_1})$ generates a commuting hierarchy $\left[K_{\theta_1}, K_{\theta_2}\right]$ with $\left\{\theta_1,\theta_2\right\}=0$. Further, $\phi_1^2 = \mathrm{id}$. and the K_I's commute with the K_θ's. Parallel statements hold with ϕ_2 replacing ϕ_1 and the J's and φ's replacing the I's and θ's respectively.

Remark 2.2: The part of example 2 which pertains to (C2) simulates the situation for the 2–soliton sector of the KdV equation. Namely, the linear momentum and the energy constitute a complete set of functionally independent constants of motion in involution. See Section 3.

Remark 2.3: It is clear how to generalize (C1) and (C2) to the case of $n \geq 3$ masses on a line. For (C1), simply add $I_j = (p_j^2 + q_j^2)/2$, and $\theta_j = \arctan(q_j/p_j)$, $j = 3,\ldots,n$. For (C2), take J_1 and φ_1 as given and set $J_j = \sum_{i=1}^{j} I_i$, and $\varphi_j = \theta_j - \theta_{j+1}$, $j = 2,\ldots,n$, where $\theta_{n+1} \equiv 0$.

Remark 2.4: The two–dimensional isotropic harmonic oscillator provides examples where conditions 1a) and 1b) of Theorem 1 hold but 2) does not. In this case we obtain hierarchies of commuting vector fields, but they are not, in general, Hamiltonian vector fields.

§3. Application to the KdV one–soliton and two–soliton sectors

Consider the KdV vector field in the following form

$$K_3(u) = \partial^3 u + 6u\, \partial u\,, \quad u \in M, \tag{3.1}$$

where $\partial = \partial/\partial x$ and $M = S(\mathbb{R})$, the Schwartz space of rapidly decreasing functions on $\mathbb{R}$. We model TM on $S(\mathbb{R})\times S(\mathbb{R})$ and let $X(u) \in S(\mathbb{R})$ denote the tangent vector at u defined by a vector field X on M.

One of the major technical problems in treating such infinite–dimensional Hamiltonian systems is the fact that T^*M is too large. Our treatment of this problem [A–D–V] differs from that in the literature. Hence, it is necessary to begin by summarizing our approach adapted to the KdV system on $S(\mathbb{R})$. Then we shall restrict it to the two and four–dimensional submanifolds corresponding to the one–soliton and two–soliton sectors, respectively.

Take the two–form ω on M defined point–wise by

$$\omega(f,g) = 1/2 \int_{\mathbb{R}} \left[f(\partial^{-1}g) - (\partial^{-1}f)g\right], \quad f,g \in S(\mathbb{R}), \tag{3.2}$$

where $[\partial^{-1}f](x) = \int_{-\infty}^{x} f(x')dx'$ (c.f. [Z–F]). Define an ω^b: $TM \to T^*M$, by $f \to \alpha = \omega(\ ,f)$. Then, the element α dual to f under ω^b is given by $\alpha = \partial^{-1}f - (1/2)\int_{\mathbb{R}} f$. It follows that ω^b is an injection, i.e., the two–form ω is weakly nondegenerate. If we restrict ourselves to the image $\omega^b(TM) = T^*_R M \subset T^*M$ defined by $T^*_R M = \left\{\alpha \mid \alpha = \omega(\ ,f),\ f \in S(\mathbb{R})\right\}$, then we can define the inverse $\omega^{\#} = (\omega^b)^{-1} = \partial$. With these conventions, we consider the KdV Hamiltonian system $(M=S(\mathbb{R}),\omega,\ H)$, where $H(u) = \int_{\mathbb{R}} (-(\partial u)^2/2 + u^3) = P_3(u)$. It follows that if the vector fields f,g are symplectic with respect to ω, then

$$\omega(f,g) = \left\{G,F\right\}, \tag{3.3}$$

where F,G are the Hamiltonians for f,g, respectively, and

$$\left\{G,F\right\} = 1/2 \int_{\mathbb{R}}((\delta G/\delta u)(\partial\, \delta F/\delta u) - (\delta F/\delta u)(\partial\, \delta G/\delta u)), \tag{3.4}$$

is the antisymmetrized Gardner bracket.

It is well known that for a solution u of the KdV equation the Lenard operator, N(u), given by

$$N(u)\, f = \partial^2 f + 4uf + 2(\partial u)\partial^{-1}f, \tag{3.5}$$

has the property that for each solution u of the KdV equation the hierarchy of vector fields $\left\{K_{2j+1} = N^j\partial\right\}_{j=0}^{\infty}$ are mutually commuting symmetries (Note: $N\partial = K_3$) . Further, they are related via the construction $K_{2j+1}(u) = \partial\,(\delta P_{2j+1}(u)/\delta u)$ with functional derivatives of an hierarchy of constants of motion (COM's) $\left\{P_{2j+1}\right\}_{j=0}^{\infty}$ such that these COM's are in involution with respect to the Gardner bracket

$$\{F,G\} = \int_{\mathbb{R}} \delta F/\delta u \; \partial \; \delta G/\delta u. \tag{3.6}$$

The operator N is not self–adjoint with respect to ω given by eq.(3.2). This fact led us to consider the operator $\phi \in \mathrm{End}(TM)$ given by

$$\phi(u) = N(u) - (\partial u)\int_{\mathbb{R}}, \tag{3.7}$$

which is symplectically self–adjoint with respect to ω given by eq.(3.2). We will now study this operator on the one–soliton and two–soliton sectors.

The one–soliton solution s_1 of the KdV equation, with the conventions established by our choice of the vector field K_3, is given by

$$s_1(x,c_1,x_1) = (c_1/2)\,\mathrm{sech}^2(\sqrt{c_1}(x+x_1)/2), \tag{3.8}$$

where $x_1 = c_1^2 t - x_{10}$, the speed $c_1 \in \mathbb{R}^+$, the time $t \in \mathbb{R}$, and $x_{10} \in \mathbb{R}$ is the initial position of the maximum of s_1. By setting $t = 0$ in x_1, we can regard s_1 in eq. (3.8) as an initial data point in $M = S(\mathbb{R})$ specified by the two parameters c_1 and x_1.

We begin by noting that $\partial/\partial x_1$ and $\partial/\partial c_1$ are two independent vector fields tangent to the two–parameter manifold of one–soliton solutions given by eq.(3.8). Define $K_1(s_1) = \partial s_1$, $K_3(s_1) = \phi(s_1)\, K_1(s_1)$ and $W_1(s_1) = s_1 + x\, \partial s_1/2$, $W_3(s_1) = \phi(s_1)\, W_1(s_1)$. Then, direct computation yields the following results:

$$\phi(s_1)\partial s_1/\partial x_1 = c_1\, \partial s_1/\partial x_1; \tag{3.9a}$$

$$\partial s_1/\partial x_1 = K_1(s_1); \tag{3.9b}$$

$$\phi(s_1)\partial s_1/\partial c_1 = c_1\, \partial s_1/\partial c_1; \tag{3.9c}$$

$$c_1 \partial s_1/\partial c_1 = W_1(s_1) + x_1\, K_1(s_1)/2; \tag{3.9d}$$

and

$$\int_{\mathbb{R}} s_1 = 2\, c_1^{1/2}; \tag{3.10}$$

$$P_1(s_1) \equiv \int_{\mathbb{R}} s_1^2/2 = c_1^{3/2}/3; \tag{3.11}$$

$$\omega(W_1(s_1),K_1(s_1)) = 3/2\; P_1(s_1). \tag{3.12}$$

Now it is straightforward to show that the two–dimensional one–soliton submanifold is symplectic and the 2–form eq.(3.2) restricts to the 2–form $\omega = dI_1 \wedge d\theta_1$, where the pair $(I_1{=}c_1, \theta_1{=}\, c_1^{1/2}x_1/2)$ are Darboux coordinates which linearize the KdV one–soliton flows.

For the 4–parameter manifold of two–soliton solutions, we take Hirota's construction, namely,

$$s_2(x,c_1,c_2,x_1,x_2) = 2\,\partial^2\ln(f_2), \tag{3.13}$$

where

$$f_2 = 1+\exp(\xi_1)+\exp(\xi_2)+((p_1-p_2)/(p_1+p_2))^2\exp(\xi_1+\xi_2), \tag{3.14}$$

and $\xi_i = p_i(x+x_i)$, $x_i = p_i^2 t - x_{i0}$, $p_i = c_i^{1/2}$, $i = 1,2$. The c's $\in \mathbb{R}^+$, with $c_2<c_1$, $t \in \mathbb{R}$, and the x_0's $\in \mathbb{R}$. Again, our starting point is $\partial/\partial x_1, \partial/\partial x_2, \partial/\partial c_1, \partial/\partial c_2$ are four linearly independent tangent vector fields to the four–parameter manifold of initial data given by eq.(3.13). Define $K_1(s_2) = \partial s_2$, $K_3(s_2) = \phi(s_2)K_1(s_2)$ and $W_1(s_2) = s_2 + x\partial s_2/2$, $W_3(s_2) = \phi(s_2)W_1(s_2)$. Direct computation, using macsyma, yields the following results:

$$\phi(s_2)\,\partial s_2/\partial x_i = c_i\,\partial s_2/\partial x_i; \tag{3.15a}$$

$$K_1(s_2) = \partial s_2/\partial x_1 + \partial s_2/\partial x_2; \tag{3.15b}$$

$$\phi(s_2)\,\partial s_2/\partial c_i = c_i\,\partial s_2/\partial c_i - c_i^{-1/2}\,\partial s_2/\partial x_j; \tag{3.15c}$$

$$W_1(s_2) = c_1\partial s_2/\partial c_1 + c_2\,\partial s_2/\partial c_2 - 1/2\,(x_1\,\partial s_2/\partial x_1 + x_2\,\partial s_2/\partial x_2); \tag{3.15d}$$

(no sum on repeated indices) $i \neq j$, $i,j = 1,2$.

(It is evident that $\partial/\partial x = \partial/\partial x_1 + \partial/\partial x_2$ on s_2, hence it is not necessary to use macsyma to obtain eq.(3.15b)) The above implies that $\phi(s_2)$ is diagonalizable on the span of the four coordinate partials $\partial/\partial x_1, \partial/\partial x_2, \partial/\partial c_1, \partial/\partial c_2$ with eigenvalues c_1, c_2, each with multiplicity two. Thus,

$$\phi^2(s_2) - (c_1 + c_2)\,\phi(s_2) + c_1c_2\,\mathrm{id}$$

is zero on this span. In particular,

$$(\phi(s_2) - c_i)\,(K_3(s_2) - c_jK_1(s_2)) = 0, \quad i,j = 1,2. \tag{3.16a}$$

$$(\phi(s_2) - c_i)\,(W_3(s_2) - c_jW_1(s_2)) = 0, \quad i,j = 1,2. \tag{3.16b}$$

It can be deduced from these formulas that ϕ is Nijenhuis on the two–soliton manifold. Let us admit the formulas $P_1(s_2) = (c_1^{3/2} + c_2^{3/2})/3$, and $P_3(s_2) = (c_1^{5/2} + c_2^{5/2})/5$ for the two–soliton COM's which heuristically follow from Lax's asymptotics [L] and the one–soliton case. This combined with the above results allows one to show that the four–dimensional two–soliton manifold is symplectic and the 2–form eq.(3.2) restricts to the 2–form

$$\omega = dI_1 \wedge d\theta_1 + dI_2 \wedge d\theta_2, \tag{3.17a}$$

where for i = 1,2,

$$I_i = c_i, \tag{3.17b}$$

$$\theta_i = c_i^{1/2} x_i/2 + \ln\{(c_1^{1/2} - c_2^{1/2})/(c_1^{1/2} + c_2^{1/2})\}, \tag{3.17c}$$

(no sum on repeated indices), are Darboux coordinates which linearize the KdV two–soliton flows. Alternatively, one can take the following as a set of linearizing Darboux coordinates

$$\mathcal{P}_1 = (c_1^{3/2} + c_2^{3/2})/3 = P_1(s_2), \tag{3.18a}$$

$$\mathcal{P}_2 = (c_1^{5/2} + c_2^{5/2})/5 = P_3(s_2), \tag{3.18b}$$

$$\mathcal{Q}_1 = 2c_1c_2(c_2 - c_1)^{-1}(c_1^{-3/2}\,\theta_1 - c_2^{-3/2}\,\theta_2), \tag{3.18c}$$

$$\mathcal{Q}_2 = 2(c_1 - c_2)^{-1}(c_1^{-1/2}\,\theta_1 - c_2^{-1/2}\,\theta_2), \tag{3.18d}$$

Thus, all the hypotheses of Theorem 1 are satisfied for the triples

$$(\phi(s_2), \partial s_2 = \partial(\delta \mathcal{P}_1/\delta s_2), \phi(s_2)\partial s_2 = \partial(\delta \mathcal{P}_2/\delta s_2))$$

and

$$(\phi(s_2), \partial(\delta \mathcal{Q}_1/\delta s_2), \phi(s_2)\partial(\delta \mathcal{Q}_1/\delta s_2) = -c_1c_2\,\partial(\delta \mathcal{Q}_2/\delta s_2)).$$

<u>Remark 3.1</u>: There is a two–soliton to one–soliton reduction given by setting $c_2 = 0$. Before implementing this reduction, eq.(3.15c) must be multiplied by $c_2^{1/2}$. After this is done, eq.(3.15) reduces to eq.(3.9) in the following way: $\partial/\partial x_1 \to \partial/\partial x_1$, $\partial/\partial x_2 \to 0$, $\partial/\partial p_1 \to \partial/\partial p_1$, and $\partial p_2 \to -2/p_1^2\,\partial/\partial x_1$, where the operators on the left of the arrows operate on the two–soliton sector and those on the right operate on the one–soliton sector.

Remark 3.2: Another two–soliton to one–soliton reduction is given by setting $p_1 = p_2$, irrespective of the values of x_1 and x_2. However, in this case, suitable care must be taken when applying the reduction to the right hand side of eq.(3.15c). In particular, the term $-(\partial s_2)\int_{\mathbb{R}} \partial s_2/\partial c_i$, which is traced to the modification of N in eq.(3.7), involves a limit which must be evaluated by L'Hospital's Rule.

Remark 3.3: Examination of the results for the 1–soliton and 2–soliton sectors leads us to predict that the following formulae hold for the N–soliton sector:

$$\phi(s_N)\,\partial s_N/\partial x_i = c_i\,\partial s_N/\partial x_i; \tag{3.19a}$$

$$K_1(s_N) = \sum_{k=1}^{N} \partial s_N/\partial x_k; \tag{3.19b}$$

$$\phi(s_N)\,\partial s_N/\partial c_i = c_i\,\partial s_N/\partial c_i - c_i^{-1/2}\,(\,\partial s_N - \partial s_N/\partial x_i); \tag{3.19c}$$

$$W_1(s_N) = \sum_{k=1}^{N} (c_k \partial s_N/\partial c_k - 1/2\, x_k\, \partial s_N/\partial x_k); \tag{3.19d}$$

(no sum on repeated indices unless indicated by Σ) $i = 1,\dots,N$, where $K_1(s_2) = \partial s_2$, $W_1(s_2) = s_2 + x\partial s_2/2$, and ϕ is given by eqs.(3.7) and (3.5). If the above holds, this implies that $\phi(s_N)$ is diagonalizable on the span of the 2N coordinate partials $\partial/x_1,\dots,\partial/\partial x_N,\partial/\partial c_1,\dots,\partial/\partial c_N$ with eigenvalues $c_1,\dots c_N$, each with multiplicity two. Therefore,

$$(\phi(s_N) - c_1\,\mathrm{id})\dots(\phi(s_N) - c_N),$$

is zero on this span. In particular,

$$\prod_{j=2}^{N} (\phi(s_N) - c_{i_j})\, K_1(s_N), \tag{3.20a}$$

and

$$\prod_{j=2}^{N} (\phi(s_N) - c_{i_j})\, W_1(s_N), \tag{3.20b}$$

where $i \in \mathscr{P}_N$, the permutation group of N objects, are eigenfunctions of $\phi(s_N)$ with eigenvalue c_{i_1}.

Eqs.(3.16) and our prediction, eqs.(3.20), agree with the results obtained by Fuchssteiner and Oevel using their master symmetry approach [F–O]. However, our eqs.(3.15) and our prediction, eqs.(3.19), are different. Further, our explicit expressions for the Darboux coordinates which linearize the flows for the two–soliton sector eqs.(3.17) and eqs.(3.18) are new.

Acknowledgements: The authors would like to thank J. Dorfmeister for helpful discussions on many topics related to this paper and B. Fuchssteiner for making available the interesting preprint [F–O]. One of the authors (RLA) would like to acknowledge stimulating discussions with I. Bialynicki–Birula in 1981 about constructing recursion operators for oscillator systems.

Note added in proof: We would like to thank Yvette Kosmann–Schwarzbach for calling to our attention that Example 1 pp.7-8 appears in reference [K–S]. We would also like to thank Stephan De Biève for a helpful comment on the action of ϕ on the W–hierarchy in §3.

References

[A–M] R. Abraham and J. Marsden, Foundations of Mechanics, Second Edition, Benjamin/Cummings, Reading, (1978).

[A] V.I Arnold, Mathematical Methods of Classical Mechanics, Springer–Verlag, New York, (1978).

[A–D–V] R.L. Anderson, J. Dorfmeister, and R. Varley, "Hierarchies of symmertries and constants of motion on symplectic spaces", unpublished manuscript (1986).

[F] B. Fuchssteiner, "Applications of hereditary symmetries to nonlinear evolution equations", Nonlinear Anal.TMA 3, 849 (1978).

[F–O] B. Fuchssteiner and G. Oevel, "Geometry and action–angle variables of multi soliton systems", Univ. of Paderborn preprint (1989).

[G–D] I.M. Gel'fand and I.Y. Dorfman, "Hamiltonian operators and algebraic structures related to them",Funct. Analysis Appl. 13, 248 (1979);"The Shouten bracket and Hamiltonian operators", Analysis Appl. 14, 223 (1980).

[H] R.Hirota, "Exact solution of the Korteweg–de Vries equation for multiple collisions of solitons", Phys. Rev. Lett. 27, 1192 (1971).

[K–S] Y. Kosmann–Schwarzbach, "Géométrie des systèmes bihamiltoniens," Séminare de Mathématiques Supérieures 102, pp.185–216, ed. P. Winternitz, University of Montréal Press (1986).

[K–S–M] Y. Kosmann–Schwarzbach and F. Magri, "Poisson–Nijenhuis Structures," to appear in Ann. Inst, Henri Poincaré, Série A.

[L] P.D.Lax, "Integrals of nonlinear equations of evolution solitary waves", Comm. Pure Appl. Math. 21, 467 (1968).

[M] F.Magri, "A simple model of the integrable Hamiltonian equation", J. Math. Phys. 19, 1156 (1978);"A geometrical approach to the nonlinear solvable equations", Lecture notes in Physics (Springer) 120, 233 (1980)(Eds. M.Boiti, F. Pempinelle, G. Soliani.).

[M–M] F. Magri and C. Morosi, "A geometrical characterization of integrable Hamiltonian systems through the theory of Poisson–Nijenhuis manifolds", Univ. of Milano preprint (1984).

[N–M–P–Z] S. Novikov, V. Manakov, L.P. Pitaevskii, and V.E. Zakharov, Theory of solitons, Consultants Bureau, New York and London (1980).

[O] P.J. Olver,"Evolution equations possessing infinitely many symmetries ", J. Math. Phys. 18, 1212 (1977).

[tE] H.M.M. ten Eikelder, Symmetries for Dynamical and Hamiltonian Systems, Mathematisch Centrum, Amsterdam, (1985).

[Z–F] V.E. Zakharov and L.D.Faddeev, "Korteweg–de Vries equation: A completely integrable Hamiltonian system", Funct. Anal. Appl. 5, 18 (1971).

INTEGRABLE HAMILTONIAN SYSTEMS ON RATIONAL COADJOINT ORBITS OF LOOP ALGEBRAS†

M.R. ADAMS,[1] J. HARNAD[2] AND J. HURTUBISE[3]

[1] Department of Mathematics, University of Georgia
Athens, GA 30602, USA
[2] Department of Mathematics, Concordia University
Montréal, Québec, Canada
and Centre de Recherches Mathématiques, Université de Montréal
C.P. 6128 – A, Montréal, Québec H3C 3J7, Canada
[3] Department of Mathematics, McGill University
Montréal, Québec H3A 2K6, Canada

ABSTRACT. This report gives an overview of results on integrable Hamiltonian systems in the context of isospectral flows in loop algebras, as developed in a recent series of papers. A moment map embedding is used to model Poisson submanifolds of the dual of a loop algebra, having symplectic leaves that are rational coadjoint orbits, in terms of reductions of symplectic vector spaces. A duality theorem explains how the flows may have two Lax pair representations in different loop algebras which give rise to the same spectral curves and ring of invariants by interchange of the spectral parameter and loop variables. Linearization of the flows on the Jacobi variety of the associated spectral curves is explained in a framework allowing for singularities in the curves due to a degenerate spectrum. A natural set of Darboux coordinates on rational coadjoint orbits is introduced and used to derive the connection between flows of line bundles and Arnold-Liouville integration in loop algebras.

INTRODUCTION

A key ingredient in the understanding of many integrable Hamiltonian systems consists of realizing them in terms of isospectral flows governed by Lax equations of the type

$$\dot{L}(\lambda) = [P(\lambda, \lambda^{-1}, L(\lambda))_+,\ L(\lambda)]\ . \tag{0.1}$$

Here $L(\lambda)$ is a Laurent polynomial or semi-infinite formal power series in (λ, λ^{-1}), with matrix coefficients, $P(\lambda, \lambda^{-1}, z)$ is a polynomial in its arguments, and the subscript $+$ denotes projection to the positive powers of λ. Our focus will be on the case when $L(\lambda)$ is rational with fixed poles in the complex λ plane. This case is of importance both in the study of finite dimensional integrable systems and special classes of quasi-periodic solutions to integrable systems of PDE's. Two aspects of such Lax equations have been exploited in relation to the complete integrability of the flows:

1. The flow described by (0.1) is isospectral for $L(\lambda)$, and hence the spectral invariants of $L(\lambda)$ give constants of motion. If $L(\lambda)$ is a rational function of λ, the characteristic polynomial

$$\det(L(\lambda) - z) = 0\ . \tag{0.2}$$

defines a spectral curve in $\mathbf{C}^2$ which is an invariant of the flow. Generically, $L(\lambda)$ defines a one dimensional eigenspace over each point of this curve and the flow for

† Research supported in part by the Natural Sciences and Engineering Research Council of Canada and by U.S army grant DAA-87-K-0110.

(0.1) is seen to map to a linear flow in the Jacobian of a compactification S of the spectral curve (see e.g. **[AvM1]**, **[vMM]**, **[RS]**, **[KN]**).

2. The flow is Hamiltonian with respect to the natural Lie-Poisson structure on the dual of the positive part of a loop algebra, and the spectral invariants are Poisson commuting first integrals arising from the Adler-Kostant-Symes (AKS) theorem (see e.g. **[AvM1]**, **[RS]**, **[FNR]**, **[AHP]**). Rational $L(\lambda)$ lie on finite dimensional coadjoint orbits in the dual of the positive part of the loop algebra (the $+/-$ splitting being understood with respect to a given domain of analyticity) and, generically, it can be shown that the AKS invariants give a completely integrable Hamiltonian system on such orbits. In theory then, the flows can be linearized on the common level surfaces of the first integrals.

The two aspects – the algebraic geometry of spectral curves and their Jacobians, and the symplectic geometry of rational coadjoint orbits of loop algebras – may be tied together by noting that the space of line bundles over a suitable family of spectral curves has a natural symplectic structure, derived from Serre duality, which coincides with the Kostant-Kirillov symplectic form on rational coadjoint orbits after reduction by global conjugation (cf. sec. 3).

The class of equations known as the generalized AKNS or Zakharov-Shabat systems can all be realized as the commutativity condition of two AKS type flows on loop algebras (see **[FNR]**, **[RS]**, **[AHP]**). Restricting these flows to finite dimensional coadjoint orbits gives families of quasi-periodic solutions that are determined by flows in stationary manifolds of higher invariants. A Hamiltonian interpretation of such "stationary" flows was developed in the framework of differential algebras in the work of Gel'fand and Dickey (**[GD]**, **[D]**).

The purpose of this report is to present an overview of an approach aimed at unifying the finite dimensional Hamiltonian framework with that of integrable systems of PDE's using moment map embeddings into the duals of loop algebras (**[AHP]**, **[AHH1]**-**[AHH5]**). In many finite dimensional examples, different approaches have led to two alternative Lax-pair representations. In one representation an n-dimensional classical system is mapped to $n \times n$ matricial polynomials of fixed low degree (usually two); in the other, the classical system is mapped to matricial polynomials whose degrees depend on n but whose size is fixed (usually 2×2). This is the case, for example, for the Neumann oscillator system (**[M]**, **[AvM1]**, **[AHP]**, **[Ra1]**), the Jacobi ellipsoid (**[M]**,**[AvM1]**) the Rosochatius system (**[Ra2]**, **[AHP]**, **[AHH1]**), the Toda systems **[AvM1]** and Moser's rank -2 reduction of generalized Euler tops (**[M]**, **[AHH1]**). Section 1 describes the relationship of classical flows on reduced symplectic vector spaces to isospectral flows on rational coadjoint orbits of a loop algebra via the moment map construction introduced in **[AHP]**, and relates the two alternative Lax representations by means of a duality theorem described in **[AHH1]**.

Section 2 discusses the algebro-geometric approach to linearization within a sheaf theoretic framework that is equally applicable in the case of singular spectral curves. These frequently arise from the structure of the moment map embeddings, as well as from degenerate spectral conditions at $\lambda = \infty$ that are required in certain applications. The formulation is based on a compactification defined, as in **[Hi]**, by embedding all spectral curves in the total space T of a line bundle $\mathcal{O}(m)$ over $\mathbf{P}_1(\mathbf{C})$.

In section 3, the relationship between the algebraic geometry of spectral curves and the symplectic geometry of rational coadjoint orbits is explained using Serre duality. A natural Darboux system consisting of the affine coordinates in T of the divisor on S determining the associated eigenvector line bundle (or the corresponding sheaf) is used to obtain a Liouville generating function for the canonical transformation to linearizing coordinates. This provides a simple Hamiltonian interpretation of the Abel map linearization, analogous to the results of Gel'fand and Dickey (**[GD]**, **[D]**), but entirely within the framework of symplectic structures on reduced coadjoint orbits of loop algebras. The formulation is also extended to include integration within an invariant symplectic submanifold of the orbit itself, without further reduction to quotient Poisson manifolds. The discussion is confined to "generic" cases only, avoiding the complications arising from degenerate spectra, singular curves or reductions under algebraic automorphisms. Such questions are addressed in the references (**[AHP]**, **[AHH1]**-**[AHH5]**), where proofs of the results cited may be found, as well as applications to numerous examples.

1. MOMENT MAP EMBEDDINGS IN LOOP ALGEBRAS AND DUALITY

In the formal series approach to loop algebras, $\widetilde{gl(r)}$ denotes the algebra of Laurent series, with coefficients in $gl(r)$, of the form

$$X(\lambda) = \sum_{i=-\infty}^{K} X_i\lambda^i , \qquad X_i \in gl(r, \mathbf{C}) , \tag{1.1}$$

and the subalgebra $\widetilde{gl(r)}^+$ consists of the polynomial elements. One has the invariant scalar product:

$$< X, Y > \equiv \mathrm{tr}[(XY)_0] , \quad X(\lambda), Y(\lambda) \in \widetilde{gl(r)}^+ , \tag{1.2}$$

(where the subscript o denotes the 0-th order term in λ), giving an identification of $\widetilde{gl(r)}$ with a dense subspace of the dual space $(\widetilde{gl(r)})^*$. Under this pairing, the dual $(\widetilde{gl(r)})^+)^*$ is identified with the subalgebra $\widetilde{gl(r)}_-$ of negative Laurent series of the form $\{\sum_{i=-\infty}^{0} X_i\lambda^i\}$.

Let $I(\widetilde{gl(r)}^*)$ denote the ring of Ad^*-invariant polynomials on $\widetilde{gl(r)}^*$. Fix $Y \in gl(r)$ and define $\mathcal{F}_Y^+$ to be the ring of functions on $(\widetilde{gl(r)}^+)^*$ given by

$$\mathcal{F}_Y^+ = \{\phi \mid \phi(X) = \Phi(\lambda Y + X),\ \Phi \in I(\widetilde{gl(r)}^*) , \quad X \in \widetilde{gl(r)}_-\} . \tag{1.3}$$

The AKS theorem states that the functions in $\mathcal{F}_Y^+$ Poisson commute on $(\widetilde{gl(r)}^+)^*$. Using the identification of $(\widetilde{gl(r)}^+)^*$ with $\widetilde{gl(r)}_-$, the Hamiltonian flow for an element ϕ of $\mathcal{F}_Y^+$ can be written in Lax form:

$$\dot{X}(\lambda) = [(d\phi(\lambda Y + X(\lambda)))_+ , \ \lambda Y + X(\lambda)] , \tag{1.4}$$

where the subscript + denotes the natural projection to $\widetilde{gl(r)}^+$.

To allow for rational orbits, the loop algebra elements will instead be viewed as maps $X : S_A^1 \to gl(r)$ from a circle S_A^1 in the complex λ - plane, centered at the origin, containing the eigenvalues of a fixed $N \times N$ matrix $A \in gl(N)$, while the subalgebra $\widetilde{gl(r)}^+$ will be the subspace consisting of $X(\lambda)$ that extend holomorphically inside S_A^1. The scalar product (1.2) is interpreted as a contour integral along S_A^1:

$$< X, Y > = \frac{1}{2\pi i} \oint_{S_A^1} \mathrm{tr}(X(\lambda)Y(\lambda)) \frac{d\lambda}{\lambda} . \tag{1.5}$$

(We note that the circle S^1_A may be replaced by any closed contour containing the eigenvalues of A.) The corresponding group $\widetilde{Gl(r)}^+$ similarly consists of maps into $Gl(r)$ extending holomorphically inside S^1_A. The dual space $(\widetilde{gl(r)}^+)^* \equiv \widetilde{gl(r)}_-$ consists of maps $X : S^1_A \to gl(r)$ extending holomorphically outside S^1_A, to $\lambda = \infty$.

Let $M_{N,r}$ denote the space of $N \times r$ matrices, with $N \geq r$. The space $M = M_{N,r} \times M_{N,r}$ has a natural symplectic structure

$$\omega(F,G) = tr(dF \wedge dG^T) \ , \quad (F,G) \in M \ . \tag{1.6}$$

In **[AHP]** a large class of integrable systems on M was derived by constructing a moment map into the dual of the positive part of a loop algebra. We define an action

$$\begin{aligned} \widetilde{Gl(r)}^+ : M &\longrightarrow M \\ g(\lambda) : (F,G) &\longmapsto (F_g, G_g) \ , \end{aligned} \tag{1.7}$$

where $(F_g, G_g) \in M$ are defined by the equations

$$(A - \lambda I)^{-1} F g^{-1}(\lambda) = (A - \lambda I)^{-1} F_g + F_{hol}(\lambda) \tag{1.7a}$$

$$g(\lambda) G^T (A - \lambda I)^{-1} = G^T_g (A - \lambda I)^{-1} + G^T_{hol}(\lambda) \ , \tag{1.7b}$$

with F_{hol} and G_{hol} holomorphic inside S^1_A.

THEOREM 1.1 (**[AHP]**, **[AHH1]**). *The $\widetilde{Gl(r)}^+$ -action (1.7a,b) preserves the symplectic form (1.6) and is generated by the equivariant moment map $\tilde{J}^A_r : M \to (\widetilde{gl(r)}^+)^*$ defined by:*

$$\tilde{J}^A_r(F,G) = \lambda G^T (A - \lambda I)^{-1} F \ , \tag{1.8}$$

thus taking its values in a finite dimensional Poisson subspace consisting of rational coadjoint orbits with poles at the spectrum of A.

Since equivariant moment maps are Poisson maps with respect to the Lie-Poisson structure, it follows that the ring $\mathcal{F}_Y$ of functions on M given by the pull-back via $\tilde{J}^A_r$ of the AKS functions $\mathcal{F}^+_Y$ on $(\widetilde{gl(r)}^+)^*$ is Poisson commutative. Thus, we obtain Hamiltonian systems on the symplectic vector space $M \sim \mathbf{R}^{2Nr}$ which have a natural Lax-pair formulation and many Poisson commuting first integrals.

To see the relationship of these systems to the flows on coadjoint orbits in $(\widetilde{gl(r)}^+)^*$ we note that there is also a symplectic action of $Gl(N)$ on M, defined by

$$g : (F,G) \longmapsto (gF, (g^T)^{-1} G), \quad g \in Gl(N), \ \ (F,G) \in M \ . \tag{1.9}$$

If the spectrum of A is not simple, we will require that the dimension k_i of the subspace $V_i \subset \mathbf{C}^N$ of generalized eigenvectors corresponding to any eigenvalue α_i of A be less than or equal to r. Let $M_0 \subset M$ denote the open, dense submanifold consisting of pairs (F,G) of rank -r matrices whose projection to each V_i is of maximal rank. Then the isotropy subgroup $Gl(N)_A$ of A in $Gl(N)$ acts freely on M_0.

LEMMA 1.2 (**[AHP]**). *The fibres of the map $\tilde{J}^A_r : M_0 \to (\widetilde{gl(r)}^+)^*$ are the orbits in M_0 of the isotropy subgroup $Gl(N)_A$. Thus $\tilde{J}^A_r$ projects to an injective Poisson map $\tilde{J}^A_{r,0} : M_0/Gl(N)_A \to (\widetilde{gl(r)}^+)^*$ from the quotient space $M_0/Gl(N)_A$ into $(\widetilde{gl(r)}^+)^*$.*

There is, moreover, a reciprocal identification, since the $Gl(N)_A$ - action is also Hamiltonian, and generated by the moment map

$$J_{Gl(N)_A} : (F,G) \longmapsto FG^T|_{gl(N)_A} \tag{1.10}$$

whose generic fibres are similarly the $\widetilde{Gl(r)}^+$ orbits. The $\widetilde{Gl(r)}^+$ and $Gl(N)_A$ actions on M form an example of a "dual pair" [W]. Since the projected map $\tilde{J}^A_{r,0}$: $M_0/Gl(N)_A \longrightarrow (\widetilde{gl(r)}^+)^*$ is injective it gives an identification between the symplectic leaves of $M_0/Gl(N)_A$ and certain rational coadjoint orbits in $(\widetilde{gl(r)}^+)^*$. The AKS flows on such orbits in $(\widetilde{gl(r)}^+)^*$ are thus modeled by reduced Hamiltonian flows on the symplectic vector space M.

There is another notion of "duality" present, however, that is more important from the viewpoint of integrable systems described by Lax equations. Choosing a matrix Y, we may define the loop group $\widetilde{Gl(N)}^+$, analogously to $\widetilde{Gl(r)}^+$, but consisting of elements $g(z)$ (z denoting the loop parameter) that are maps from a circle S^1_Y in the complex z-plane containing the eigenvalues of Y, with $g(z)$ extending holomorphically inside S^1_Y, and similar definitions for the Lie algebra $\widetilde{gl(N)}^+$ and its dual $(\widetilde{gl(N)}^+)^* \equiv \widetilde{gl(N)}_-$. We also have a Hamiltonian action of $\widetilde{Gl(N)}^+$ on M defined, similarly to (1.7), by

$$\begin{aligned} \widetilde{Gl(N)}^+ : M &\longrightarrow M \\ g(z) : (F, G) &\longmapsto (F_g, G_g) \end{aligned} \tag{1.11}$$

where

$$g(z)F(y - zI)^{-1} = F_g(Y - zI)^{-1} + F_{hol}(z) \tag{1.11a}$$
$$(Y - zI)^{-1}G^T g^{-1}(z) = (Y - zI)^{-1}G^T_g + G^T_{hol}(z)\ , \tag{1.11b}$$

and $F_{hol}(z), G_{hol}(z)$ are holomorphic inside S^1_Y. The moment map $\tilde{J}^Y_N : M \rightarrow (\widetilde{gl(N)}^+)^*$ generating this action is given by

$$\tilde{J}^Y_N(F, G) = -zF(Y - z)^{-1}G^T\ . \tag{1.12}$$

Just as above, it can be shown that the generic fibres of the map $\tilde{J}^Y_N$ are given by the orbits of the isotropy subgroup, $Gl(r)_Y \subset Gl(r)$, of Y , where $Gl(r)$ acts on M by

$$g(F, G) \rightarrow (Fg^{-1}, Gg^T)\ . \tag{1.13}$$

The functions in the ring $\mathcal{F}_A$ of functions on M given by the pullback, via $\tilde{J}^Y_N$, of the AKS ring $\mathcal{F}^+_A$ defined by

$$\mathcal{F}^+_A = \{\phi \mid \phi(X) = \Phi(-zA + X),\ \Phi \in I(\widetilde{gl(N)}^*)\} \tag{1.14}$$

are similarly mutually Poisson commutative. Now, let

$$\mathcal{N}(\lambda) \equiv \lambda Y + \lambda G^T(A - \lambda I)^{-1}F \tag{1.15a}$$
$$\mathcal{M}(z) \equiv -zA - zF(Y - zI)^{-1}G^T \tag{1.15b}$$

be viewed as elements of $\widetilde{gl(r)}^*$ and $\widetilde{gl(N)}^*$, respectively.

THEOREM 1.3 (**[AHH1]**). *The two rings of functions on M, $\mathcal{F}_Y$ and $\mathcal{F}_A$, are equal. Their Hamiltonian flows preserve the spectrum of both $\mathcal{N}(\lambda)$ and $\mathcal{M}(z)$. If $\psi \in \mathcal{F}_Y = \mathcal{F}_A$ is of the form:*

$$\psi(F, G) = \phi_1(\mathcal{N}) = \phi_2(\mathcal{M})\ , \tag{1.16}$$

with

$$\phi_1 \in I(\widetilde{gl(r)}^*) \ , \qquad \phi_2 \in I(\widetilde{gl(N)}^*) \ , \tag{1.17}$$

the integral curves $(F(t), G(t))$ of the corresponding Hamiltonian flows satisfy both the Lax equations:

$$\frac{d\mathcal{N}}{dt} = [(d\phi_1(\mathcal{N}))_+, \mathcal{N}] \tag{1.18a}$$

$$\frac{d\mathcal{M}}{dt} = [(d\phi_2(\mathcal{M}))_+, \mathcal{M}] \ , \tag{1.18b}$$

where $(d\phi_1)_+$ and $(d\phi_2)_+$ denote, respectively, the projections to the parts of $\widetilde{gl(r)}^*$ and $\widetilde{gl(N)}^*$ that are holomorphic inside the circles S^1_A and S^1_Y in the complex λ and z - planes. Moreover, the invariant spectral curves defined by

$$\det(\mathcal{N}(\lambda) - z\lambda I) = 0 \tag{1.19a}$$

$$\det(\mathcal{M}(z) - z\lambda I) = 0 \tag{1.19b}$$

are biholomorphically equivalent.

This theorem gives the notion of duality of moment maps discussed in **[AHH1]**. The collective Hamiltonian systems on M obtained by pulling back the AKS ring $\mathcal{F}^+_Y$ under the moment map $\tilde{J}^A_r$ and the corresponding ring $\mathcal{F}^+_A$ under the moment map $\tilde{J}^Y_N$ are identical, and have Lax pair representations in both $\widetilde{gl(r)}$ and $\widetilde{gl(N)}$. The linearization of their flows, as discussed in the following sections, may be accomplished in the Jacobi variety of the spectral curve associated to either representation. Reducing these maps by $Gl(N)_A$ and $Gl(r)_Y$, respectively, gives injective moment maps

$$\tilde{J}^A_{r,0} : M_0/Gl(N)_A \hookrightarrow (\widetilde{gl(r)}^+)^* \tag{1.20}$$

and

$$\tilde{J}^Y_{N,0} : M_0/Gl(r)_Y \hookrightarrow (\widetilde{gl(N)}^+)^* \ . \tag{1.21}$$

The first of these is equivariant with respect to the action of $Gl(r)_Y$ on $M_0/Gl(N)_A$ inherited from the action on M and the natural action of $Gl(r)_Y$ as "constant" loops on $(\widetilde{gl(r)}^+)^*$. Similarly, the second map (1.21) is equivariant with respect to the corresponding actions of $Gl(N)_A$ on $M_0/Gl(r)_Y$ and $(\widetilde{gl(N)}^+)^*$. Reducing (1.20) by the $Gl(r)_Y$ action and (1.21) by the $Gl(N)_A$ action, gives the following commutative diagram, yielding an identification of reduced symplectic leaves in $(\widetilde{gl(r)}^+)^*$ with reduced symplectic leaves in $(\widetilde{gl(N)}^+)^*$.

$$\begin{array}{ccccccc}
\widetilde{gl(N)}_- & \xleftarrow{\ \tilde{J}^Y_N\ } & & M_0 & & \xrightarrow{\ \tilde{J}^A_r\ } & \widetilde{gl(r)}_- \\
& \nwarrow \tilde{J}^Y_{N,0} & \swarrow & & \searrow & \tilde{J}^A_{r,0} \nearrow & \\
\downarrow & & M_0/Gl(r)_Y & \downarrow & M_0/Gl(N)_A & & \downarrow \\
& & \searrow & & \swarrow & & \\
\widetilde{gl(N)}_-/Gl(N)_A & \longleftarrow & & M_0/(Gl(r)_Y \times Gl(N)_A) & & \longrightarrow & \widetilde{gl(r)}_-/Gl(r)_Y
\end{array}$$

The space $M_0/Gl(r)_Y$ also projects to the set of rank - r perturbations of A of the form

$$L_A = A + FaG^T \ , \tag{1.22}$$

where

$$a \equiv (Y + I)^{-1} \ , \tag{1.23}$$

(assuming $z = -1$ is not an eigenvalue of Y; otherwise, rescale). From the Lax equation (1.18b), we see that L_A undergoes isospectral flow governed by the equation:

$$\frac{dL_A}{dt} = [B, L_A] \ , \tag{1.24}$$

where

$$B = (d\phi_2(L_A))_+|_{z=-1} \ . \tag{1.25}$$

This generalizes a formulation of integrable systems in terms of rank - 2 perturbations of a diagonal matrix A studied earlier by Moser **[M]**.

The embedding of a large number of finite dimensional Hamiltonian systems in loop algebras via one or another, or both of these moment maps is discussed in **[AHP]**, **[AHH1]**, **[HH]**, **[HHM]**. This includes all of Moser's rank - 2 systems, such as the Neumann oscillator, the rank -2 reduction of N -dimensional Euler tops and the Rosochatius system, as well as the generalized tops treated in **[AvM1]**, **[AvM2]**, **[RS2]**, **[BRS]**, and the Toda lattices. To formulate many of these systems in terms of Lax pair flows it is necessary to restrict the generic flows to invariant symplectic subspaces or reduce them to subquotients determined either by discrete or continuous Hamiltonian symmetry groups. Those reductions involving fixed point sets under symplectic automorphisms also imply reductions of the spectral data. A discussion of discrete reductions in a general context as well as a number of detailed examples may be found in **[AHP]**, **[AHH2]**, **[HH]** and **[HHM]**.

Flows on rational orbits also give rise to finite dimensional families of quasi-periodic solutions of integrable systems of PDE's, such as the KdV and MKdV families of equations, the cubically nonlinear Schrödinger equations and other systems admitting a Lax pair formulation. These may also be analyzed in terms of the coordinates provided by the space M and its reductions (**[AHP]**, **[AHH2]**, **[HW]**). The embeddings (1.20), (1.21) show how collective coordinates for finite dimensional systems provide such special classes of solutions to the systems of PDE's implied by the commutativity of the AKS flows.

2. LINEAR FLOWS OF LINE BUNDLES

In this section we describe briefly the algebro-geometric integration of the Lax equation (0.1) for $L(\lambda)$ an $r \times r$ matricial polynomial of degree m in λ. Such a system may be obtained from the equivalent Lax equation (1.18a) for rational $\mathcal{N}(\lambda)$ by defining

$$L(\lambda) = \lambda^{-1} a(\lambda) \mathcal{N}(\lambda) \tag{2.1}$$

where $a(\lambda)$ is the minimal polynomial for A. The approach used here is detailed in **[AHH2]**, along similar lines to **[AvM]**, **[vMM]**, **[KN]**, **[Du]**, **[RS1]**, but in a formulation, as in **[Hi]**, that is adapted so as to allow singularities in the spectral curve at finite points or over $\lambda = \infty$. Consider the plane curve defined by

$$f(\lambda, z) \equiv \det(L(\lambda) - z\mathbf{I}) = 0 \ . \tag{2.2}$$

We compactify this curve by considering $\mathbf{C}^2$ as an open set U_0 in the total space T of the line bundle $\mathcal{O}(m)$ over $\mathbf{P}_1(\mathbf{C})$, with λ thought of as an affine coordinate lifted from $\mathbf{P}_1(\mathbf{C})$ and z as the fibre coordinate. T is covered by two coordinate patches U_0 and U_1, with coordinates (z,λ), $(z',\lambda') = (z\lambda^{-m}, \lambda^{-1})$, respectively. Let $S \subset T$ denote the compactified spectral curve obtained by extending (2.2) to T. The projection $T \to P_1(\mathbf{C})$ determines S as an r - sheeted branched covering of the Riemann sphere.

From $L(\lambda)$, one can define a sheaf E supported over S by

$$0 \to \mathcal{O}(-m)^{\oplus r} \stackrel{zI-L(\lambda)}{\longrightarrow} \mathcal{O}^{\oplus r} \to E \to 0 \ . \tag{2.3}$$

where $\mathcal{O}(j)$ is the pull-back to T of the unique line bundle of degree j over $\mathbf{P}_1(\mathbf{C})$. Under suitable genericity assumptions, e.g. when S is smooth and reduced, E is a line bundle of degree $(g+r-1)$, where

$$g \ = \ \text{genus}(S) \ = \tfrac{1}{2}[(r-1)(rm-2)] \ . \tag{2.4}$$

The sequence (2.3) imples

$$H^0(S,E) = \mathbf{C}^r \ , \ H^0(S, E\otimes \mathcal{O}(-1)) = 0 \ . \tag{2.5}$$

It is shown in **[AHH2]** how the solution $L(\lambda,t)$ of the Lax equation (0.1) corresponds to a linear flow of the sheaves E_t. More precisely, if $P(\lambda,\lambda^{-1},L(\lambda))$ is the polynomial which appears in (0.1), then

$$E_t = E_0 \otimes F_t \tag{2.6}$$

where F_t is the line bundle on T whose transition function from U_0 to U_1 is $\exp(tP(\lambda,\lambda^{-1},z))$.

Given the curve S and the flow E_t, one can obtain $L(\lambda,t)$ as follows. Let $\{\psi_i(t)|\}_{i=1,\ldots,r}$ be a basis of sections of E_t. Note that eq. (2.5) is satisfied for generic E. In the trivialisations of E_t corresponding to the decomposition (2.6), $\psi_{i,t}$ are essentially generalized Baker-Akhiezer functions (**[KN]**, **[Du]**). Let $\{z_j(\lambda)\}_{j=1,\ldots,r}$ be the z-coordinates of the points $\{s_j(\lambda)\}$ of S over λ; i.e., the eigenvalues of $L(\lambda)$. Assume for simplicity that the curve S has r points $\{s_j(\infty)\}$ over $\lambda=\infty$. Choose a trivialization of E_0 over U_1 and take a trivialization of F_t corresponding to the transition function $e^{tP(\lambda,\lambda^{-1},z)}$. This induces a trivialization of E_t allowing us to represent sections as functions on U_1. We define the $\psi_i(t)$, which are determined by their values at the r points over $\lambda=\infty$, by setting

$$\psi_i(t)(s_j(\infty)) = \delta_{i,j} \ . \tag{2.7}$$

Defining the matrix

$$\Psi_{ij}(\lambda,t) = \psi_{i,t}(s_j(\lambda)) \ , \tag{2.8}$$

we have

$$L(\lambda,t) = \Psi(\lambda,t)\text{diag}(z_j(\lambda)) \ \Psi(\lambda,t)^{-1} \ . \tag{2.9}$$

Note that the pair (S,E) only determines $L(\lambda,t)$ up to conjugation by a diagonal matrix since there is implicit in the solution (2.9) a choice of trivialisation of E_0.

The line bundle E_t can be represented by the divisor of zeroes of one of its sections. These sections are projections of the sections of $\mathcal{O}^{\oplus r}$ under (2.3). For example, we could choose a section ψ such that $\psi(s_j(\infty)) = \delta_{1j}$. If $L(\lambda,t)$ can be conjugated so that its highest order term in λ is $\text{diag}(z_j(\infty))$, then by (2.3) the divisor of ψ over the finite portion S_0 of S is cut out by the constraint

$$(L(\lambda,t) - zI)_{adj} \cdot (1,0,..,0)^T = 0 \ , \tag{2.10}$$

where M_{adj} denotes the matrix of cofactors of M, so that

$$M \cdot M_{adj} = \det(M) \cdot I \tag{2.11}$$

This condition cuts out g points $\{p_1(t), \dots, p_g(t)\}$ on S_0, and the linearity of the flow is expressed in the fact that on S

$$\sum_{\nu=1}^{g} \int_{p_0}^{p_\nu(t)} \omega_j = k_j t + c_j \ , \tag{2.12}$$

where p_0 is a base point, the ω_j form a basis for the holomorphic differentials on S, and the k_j, c_j are constants.

3. DARBOUX COORDINATES AND LIOUVILLE-ARNOLD INTEGRATION

The method of integration presented in section 2 was purely algebro-geometric in content. The Lax equations (0.1), (1.18a), however, are Hamiltonian with respect to the Lie-Poisson structure on $(\widetilde{gl(r)}^+)^*$. We now show how to derive the above results using this Hamiltonian structure.

To simplify the discussion, we only consider the case when $Y = 0$ and A is diagonalizable in eq.(1.18a), although the general case may also be treated with minor modifications. The flow then takes place on a coadjoint orbit $\mathcal{O}_\mathcal{N}$ of a rational element $\mathcal{N}(\lambda)$ of $(\widetilde{gl(r)}^+)^* = \widetilde{gl(r)}_-$ with simple poles at the eigenvalues $\{\alpha_i\}_{i=1,\dots n}$ of A:

$$\mathcal{N}(\lambda) = \lambda \sum_{i=1}^{n} \frac{R_i}{(\lambda - \alpha_i)} \ . \tag{3.1}$$

Proposition 3.1 ([**AHP**]). *The coadjoint orbit in* $\widetilde{gl(r)}_- \equiv (\widetilde{gl(r)}^+)^*$ *through* $\mathcal{N}(\lambda)$ *is given by*

$$\mathcal{O}_\mathcal{N} = \left\{ \lambda \sum_{i=1}^{n} \frac{g_i R_i g_i^{-1}}{\lambda - \alpha_i}, \quad g_i \in Gl(r, \mathbf{C}) \right\} . \tag{3.2}$$

We first consider the flow on a reduction of $\mathcal{O}_\mathcal{N}$ by the action of conjugation by $Gl(r, \mathbf{C})$. This corresponds to reducing the arbitrariness in the choice of basis at $\lambda = \infty$ implicit in the normalization of E_0 at ∞. The action is Hamiltonian with respect to the Kostant - Kirillov symplectic form ω_{orb} and the associated moment map is just $\mathcal{N}(\infty)$. The symplectic reduction of the orbits by this action consists of fixing the Jordan form of $\mathcal{N}(\infty)$, and quotienting by $Gl(r, \mathbf{C})$, giving a reduced space $(\mathcal{O}_\mathcal{N})_{red}$ with corresponding reduced symplectic form ω_{red}. For simplicity, we make the genericity assumption that the eigenvalues of $\mathcal{N}(\infty)$ are distinct. (See [**AHH4**]-[**AHH5**] for a discussion of how the approach may be generalized to allow for cases where this condition is not satisfied, e.g. in the coupled non-linear Schrödinger equation.)

Define the matricial polynomial

$$L(\lambda) = \lambda^{-1} a(\lambda) \mathcal{N}(\lambda) \ , \tag{3.3}$$

where

$$a(\lambda) = \prod_{i=1}^{n} (\lambda - \alpha_i) \ . \tag{3.4}$$

As in (2.2), let S be the spectral curve of $L(\lambda)$. We simplify the discussion further by supposing that the R_i are chosen so that the generic S on $\mathcal{O}_{\mathcal{N}}$ are smooth. Let us consider the divisor $\sum_{\nu=1}^{g} p_\nu$ of E determined by (2.10). Let $\{(z_\nu, \lambda_\nu)\}$ denote the coordinates in T of the points $\{p_\nu\}$, viewed as functions on the orbit $\mathcal{O}_{\mathcal{N}}$ and set

$$\zeta_\nu \equiv z_\nu / a(\lambda_\nu) \ . \tag{3.5}$$

THEOREM 3.2 (**[AHH3]**-**[AHH5]**). *The functions $(\lambda_\nu, \zeta_\nu)_{\nu=1,\dots,g}$ project to a set of Darboux coordinates for the reduced Kostant-Kirillov symplectic form ω_{red} on $(\mathcal{O}_{\mathcal{N}})_{red}$.*

$$\omega_{red} = \sum_{\nu=1}^{g} d\lambda_\nu \wedge d\zeta_\nu \ . \tag{3.6}$$

The reduced symplectic form ω_{red} has an algebro-geometric interpretation in terms of curves and line bundles. If one defines the weighted degree δ of a monomial $\lambda^j z^k$ by the rule

$$\delta(\lambda^j z^k) = j + k(n-1) \ , \tag{3.7}$$

and extends this in the natural way to polynomials, the polynomial f in the defining equation (2.2) of the spectral curve S satisfies

$$\delta(f) = r(n-1) \ . \tag{3.8}$$

Let $\mathcal{C}$ be the set of smooth compact curves S' in the linear system $|\mathcal{O}(r(n-1))|$ of S (i.e. cut out by polynomials with same degree δ), which coincide with S on the $r(n+1)$ points over $\{\lambda = \alpha_i, \ \infty\}$. By results of **[AHH2]**, **[AHH3]**, **[AHH5]**, the map $L(\lambda) \to (S, E)$ gives an isomorphism between an open dense set of the reduced coadjoint orbit $(\mathcal{O}_{\mathcal{N}})_{red}$ and the variety V of pairs:

$V = \{\ (S', E')$ with $S' \in \mathcal{C}$ and E' a line bundle of degree $g+r-1$ over S' satisfying (2.5)$\}$.

The space V is $2g$-dimensional, and $\mathcal{C}$ is g-dimensional (see **[AHH2]**, **[B]**). The map $\pi : V \to \mathcal{C}$ induces an exact sequence of tangent bundles:

$$0 \to TF \to TV \to \pi^* T\mathcal{C} \to 0 \ . \tag{3.9}$$

The fibres $\pi^{-1}(S)$ are naturally isomorphic to an open subvariety of $Pic^{g-1}(S)$, which is essentially the complement of the theta-divisor. Infinitesimal deformations of these line bundles are parametrised by $H^1(S, \mathcal{O})$. On the other hand, infinitesimal deformations of the curve S in T are given by elements of $H^0(S, \mathbf{N})$, where $\mathbf{N}$ is the normal bundle. These sections of N, however, are constrained to be zero at $\{\lambda = \alpha_i, \infty\}$, so the deformations of S correspond to elements of $H^0(S, \mathbf{N}(-n-1))$. By the adjunction formula **[GH]**, this is isomorphic to $H^0(S, K_S)$ where K_S is the canonical bundle of S. Thus, at a point (S, E) of V, (3.9) becomes

$$0 \to H^1(S, \mathcal{O}) \to TV \to H^0(S, K_S) \to 0 \ . \tag{3.10}$$

Now, suppose that we have extended E to a neighbourhood of S in T. This gives an ambient space line bundle whose restriction to any variation of S in T determines a cross section of $\pi : V \to \mathcal{C}$, and hence a splitting of (3.10):

$$TV \cong H^0(S, K_S) \oplus H^1(S, \mathcal{O}) \ . \tag{3.11}$$

Since these two summands are Serre duals, we can define a natural skew form W on TV by

$$W((a,v),(b,w)) = a(w) - b(v) \ , \tag{3.12}$$

where $a, b \in H^0(S, K_S)$ and $v, w \in H^1(S, \mathcal{O})$. Although the extension of E to a neighbourhood of S in T is not unique, all such extensions give the same form W. The relationship between the symplectic geometry of coadjoint orbits and the algebraic geometry of spectral curves is given by the following.

THEOREM 3.3 (**[AHH3]-[AHH5]**). *Under the identification between the reduced coadjoint orbit $(\mathcal{O}_{\mathcal{N}})_{red}$ and V, W coincides with the reduced Kostant-Kirillov form ω_{red}.*

The Darboux coordinates of Theorem 3.2 can be used to obtain the linearization (2.12) for the AKS flows. Let f be the polynomial in eq.(2.2) defining one of the curves in $\mathcal{C}$. The other curves in $\mathcal{C}$ form an open dense set within the space of curves defined by equations of the form

$$f_R(\lambda, z) = f(\lambda, z) + a(\lambda)R(\lambda, z) = 0 \ , \tag{3.13}$$

where the polynomial $R(\lambda, z)$ satisfies

$$\delta(R) \leq (r-1)(n-1) - 2 \ . \tag{3.14}$$

There is a g-dimensional linear space of such polynomials $R(\lambda, z)$. Let $\{P_1, ..., P_g\}$ denote a choice of independent coefficients determining R. The AKS Hamiltonians on the orbit $\mathcal{O}_{\mathcal{N}}$ giving Lax pairs (1.18a) generating nontrivial flow on the reduced space $(\mathcal{O}_{\mathcal{N}})_{red}$ are functions $h(P_1, .., P_g)$, and hence the $\{P_\mu\}$ provide a complete set of constants of motion.

One can obtain the coordinates $\{Q_1, .., Q_g\}$ conjugate to the $\{P_\mu\}$ by constructing a Liouville generating function for the appropriate canonical transformation, starting from the coordinates (λ_ν, ζ_ν). Let $\zeta(\lambda, P_1, \cdots, P_g)$ be the algebraic function defined implicitly by eq.(3.13). From Theorem 3.2 it follows that the appropriate generating function G may be taken as

$$G(\lambda_\nu, P_\mu) = \sum_{i=1}^{g} \int_{\lambda_0}^{\lambda_\nu} \zeta(\lambda, P_1, \cdots, P_g) d\lambda \ . \tag{3.15}$$

Thus,

$$Q_\mu \equiv \frac{\partial G}{\partial P_\mu} = -\sum_{i=1}^{g} \int_{\lambda_0}^{\lambda_\nu} (\frac{\partial f_R/\partial P_\mu)}{(\partial f_R/\partial \zeta)} d\lambda \tag{3.16}$$

gives us the required conjugate coordinates in which the flow becomes linear. As a consequence of the Poincaré residue formula **[GH]**, the integrands in (3.16) form a basis of the holomorphic differentials on the curve. We have therefore rederived formula (2.12) and obtained the integration of the Lax equation in purely Hamiltonian terms, at least on the reduced space.

This derivation of the flow can also be extended to the non-reduced space as follows. Let

$$L(\lambda) = L_0\lambda^{n-1} + L_1\lambda^{n-2} + ... + L_{n-1} \tag{3.17}$$

be the polynomial corresponding under (3.3) to the element $\mathcal{N}(\lambda)$ of the coadjoint orbit. We can define an invariant submanifold $\mathcal{O}_d \subset \mathcal{O}_{\mathcal{N}}$ of the coadjoint orbit by the constraint:

$$(L_0)_{ij} = 0 \quad \text{if} \quad i \neq j \ . \tag{3.18}$$

This submanifold is symplectic over the open set of L_0's with distinct eigenvalues $(L_0)_{ii}$. Since L_0 is a constant of motion and the equation (1.18a) determining the flow is invariant under conjugation, the restriction to $\mathcal{O}_d$ involves very little loss of generality. In fact, cases when some eigenvalues are equal can also be dealt with (**[AHH4]**, **[AHH5]**).

We define functions $\{p_i, q_i\}_{i=1,\dots,r}$ on $\mathcal{O}_d$ by

$$\begin{aligned} p_i &\equiv (L_0)_{ii} \\ q_i &\equiv \ln((L_1)_{1i} + \frac{1}{2}\sum_{j\neq 1,i} ln(p_j - p_i)\ . \end{aligned} \tag{3.19}$$

Since the trace of L_0 is constant over the orbit $\mathcal{O}_{\mathcal{N}}$, p_1 is determined by $p_2, .., p_r$.

THEOREM 3.4 (**[AHH5]**). *The restriction of the Kostant-Kirillov form ω_{orb} to $\mathcal{O}_d$ is*

$$(\omega_{orb})|_{\mathcal{O}_d} = \sum_{\nu=1}^{g} d\lambda_\nu \wedge d\zeta_\nu \ + \ \sum_{i=2}^{r} dq_i \wedge dp_i\ . \tag{3.20}$$

Thus, the functions $\{\lambda_\nu, \zeta_\nu, p_i, q_i\}_{\nu=1,\dots g, i=2,\dots r}$ form a Darboux coordinate system on $\mathcal{O}_d$. Again, we can give an algebro-geometric interpretation of the symplectic form (3.20). Let $\mathcal{C}'$ be the set of deformations in T of one of the spectral curves S_0 of the orbit which coincide with S_0 at $\lambda = \alpha_i$ (but not necessarily at $\lambda = \infty$) and which are smooth. $\mathcal{O}_d$ has an open dense submanifold isomorphic to the set:

> $V' = \{\ (S', E')$ with $S' \in \mathcal{C}'$ and E' an equivalence class of pairs (E, τ) , E a line bundle of degree $g + r - 1$ over S' satisfying (3.2) and τ a trivialisation of E over $\lambda = \infty\}$.

As above, the tangent space can be decomposed as

$$TV' \cong H^0(S, K_S(1)) \oplus H^1(S, \mathcal{O}(-1))\ , \tag{3.21}$$

the summands corresponding to infinitesimal deformations of S' and E', respectively. The two terms are again Serre duals, so we may define, as in eq. (3.12), a 2-form W' on V' by

$$W'((a, v), (b, w)) = a(w) - b(v)\ , \tag{3.22}$$

where $a, b \in H^0(S, K_S(1))$ and $v, w \in H^1(S, \mathcal{O}(-1))$. We then have:

THEOREM 3.5 (**[AHH5]**). *Under the identification betweeen the invariant submanifold $\mathcal{O}_d \subset \mathcal{O}_{\mathcal{N}}$ and V', we have*

$$W' = (\omega_{orb})|_{\mathcal{O}_d} \ + \ \frac{1}{2}\sum_{\substack{i,j=1 \\ i\neq j}}^{r} \frac{dp_i \wedge dp_j}{p_i \ - \ p_j}\ . \tag{3.23}$$

Thus, W' and the restricted symplectic form $\omega_{orb}|_{\mathcal{O}_d}$ differ only by a form involving constants of motion. For the AKS class of Hamiltonians they therefore give the same flows, up to a choice of integration constants.

It follows from Theorem 3.4 that we can also obtain the Darboux coordinates conjugate to $\{P_\nu, p_i\}_{\nu=1,\dots g, i=2,\dots r}$ from the Liouville generating function

$$G'(\lambda_\nu, q_i, P_\mu, p_j) = \sum_{\nu=1}^{g} \int_{\lambda_0}^{\lambda_\nu} \zeta(\lambda, P_\mu, p_i) d\lambda \ + \sum_{i=2}^{r} p_i q_i\ . \tag{3.24}$$

Thus,

$$Q_\mu \equiv \frac{\partial G}{\partial P_\mu} = \sum_{i=1}^{g} \int_{\lambda_0}^{\lambda_\nu} \frac{\partial f_P / \partial P_\mu}{(\partial f_P / \partial z)} \, d\lambda \tag{3.25}$$

$$\tilde{q}_i \equiv \frac{\partial G}{\partial p_i} = \sum_{i=1}^{g} \int_{\lambda_0}^{\lambda_\nu} \frac{\partial f_P / \partial p_i}{(\partial f_P / \partial z)} \, d\lambda \ + \ q_i \tag{3.26}$$

defines our conjugate coordinates, in which the flows generated by the AKS ring of Hamiltonians linearize. The integrands in (3.25) are holomorphic differentials on the curve, while those in (3.26) have simple poles over ∞. The integrals in (3.26) can be computed from the ones in (3.25) in terms of quotients of Θ - functions by use of a standard reciprocity formula (**[GH]**, **[AHH5]**).

The application of these results, both to the integration of finite dimensional systems and to the finite dimensional quasi-periodic solutions of integrable PDE's such as the coupled cubically nonlinear Schrödinger equation may be found in refs. **[AHH4]**, **[AHH5]**.

References

[AHH1] Adams, M., Harnad, J. and Hurtubise, J., "Dual Moment Maps to Loop Algebras", *Lett. Math. Phys.* (in press, 1990)

[AHH2] Adams, M., Harnad, J. and Hurtubise, J., "Isospectral Hamiltonian Flows in Finite and Infinite Dimensions II. Integration of Flows", (CRM Preprint 1635 (1989)), *Commun. Math. Phys.* (in press, 1990) .

[AHH3] Adams, M., Harnad, J. and Hurtubise, J., "Coadjoint Orbits, Spectral Curves and Darboux Coordinates", preprint CRM-1638 (1989), to appear in: *MSRI Workshop on the Geometry of Hamiltonian Systems*, MSRI Publ., (ed. T. Ratiu), Berkeley, (1990)

[AHH4] Adams, M., Harnad, J. and Hurtubise, J., "Liouville Generating Function for Isospectral Flow in Loop Algebras", preprint CRM-1652 (1990), to appear in: *Integrable and Superintegrable Systems*, (ed. B. Kupershmidt), World Scientific, Singapore (1990)

[AHH5] Adams, M., Harnad, J. and Hurtubise, J., "Darboux Coordinates and Liouville-Arnold Integration in Loop Algebras", (CRM preprint, in preparation).

[AHP] Adams, M., Harnad, J. and Previato, E., "Isospectral Hamiltonian Flows in Finite and Infinite Dimensions I. Generalised Moser Systems and Moment Maps into Loop Algebras", *Commun. Math. Phys.* **117**, 451-500 (1988).

[AvM1] Adler, M. and van Moerbeke, P., "Completely Integrable Systems, Euclidean Lie Algebras, and Curves", *Adv. Math.* **38**, 267-317 (1980); "Linearization of Hamiltonian Systems, Jacobi Varieties and Representation Theory", *ibid.* **38**, 318-379 (1980).

[AvM2] Adler, M. and van Moerbeke, P., "The Kowalewski and Hénon-Heiles Motions as Manakov Geodesic Flows on $SO(4)$ - A Two Dimensional Family of Lax Pairs", *Commun. Math. Phys* **113**, 659-700 (1988).

[B] Beauville, A., "Jacobiennes des courbes spectrales et systèmes Hamiltoniens complètement intégrables", (Preprint, 1989).

[BRS] Bobenko, A.I., Reyman, A.G., and Semenov-Tian-Shansky, M.A., "The Kowalewski Top 99 Years Later: A Lax Pair, Generalizations and Explicit Solutions", *Commun. Math. Phys.* **122**, 321-354 (1989).

[D] Dickey, L.A., "Integrable Nonlinear Equations and Liouville's Theorem I, II", *Commun. Math. Phys.*, **82**, 345-360 (1981); *ibid.* **82**, 360-375 (1981).

[Du] Dubrovin, B.A., "Theta Functions and Nonlinear Equations", *Russ. Math. Surv.*, **36**, 11-92 (1981).

[FNR] Flaschka, H., Newell, A.C. and Ratiu, T., "Kac-Moody Algebras and Soliton Equations II. Lax Equations Associated to $A_1^{(1)}$", Physica **9D**, 300 (1983).

[GD] Gel'fand. I.M. and Dik'ii, L.A., " Integrable Nonlinear Equations and the Liouville Theorem", Funct. Anal. Appl. **13**, 8-20 (1979).

[GH] Griffiths, P. and Harris, J., *Principles of Algebraic Geometry*, Wiley, New York, 1978.

[Hi] Hitchin, N., "On the Construction of Monopoles" *Commun. Math. Phys.* **89**, 145-190 (1983).

[HH] Harnad, J. and Hurtubise, J., "Generalized Tops and Moment Maps into Loop Algebras", preprint CRM (1990).

[HHM] Harnad, J., Hurtubise, J. and Marsden, J., "Reduction of Hamiltonian Systems with Discrete Symmetry", (preprint CRM (1990), in preparation).

[HW] Harnad, J. and Wisse, A., "Darboux Coordinates for Stationary Manifolds of Matrix Nonlinear Schrödinger Equations", (preprint CRM (1990), in preparation)

[KN] Krichever, I.M., Novikov, S.P., "Holomorphic Bundles over Algebraic Curves and Nonlinear Equations", *Russ. Math. Surveys* **32**, 53-79 (1980).

[M] Moser, J., "Geometry of Quadrics and Spectral Theory", *The Chern Symposium, Berkeley, June 1979*, 147-188, Springer, New York, (1980).

[Ra1] Ratiu, T. "The C. Neumann Problem as a Completely Integrable System on an Adjoint Orbit", Trans. AMS **439**, 321-329 (1981).

[Ra2] Ratiu, T. "The Lie Algebraic Interpretation of the Rosochatius System" in: Mathematical Methods in Hydrodynamics and Integrability in Dynamical Systems, AIP Conf. Proc. **88** , La Jolla (1981).

[RS1] Reiman, A.G., and Semenov-Tian-Shansky, M.A., "Reduction of Hamiltonian systems, Affine Lie algebras and Lax Equations I, II", *Invent. Math.* **54**, 81-100 (1979); *ibid.* **63**, 423-432 (1981).

[RS2] Reyman, A.G., and Semenov-Tian-Shansky, M.A., "A New Integrable Case of the Motion of the 4-Dimensional Rigid Body", *Commun. Math. Phys.* **105**, 461-472 (1986).

[vMM] van Moerbeke, P. and Mumford, D., "The Spectrum of Difference Operators and Algebraic Curves" *Acta Math.* **143**, 93-154 (1979).

[W] Weinstein, A., "The Local Structure of Poisson Manifolds", *J. Diff. Geom.* **18**, 523-557 (1983).

AN EXOTIC SYMPLECTIC R^4

L. M. BATES

Department of Mathematics, University of Calgary
Calgary, Alberta T2N 1N4, Canada

ABSTRACT

There once was a man named Darboux
Who asked 'What can symplectic forms do?'
In the small he could see
How to find q and p
But globally it need not be true.

In this article we describe a method for constructing an exotic symplectic geometry, and then use the resulting symplectic manifold as the phase space of a completely integrable Hamiltonian system. The main problem that we are trying to address is how to understand just what it means for a symplectic geometry to be exotic. So our point of view is to try to use the completely integrable Hamiltonian system as a tool to understand exotic geometry. For example, when an integrable equation, containing a homoclinic orbit, is perturbed, generally speaking, the homoclinic orbit is broken. This allows frequent crossings of the homoclinic orbit of the unperturbed system which, in turn, may lead to horseshoes and chaos. In two dimensional systems a homoclinic orbit naturally divides phase space into two regions, inside and outside the homoclinic orbit. A crossing of the homoclinic orbit is identified as an orbit passing from inside the homoclinic orbit to its outside, or vice versa. Melnikov's method allows one to measure the distance between the stable and unstable manifolds, broken under the perturbation and gives precise information regarding homoclinic crossings.

It all begins with a theorem of Darboux that appeared in 1882.

THEOREM. About each point x in a symplectic manifold (M,ω) there exists a local coordinate chart $\{q^a, p_b\}$ so that the local expression of ω is

$$\omega = dq^a \wedge dp_a.$$

This theorem follows immediately from his results on the local canonical form for one-forms. Probably the nicest proof of this theorem is to use the method of Lie transforms, as in Moser [4].

This was pretty much the state of affairs until the following fantastic theorem of Gromov [3].

THEOREM. There exist exotic symplectic geometries on R^{2n} for each $n \geq 2$.

Let us say that a symplectic geometry is exotic if it does not embed in the standard one of the same dimension. In particular, this tells us that we can not find a global canonical chart for an exotic symplectic geometry. This is fascinating because the obstruction sits in the geometry of the symplectic form, and not in the topology of the underlying manifold.

Gromov proves this theorem by using his h-principle. Roughly speaking, one first constructs an algebraic section of the jet bundle that one is trying to solve the partial differential relation in, and then homotopes the algebraic section to a holonomic one. The drawback of such a technique is that it is non-constructive, so one doesn't know what the solution looks like.

Now we try to be a little more precise and so we have the following

DEFINITION. Let L be a closed (compact, no boundary) Lagrangian submanifold of (R^{2n}, ω). L is an exact Lagrangian submanifold if

$$0 = [\omega] \in H^2(R^{2n}, L; R).$$

This means that if we pick any one-form θ so that $d\theta = \omega$, then $[\theta|_L] \in H^1(L; R)$ is well-defined and it vanishes. Gromov has proved the following

THEOREM. There are no closed exact Lagrangian submanifolds in the standard symplectic R^{2n}.

In the plane this is easy to see because

$$\begin{aligned} \int_L \theta_o &= \int_{\text{int}\, L} \omega_o \\ &= \text{area enclosed by } L \\ &\neq 0 \end{aligned}$$

Here we are using θ_o and ω_o for the canonical one- and two-forms on R^{2n}.

Gromov proves this theorem from general results on pseudo-holomorphic curves, but it can be proved by the following approach of Sikorav [6].

THEOREM. Let L be exact Lagrangian in (R^{2n}, ω_o). Let g be defined on $D^2 = \{z \mid |z| < 1\}$ so that $\partial g = 0$ and g has boundary values. Then there exists an f such that

$$f : (D^2, \partial D) \longrightarrow (R^{2n}, L)$$

and $\bar{\partial} f = g$.

The proof of this theorem is actually very readable.

Note that $\partial\bar{\partial} f = 0$ so that f is harmonic. Now estimate

$$||g(0)|| = ||\bar{\partial} f(0)|| \leq 2 \cdot \text{diam}\, L.$$

The inequality follows from using the Poisson formula and differentiating under the integral sign. This gives us that there are no exact Lagrangian submanifolds in the standard R^{2n} because $g(0)$ is arbitrary.

We should also mention an approach due to Viterbo [7]. He has shown that the non-existence of closed exact Lagrangian submanifolds implies the existence of an exotic symplectic geometry with infinite volume. The approach is to map $S^1 \times S^1$ in R^4 to a three-lobed figure eight crossed with itself. The three-lobed figure eight crossed with itself is designed so that it is an immersed exact Lagrangian torus. Viterbo then shows that this immersion can be extended to a surjective immersion of all of R^4 to itself. The required symplectic form is then obtained by pulling back the standard one via the immersion. The only drawback of this approach is that it relies on the Hirsch-Smale immersion theory, and so is not as constructive as we would wish.

About the only other existence result on exotic symplectic geometries is the following theorem of Muller [5].

THEOREM. There exists an exotic symplectic R^6 with a Lagrangian S^3.

The proof of this result also uses the h-principle.

Our plan is to produce an explicit example of a symplectic form ω on a manifold diffeomorphic to R^4 so that

$$0 = [\omega] \in H^2(R^4, T^2; R).$$

Here T^2 is a two-torus.

The idea is to think of R^3 sitting as a plane in R^4 as in the picture.

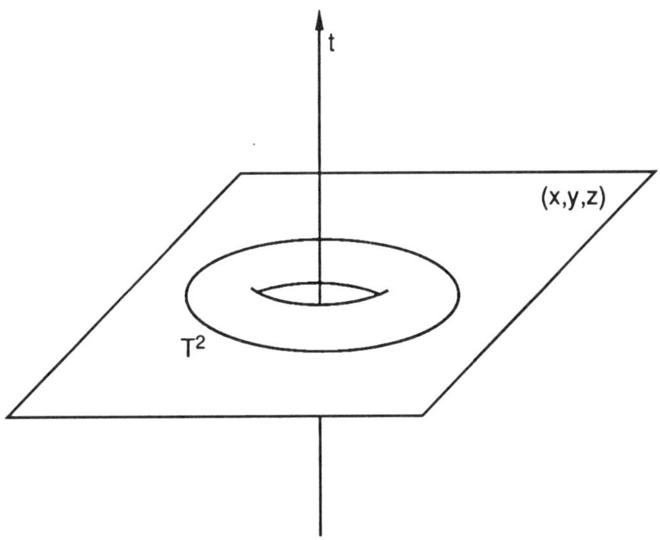

We also think of the torus sitting in the three-plane. Let (x, y, z) be the usual coordinates on R^3 and t be the fourth coordinate. Let

$$\theta = \alpha + \beta t$$

be a one-form with α and β one-forms on R^3. So we may think of θ as a one-form on R^4. We would like θ to be a potential for an exotic symplectic structure. So we compute

$$d\theta = d\alpha + td\beta + \beta \wedge dt,$$

so that

$$d\theta \wedge d\theta = 2d\alpha \wedge \beta \wedge dt + O(t).$$

We get an exotic symplectic structure in the neighbourhood of the $t = 0$ plane if we can make $d\alpha \wedge \beta \neq 0$ on R^3. Set $\beta = *d\alpha$ where $*$ is the Hodge dual for the usual Euclidean inner product. So our problem is solved if we can find an α so that $\alpha|_{T^2} = 0$

and $d\alpha \neq 0$. Note that this construction is just the co-isotropic embedding theorem of Gotay, Nester and Hinds [2].

We leave the construction of such an α as an exercise in vector calculus for the reader. However, we will provide a hint. The problem is easy to solve in toroidal coordinates if one tries to write down a formula using the picture as a guide.

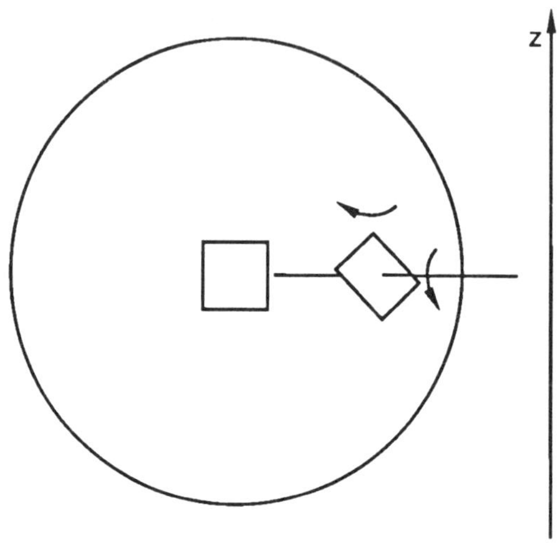

The square elements in the picture represent the kernel of the one-form.

Curiously enough, the analytic formulae are actually easier to write down if we adopt a slightly different approach and instead of working on R^3 work on S^3 instead. This is alright as when we are finished we can push everything over to R^3 by stereographic projection (see Bates and Peschke [1]).

As an example, consider the one-form defined by

$$\alpha = \cos r^2 d\theta + \cos s^2 d\phi$$

where (r, θ) and (s, ϕ) are two sets of polar coordinates in the plane. Then it is a straightforward calculation to show that $d\alpha$ is a presymplectic form on the three-sphere defined by $r^2 + s^2 = \pi$. Furthermore, α vanishes on the torus defined by $r^2 = \pi/2$, $s^2 = \pi/2$. This gives, via the co-isotropic embedding, a symplectic form on $S^3 \times I$ with an exact Lagrangian torus.

It is again a simple calculation to show that both the symplectic form and the potential just constructed are invariant under the action of the torus group given by rotations in the (r, θ) plane and rotations in the (s, ϕ) plane. Hence the momentum map for the action is computed by contracting the potential by the vector fields

$$\partial/\partial\theta \quad \text{and} \quad \partial/\partial\phi.$$

This gives us our symplectic manifold as a pinched two-torus bundle over the image of the momentum map. In this way we have an exotic model of the harmonic oscillator if we restrict ourselves to an energy surface given by $r^2 + s^2 = \text{const}$. It is far from clear, however, at this stage just what we should be looking for to see just what exotic means.

Several phenomena can be seen by constructing examples along this line. They can all be produced so that they are invariant under the same action of the torus group. To explain this, recall that the usual action of the torus on the three-sphere has two singular fibres, which are linked. The foliation by tori is regular away from these singular fibres, and can be parametrized by an interval. Away from the singular fibres of the foliation, the tori are framed by the action of the torus group. One thing that can happen is that the Hamiltonian vector field can be made to wind around this framing as we move from one singular fibre to another, in contrast to the standard case where it is fixed. Another thing that can happen is the the linking number of the two degenerate fibres can be changed from $+1$ to -1. So one can see that much work remains to understand the connections between the blow-up of Darboux charts, the gross features of symplectic geometry and the geometry of integrable Hamiltonian systems.

REFERENCES

1. L. Bates and G. Peschke, *A remarkable symplectic structure*, J. Diff. Geom., (to appear)

2. M. Gotay, J. Nester and G. Hinds, *Presymplectic manifolds and the Dirac-Bergmann theory of constraints*, J. Math. Phys. **19** (11) 2378-2399, 1978.

3. M. Gromov, *Pseudo-holomorphic curves in symplectic manifolds*, Invent. math. **82** (1985),307-347.

4. J. Moser, *On the volume elements on a manifold*, Trans. Am. Math. Soc. **120**: 296-294,1965.

5. M.-P. Muller, *une structure symplectique complete sur R^6 avec une sphere Lagrangienne plongee*, thesis, Strasbourg, 1989.

6. J.-C. Sikorav,*Non-existence de sous-variété Lagrangienne exacte dans C^n (d'après Gromov)*, Travaux en cours 25, Hermann, Paris, 1987.

7. C. Viterbo *Lagrangian embeddings and symplectic structures*, Actes du colloque en l'honneur de R. Couty, Limoges, 1984.

ANOTHER EXAMPLE OF A τ – FUNCTION

L.A. DICKEY

Department of Mathematics, University of Oklahoma
Norman, OK 73019, USA

A remarkable topic in the theory of integrable systems is its relation to the theory of representations of Kac-Moody Lie algebras. The central point in this connection is the τ-function discovered by Japanese mathematicians Hirota, Sato and others (see [1] and an important development of this theory by Segal and Wilson [2]). However, the τ-function was introduced not for every integrable system but only for the KP hierarchy (and its multi-component variant) and for its reductions. Not less important is a hierarchy generated by a matrix first-order differential operator

$$L = \partial + U - zA, \ \ \partial = d/dx.$$

This hierarchy is in some way related to the multi-component KP but does not coincide with it. For $n = 2$ (where n is the matrix dimension) this hierarchy is known as AKNS ([3]). A most important contribution to its theory was made by Dubrovin [4]; therefore we shall call it AKNS-D.

This work does not have a final character, we try here to describe what is clear now and what remains an open problem so far. First we summarize well-known facts about the KP hierarchy.

1. KP-HIERARCHY

1a. Let

$$L = \partial + u_1\partial^{-1} + u_2\partial^{-2} + \cdots, \ \ \partial = d/dx$$

be a pseudo differential operator (PDO). Put $B_k = L^k_+$ where $+$ means the positive part of a PDO). The KP hierarchy is the set of equations

$$\partial_k L = [B_k, L], \ \ \partial_k = \partial/\partial t_k$$

and t_k are some variables. It can be proved that these equations commute, $\partial_k\partial_l = \partial_l\partial_k$ which implies an equivalent set of equations

$$\partial_k B_l - \partial_l B_k + [B_l, B_k] = 0.$$

The variable x is one of t_i, namely, $x = t_1$:

$$\partial = \partial_1.$$

1b. *Baker (wave) function.* Let

$$L = \hat{w}\partial\hat{w}^{-1}$$

where $\hat{w} = \sum_0^\infty w_j\partial^{-j}, w_0 = 1$. The Baker function is

$$w = \hat{w}\exp\sum_1^\infty z^k t_k = \sum_1^\infty w_j z^{-j}\exp\sum_1^\infty z^k t_k.$$

This function satisfies equations

$$Lw = zw, \quad \text{and} \quad \partial_k w = B_k w$$

(the latter can also be written as $\partial_k \hat{w} = -L_-^k \hat{w}$).

Put $w^* = (\hat{w}^*)^{-1} \exp(-\sum_1^\infty z^k t_k)$ where $\hat{w}^*$ is a formal adjoint operator, then $L^* w^* = zw^*$, w^* is called the *adjoint Baker function.*

1c. *Bilinear identity.* The relation

$$\text{res}_{z=\infty} w(t', z) \cdot w^*(t, z) = 0$$

holds. This can also be written as

$$\text{res}_{z=\infty} \partial_{k_1} \cdots \partial_{k_s} w(t, z) \cdot w^*(t, z) = 0.$$

This identity is characteristic for Baker functions: if two formal series w and w^* satisfy this equality for any $t_1, t_2, ...$ then they are the Baker function and the adjoint one for some L that is a solution to the KP hierarchy.

1d. *τ-function.* A function $\tau(t)$ of $t_1, t_2, ...$ exists such that

$$\hat{w}(t_1, t_2, ...) = \tau(t_1 - 1/z, t_2 - 1/2z^2, t_3 - 1/3z^3, ...)/\tau(t_1, t_2, ...).$$

2. MULTI-COMPONENT KP HIERARCHY

2a. Now

$$L = A\partial + u_1 \partial^{-1} + \cdots$$

where u_i are $n \times n$ matrices, $A = \text{diag}(a_1, ..., a_n)$, a_i are distinct non-zero constants (Japanese authors take $A = I$ which seems to be an unfortunate choice). Let $R_\alpha = \sum_{j=0}^\infty R_{j\alpha} \partial^{-j}, \alpha = 1, ..., n$; where $R_{0\alpha} = E_\alpha$ which is a matrix with only non-vanishing element 1 on the (α, α) place, and

$$[L, R_\alpha] = 0.$$

It can be shown that such matrices exist and their elements are differential polynomials in elements of u_i being

$$[R_\alpha, R_\beta] = 0, \quad R_\alpha R_\beta = \delta_{\alpha\beta} R_\alpha, \quad \sum R_\alpha = I.$$

Let $B_{k\alpha} = (L^k R_\alpha)_+$. The mcKP hierarchy (multicomponent KP) is

$$\partial_{k\alpha} L = [B_{k\alpha}, L], \quad \partial_{k\alpha} R_\beta = [B_{k\alpha}, R_\beta]$$

which is equivalent to the system

$$\partial_{k\alpha} B_{l\beta} - \partial_{l\beta} B_{k\alpha} + [B_{l\beta}, B_{k\alpha}] = 0.$$

The variables x and $t_{k\alpha}$ are not independent:

$$\partial = \sum_\alpha a_\alpha^{-1} \partial_{1\alpha}$$

2b. *Baker function.* Let

$$L = \hat{w} A \partial \hat{w}^{-1},\ \hat{w} = \sum_0^\infty w_i \partial^{-i}, w_0 = I;$$

Then $R_\alpha = \hat{w} E_\alpha \hat{w}^{-1}$. Put

$$w = \hat{w} \exp(\sum_{k=1}^{\infty} \sum_{\alpha=1}^{n} z^k E_\alpha t_{k\alpha}).$$

This Baker function satisfies the equations

$$Lw = zAw, \quad \text{and} \quad \partial_{k\alpha} w = B_{k\alpha} w.$$

The latter equation is equivalent to

$$\partial_{k\alpha} \hat{w} = -(L^k R_\alpha)_- \hat{w}.$$

The adjoint Baker function is $w^* = (\hat{w}^*)^{-1} \exp(-\sum\sum z^k E_\alpha t_{k\alpha})$. It satisfies the equations

$$L^* w^* = zAw^*, \qquad \partial_{k\alpha} w^* = B^*_{k\alpha} w^*.$$

2c. *Bilinear identity.* In this case it has a form:

$$\text{res}\, \partial_{k_1\alpha_1} \cdots \partial_{k_s\alpha_s} w \cdot w^{*t} = 0$$

where the superscript t means the transpose.

2d *τ-function.* Here the τ-function depends on indices:

$$\hat{w}_{\alpha\beta}(t, z) = \tau_{\alpha\beta}(..., t_{k\beta} - 1/kz^k, ...)/\tau(t)$$

(we emphasize that $1/kz^k$ is subtracted only from $t_{k\beta}$ with the same β as in $\tau_{\alpha\beta}$). Moreover, $\tau = \tau_{\alpha\alpha}$ which is independent of α.
As to this formula, we refer to an article by K.Ueno and K. Takasaki [5] who in their turn refer to M.Sato [6]. In that paper the formula was not written but it was said that it could be derived similarly to the case of the simple KP, ***mutatis mutandi***. The latter must probably mean the dependence of the τ-function on two indices.

3. AKNS-D HIERARCHY

3a. We have

$$L = \partial + U - zA, \qquad A = \text{diag}\,(a_1, ..., a_n) = const,\ U_{\alpha,\alpha} = 0$$

with distinct a_i. *Resolvents* are series $R = \sum_{i_0}^\infty R_j z^{-j}$ that commute with L: $[L, R] = 0$. Their elements are differential polynomials in elements of the matrix U. Resolvents form a n-dimensional commutative algebra over the field of constant diagonal series $c(z) = \sum_{i_1}^\infty c_i z^{-i}$. Basis is formed of n resolvents $R_\alpha = \sum_0^\infty R_{j\alpha} z^{-j}$ where $R_{0\alpha} = E_\alpha$. They satisfy relations $R_\alpha R_\beta = \delta_{\alpha\beta} R_\alpha$.
Now let

$$B_{k\alpha} = (z^k R_\alpha)_+ = \sum_{j=0}^{k} R_{j\alpha} z^{k-j}.$$

Here the subscript + means taking positive powers of z. The hierarchy is the set of equations:

$$\partial_{k\alpha} L = [B_{k\alpha}, L], \qquad \partial_{k\alpha} R_\beta = [B_{k\alpha}, R_\beta]$$

which yields

$$\partial_{k\alpha} B_{l\beta} - \partial_{l\beta} B_{k\alpha} + [B_{l\beta}, B_{k\alpha}] = 0.$$

Here we also have a dependence between x and $(t_{1\alpha})$:

$$\partial = -\sum_\alpha a_\alpha \partial_{1\alpha}.$$

We see that these formulas are very similar to those in the case of mcKP (sect.2). But there is also an important distinction: in that case R_α were PDO, so were $B_{k\alpha}$, now they are matrix series in z.

3b. *Baker function.* Let

$$L = w\partial w^{-1}, \qquad R_\alpha = wE_\alpha w^{-1}$$

which implies $L(w) = 0$. Put

$$w = \hat{w}(z)\exp(\sum_{k=0}^{\infty}\sum_{\alpha=1}^{n} z^k E_\alpha t_{k\alpha}), \qquad \hat{w}(z) = \sum_0^\infty w_j z^{-j}.$$

The equations of the hierarchy are equivalent to

$$\partial_{k\alpha} w = B_{k\alpha} w, \qquad \text{or} \qquad \partial_{k\alpha}\hat{w} = -(z^k R_\alpha)_- \hat{w}.$$

This time the adjoint Baker function is $w^* = (w^{-1})^t$ and satisfies the equation $L^* w^* = 0$ where $L^* = -\partial + (U + zA)^t$.

3c. *Bilinear identity.*

$$\text{res}\, z^l (\partial_{k_1\alpha_1} \cdots \partial_{k_s\alpha_s} w \cdot (w^*)^t) = 0, \quad l = 0, 1, \ldots$$

This differs from the corresponding relation in sect.2c by z^l which implies an important corollary. Every Baker function of the AKNS-D hierarchy is at the same time a Baker function of the mcKP hierarchy. However operators of these hierarchies are constructed in different ways: for the mcKP this is $L = \hat{w}(\partial)A\partial\hat{w}^{-1}(\partial)$ where $\partial = \sum a_\alpha^{-1}\partial_{1\alpha}$, and for the AKNS-D this is $L = w\partial w^{-1}$ where $\partial =$
$suma_\alpha \partial_{1\alpha}$.
In fact an alternative holds: if a Baker function satisfies the strong bilinear identity, with z^l, then it yields a solution to AKNS-D but the solution to mcKP which can be constructed with the help of this function is trivial, $L = \partial$. Thus, either a solution of AKNS-D or a non-trivial solution of mcKP; both of them cannot exist together (it may happen that neither of them exist: for triviality of solution for mcKP it suffices that the bilinear identity holds for solely one value of l, namely, $l = 1$).

3d. *τ-function.* There are various possibilities to define a τ-function.

I. Since a Baker function for this hierarchy is at the same time a Baker function for the mcKP the formula of Sato-Ueno must hold. Of course, we have an additional condition for τ-functions which follow from the stronger bilinear identity.

II. There is a τ-function suggested by Flaschka, Newell, and Ratiu [7] (in the case $n=2$, but we can formulate this in general case). Let

$$F_{k_1\alpha_1,k_2\alpha_2} = \text{tr res}\,(\partial_z B_{k_1\alpha_1})z^{k_2}R_{\alpha_2}$$

Then it is easy to check that the quantity $\partial_{k_3\alpha_3}F_{k_1\alpha_1,k_2\alpha_2}$ is symmetrical with respect to (k_i,α_i). This implies the existence of a function $f(t)$ such that

$$F_{k_1\alpha_1,k_2\alpha_2} = \partial_{k_1\alpha_1}\partial_{k_2\alpha_2}f(t).$$

It remains to denote $f(t) = \log\tau(t)$.
The existence of such a function is a very interesting fact that should be studied in detail. Using reasoning close to that in [9] it can be proved that it is the same τ–function as in sect. 2d ($\tau = \tau_{\alpha\alpha}$)

III. There is another possibility: to construct a τ-function in the Segal-Wilson style.

IIIa. Let H be $L^2(S,\mathbf{C}^n)$, a space of series $v(z) = \sum_{-\infty}^{\infty} v_k z^k$ where $v_k \in \mathbf{C}^n$, $|z|=1$, let H_+ and H_- be spaces of truncated series: $H_+ = \{\sum_0^\infty\}$, $H_- = \{\sum_{-\infty}^{-1}\}$; $H = H_+ \oplus H_-$. Let $p_\pm$ be natural projections of H onto these subspaces. We shall think of vectors as vector-rows.
The Grassmanian, Gr, will be the set of all subspaces $W \subset H$ satisfying properties (oversimplifying a little bit; however, generically it is true): i) $p_+|_W$ is one-to-one correspondence, ii) $zW \subset W$. (Following to Wilson [8] we can prove that this Grassmanian is isomorphic to the Grassmanian $\text{Gr}^{(n)}$ of scalar functions, we shall not use this now). We shall say that a matrix function belongs to W if all its rows do. Let $g(t,z) = \exp\sum_{k=0}^{\infty}\sum_{\alpha=1}^{n} z^k E_\alpha t_{k\alpha}$. We consider a transformation $\mathbf{g}$ of the space H:

$$\mathbf{H} : H \to H,\ v \mapsto vg.$$

IIIb. *Baker function.* For a $W \in$Gr $w_W(t,z)$ is a Baker function if i) for any $t = (t_1,t_2,\ldots)$ $w_W \in W$ as a function of z, ii)$p_+(wg^{-1}) = 1$. The both conditions mean that $w_W(t,z)$ is the only element of W of the form

$$w_W(t,z) = (1 + \sum_1^\infty w_i z^{-i})g(t,z).$$

IIIc. *τ-function.* Let W be given. We consider mappings:

$$H_+ \overset{(p_+|_W)^{-1}}{\longrightarrow} W \overset{\mathbf{g}^{-1}}{\longrightarrow} Wg^{-1} \overset{p_+}{\longrightarrow} H_+ \overset{\mathbf{g}}{\longrightarrow} H_+.$$

The resulting mapping we denote as $\mathrm{T}_W(t) = \mathbf{g}p_+\mathbf{g}^{-1}(p_+|_W)^{-1}$. Finally, let $\tau_W(t) = \det \mathrm{T}_W(t)$. This scalar function is exactly the same as in the Sato-Ueno-Takasaki formula ($\tau = \tau_{\alpha\alpha}$). A natural question arises: how to obtain in the same way all the $\tau_{\alpha\beta}$. I do not know the answer, and this is an important problem. But, meanwile, a very interesting formula can be obtained for the mapping T itself (not its determinant). Namely,

$$\begin{aligned} &\hat{w}(t,\zeta)\exp(\sum_{k=1}^{\infty}\sum_{\alpha=1}^{n} z^k E_\alpha t_{k\alpha}) \\ = \quad &\mathrm{T}_W(t_{k\alpha} - 1/k\zeta^m)\mathrm{T}_W^{-1}(t)\exp(\sum_{k=1}^{\infty}\sum_{\alpha=1}^{n} z^k E_\alpha t_{k\alpha}) \end{aligned}$$

which has an ordinary form of a Sato relation. (The action of T on matrices is defined as its action on each row separately). Thus, in a sense, the mapping T can be considered as an analog of a τ-function. But how to use it remains unclear.

REFERENCES

1. E.Dato, M.Jimbo, M.Kashiwara, and T.Miwa, Transformation groups for soliton equations, in: Jimbo and Miwa (ed), Non-linear Integrable Systems - Classical Theory and Quantum Theory, Proc. RIMS symposium, Singapore, 1983.

2. G.Segal, and G.Wilson, Loop groops and equations of KdV-type, Publ.Math. IHES, 63, 1-64, 1985.

3. M.J.Ablowitz, D.J.Kaup, A.C.Newell, and H.Segur, Nonlinear evolution equations of physical significance, Phys.Rev.Letters, **31**, 125-7, 1973.

4. B.A.Dubrovin, Completely integrable Hamiltonian systems connected with matrix operators and Abel manifolds, Funkts. Anal. Prilozh., **11**, #4, 28-41, 1977.

5. K. Ueno and K. Takasaki, Toda lattice hierarchy, in: Advanced Studies in Pure Mathematics 4, 1984, 1-95.

6. M.Sato, Soliton equations as dynamical systems on an infinite-dimensional Grassman manifold, RIMS Kokyuroku, **439** (1981), 30-46.

7. H.Flaschka, A.C.Newell, and T.Ratiu, Kac-Moody Lie algebras and soliton equations, Physica **9D**, 300-323, 1983.

8. G.Wilson, Infinite-dimensional Lie groups and algebraic geometry in soliton theory, Phil.Trans.Royal Soc.London **A315**, 393-404, 1985.

9. M.J. Bergvelt and A.P.E. ten Kroode, τ-function and zero curvature equations of Toda-AKNS type, J. Math. Phys. 29 (6), 1988, 1308-1320.

N=1,2 SUPERSYMMETRIC KORTEWEG-DE VRIES EQUATIONS*

PIERRE MATHIEU

Département de physique, Université Laval
Québec, Québec G1K 7P4, Canada

ABSTRACT. We review known results on integrable N=1 and 2 space supersymmetric extensions of the Korteweg- de Vries equation, with an emphasis on their super (second) Hamiltonian structure and related canonical maps (Miura and Schwarzian derivative). We also identify a certain number of open problems.

1. MOTIVATION

Let us first start by motivating briefly the study of superintegrable systems. In physics the word "super" is normally an abreviation for supersymmetry and it is understood as such here. Implementing supersymmetry is a natural way of increasing the symmetry of a system. This extra symmetry, which relates bosonic (= even = commuting) fields to fermionic (= odd = anticommuting) fields, is a space-time symmetry: heuristically a supersymmetric transformation is just the square root of a space-time translation. The concept of supersymmetry is crucial in today's scenarios for the unification of all forces [1]. Quite interestingly, supersymmetry also appeared recently in statistical mechanics. In fact it is in statistical mechanics that the first genuine supersymmetric system has been identified: adsorbed helium on Kripton-plated graphite, which is a physical realization of the tricritical Ising model. The tricritical Ising model is the simplest example of a minimal (= completely solvable) two-dimensional superconformal field theory [2]. (Conformal field theory describes statistical systems at their critical points [3]).

This physical example of a supersymmetric system provides clear motivation for the study of superintegrable systems and more particularly super extensions of the Korteweg- de Vries (KdV) equation. Indeed the theory of the KdV equation turns out to be closely related to two-dimensional conformal field theory. The basic connection is the

---

* Work supported by NSERC and FCAR.

following: the KdV second Poisson structure is equivalent to the global form of the Virasoro (= conformal) algebra realized in terms of Poisson brackets [4]. A similar statement also holds for the super KdV equations considered here. Therefore, we might expect these systems to be relevant to superconformal field theories. In fact one such application has already been found. It was recently shown by Zamolodchikov [5] that there exists particular perturbed conformal field theories which are completely integrable systems. (The perturbation breaks the conformal invariance and can drive the system off criticality). For minimal superconformal models there seems to be a unique generic way of perturbing the systems such that it preserves integrability. The conservation laws of this off-critical field theory turn out to be identical to those of the (quantum form of the) N=1 supersymmetric KdV equation described below [6]. (For related results in the non supersymmetric case, see [5,7].)

2. THE KdV EQUATION - A BRIEF REVIEW

The KdV equation,

$$u_t = u_{xxx}+6uu_x \tag{1}$$

can be written using two distinct Hamiltonian structures:

$$u_t = \partial \frac{\delta}{\delta u} \int (u^3 - u_x^2)\, dx = \mathcal{D}_u \frac{\delta}{\delta u} \int u^2\, dx$$

where

$$\mathcal{D}_u = [\tfrac{1}{2}\, \partial^3 + 2u\partial + u_x]$$

($\partial=\partial_x$). That the KdV equation is a bi-Hamiltonian system turns out to be sufficient to guarantee its complete integrablility [3]. This also follows from its Lax representation,

$$L_t = [-4L^{3/2}{}_+, L] \tag{2}$$

with

$$L = \partial^2 + u$$

where the + indicates the differential part of a pseudo differential operator. The Lax representation yields a formula for the conservation laws [9]

$$\int \text{Res}\,(L^{k/2})\,dx$$

where

$$\text{Res}\,\Sigma(a_i\partial^i) = a_{-1}.$$

The KdV equation is closely related to two other integrable systems: the modified KdV equation,

$$v_t = v_{xxx} - v^2 v_x$$

and the Schwarzian KdV equation,

$$q_t = q_{xxx} - \frac{3}{2}\frac{q_{xx}^2}{q_x}.$$

Both can be written as Hamiltonian systems in the form

$$f_t = \mathcal{D}_f\, \delta\mathcal{H}_f/\delta f$$

with

$$\mathcal{D}_v = \frac{1}{2}\partial$$

$$\mathcal{D}_q = -2\,\partial^{-1} q_x\, \partial^{-1} q_x\, \partial^{-1}$$

and $\mathcal{H}_f$ given by $\int u^2\,dx$ where the $u(f)$ are

$$u(v) = v_x - v^2$$

$$u(q) = \frac{1}{2}\left[\frac{q_{xxx}}{q_x} - \frac{3}{2}\left(\frac{q_{xx}}{q_x}\right)^2\right].$$

The first of these relations is the famous Miura transformation [10]. It maps solutions of the mKdV equation into solutions of the KdV equation. More precisely, it is a canonical map between $\mathcal{D}_v$ and $\mathcal{D}_u$, i.e. it satisfies [11]

$$\left[\frac{du}{dv}\right] \mathcal{D}_v \left[\frac{du}{dv}\right]^{+} = \mathcal{D}_u$$

where du/dv is the Fréchet derivative of u with respect to v and the dagger denotes its formal adjoint. Relation u(q) shows that 2u is equal to the Schwarzian derivative of q with respect to x and it is also easily checked to be a canonical map [12]. Notice that the Schwarzian KdV equation has the same invariance property as the Schwarzian derivative: both are left invariant by a Mobius transformation [12]

$$q \to \frac{aq+b}{cq+d}$$

Finally we recall the relation with the sine-Gordon equation:

$$p_{xt} = \sin p$$

By setting $v=ip_x$, one finds that the polynomial conservation laws of this equation are identical to those of the mKdV equation.

3. N=1 SUPERSYMMETRIC KdV EQUATION

Here we consider only a space supersymmetric extension, where the space variable x is extended to a superspace doublet, (x,θ) with $\theta^2=0$ (i.e. θ is anticommuting). For space supersymmetry, we require invariance under the transformation $x \to x - \eta\theta$ and $\theta \to \theta + \eta$ where η is an anticommuting constant parameter. The fundamental differential operator is now the super derivative, $D = \theta\partial + \partial_\theta$, which satisfies $D^2 = \partial$, and fields are replaced by superfields, functions of both x and θ. Notice that due to the anticommuting character of θ, the Taylor expansion in θ of a superfield contains only two terms: one independent of θ and a linear term.

Let us consider the direct supersymmetrization of the KdV equation. If we do not introduce inverse powers of the superfield, it turns out that this extension is trivial for an even superfield of the form $u(x) + \theta\sigma(x)$ (σ is anticommuting; in this section odd fields are always denoted by Greek letters).

On the other hand for an odd superfield of the form $\Phi(x,\theta) = \theta u(x) + \xi(x)$ it yields a one-parameter family of equations:

$$\Phi_t = \Phi_{xxx} + c(\Phi D\Phi)_x + (6-2c)\Phi_x D\Phi$$

Different analyses (e.g. Painlevé test [13], integrable deformation [14,15], Lax representation [15]) show that this equation is completely integrable if and only if c=3. The corresponding equation will be called the supersymmetric KdV equation. In component fields it is

$$u_t = u_{xxx} + 6uu_x - 3\xi\xi_{xx}$$

$$\xi_t = \xi_{xxx} + 3(\xi u)_x \quad . \tag{3}$$

This is the Manin-Radul super KdV equation, obtained by reduction from their super KP hierarchy [16], but it does not correspond to the system found by Kupershmidt [17]. Eq. (3) has a Lax representation of the form (2) but with [13,16]

$$L = \partial^2 - \Phi D \quad \text{or} \quad L = \partial^2 + \Phi D - (D\Phi).$$

It has an infinite number of conservation laws given by

$$\int \mathrm{Res}^{(1)} \, (L^{k/2}) \, dxd\theta$$

(recall that $\int d\theta = 0$ and $\int \theta d\theta = 1$) where the super residue is defined by

$$\mathrm{Res}^{\,(1)} \Sigma(\alpha_i D^i) = \alpha_{-1}$$

(i.e. it is the coefficient of $D\partial^{-1}$). Furthermore, the super KdV equation can be written as a Hamiltonian system with [14,15,18]

$$\mathcal{D}_\Phi = \frac{1}{2} (D\partial^2 + 3\Phi\partial + (D\Phi)D + 2\Phi_x)$$

and

$$\mathcal{H}_\Phi = \int \Phi(D\Phi) \, dxd\theta.$$

The operator $\mathcal{D}_\Phi$ is the supersymmetric form of $\mathcal{D}_u$ and the Poisson bracket it defines is the global form of the superconformal algebra [14,18]. Notice on the other hand that (3) is not Hamiltonian with respect to the super extension of the KdV first Hamiltonian structure.

The super KdV equation is found to be related to the following equations (the super versions of the mKdV and the Schwarzian KdV equations) [19]:

$$\Psi_t = \Psi_{xxx} - 3(D\Psi)(\Psi D\Psi)_x$$

$$\Lambda_t = \Lambda_{xxx} - 3\frac{\Lambda_{xx}\,(D\Lambda_x)}{(D\Lambda)}$$

via the following maps

$$\Phi = \Psi_x - \Psi(D\Psi)$$

$$\Phi = \frac{\Lambda_{xx}}{(D\Lambda)} - 2\,\frac{\Lambda_x(\,D\Lambda_x)}{(D\Lambda)^2}\ .$$

These last two relations are the super Miura transformation and the super Schwarzian derivative respectively. Both are canonical maps from

$$\mathcal{D}_\Psi = -\tfrac{1}{2}D$$

or

$$\mathcal{D}_\Lambda = \tfrac{1}{2}\ D^{-1}(D\Lambda)\ D^{-1}(D\Lambda)\ D^{-1}$$

to $\mathcal{D}_\Phi$. The super Schwarzian KdV equation (like the super Schwarzian derivative) is invariant with respect to the super Mobius transformation [19]

$$\Lambda \to \frac{\Lambda + \mu D^{-1}(\Lambda D\Lambda) + \delta}{cD^{-1}(\Lambda D\Lambda) + d}$$

where c,d (μ,δ) are even (odd) constants. Finally, the connection with the (space-time) supersymmetric sine-Gordon equation

$$D_t DF = \sin F$$

(where $D_t = \tau\partial_t + \partial_\tau$, τ being the super partner of the time variable) is obtained by setting $\Psi = iDF$ [18]. All in all, one sees that most results of the KdV theory have a super analogue.

4. N=2 SUPERSYMMETRIC KdV EQUATIONS

Consider now the introduction of an additional space supersymmetry. The space variables are then x, θ_1, and θ_2 and there are two superderivatives

$$D_i = \theta_i\partial + \partial_{\theta_i} \quad , \qquad i=1,2.$$

The N=2 (where N refers to the number of space supersymmetries) extension of the N=1 KdV odd superfield $\Phi = \theta u + \ldots$ will then be an even superfield of the form $\theta_1\theta_2 u + \ldots$. and it will also be denoted by Φ. Notice,however,that it is an even superfield.

By brute force, we found three integrable N=2 supersymmetric KdV equations [20,21]

$$\Phi_t = -\Phi_{xxx} + 3(\Phi\, D_1 D_2\Phi)_x + \frac{(a-1)}{2}(D_1 D_2\Phi^2)_x + 3a\Phi^2\Phi_x \tag{4}$$

with a=-2, 1 and 4. For a=-2 and 4 we have obtained the following Lax operators [20]

$$L(a=-2) = -\partial^2 + \varepsilon_{ik} D_i(D_1 D_2 + \Phi)\, D_k \qquad (\varepsilon_{12} = 1)$$

$$L(a=-2) = -(D_1 D_2 + \Phi)^2.$$

For these cases eq. (4) is equivalent to (2) and integrability is established by standard arguments. The conservation laws are of the form

$$\int \mathrm{Res}^{(2)}\,(L^{k/2})\; dx d\theta_1 d\theta_2$$

where

$$\text{Res}^{(2)}\ \Sigma(g_i + \alpha_i D_1 + \beta_i D_2 + f_i D_1 D_2)\partial^i = f_{-1}\ .$$

(This corrects a misprint in [20]). For a=1, the integrability of (4) is only a conjecture, based on the existence of two nontrivial conservation laws [21].

Quite interestingly, eq. (4) can be written as a Hamiltonian system in the following way [20,22]:

$$\Phi_t = \mathcal{D}_\Phi \frac{\delta}{\delta\Phi} \int [\ \Phi\ D_1 D_2\Phi + (a/3)\Phi^3\ ]\ dx d\theta_1 d\theta_2$$

with

$$\mathcal{D}_\Phi = \tfrac{1}{2}(D_1 D_2\partial + 2\Phi\partial - (D_1\Phi)D_1 - (D_2\Phi)D_2 + 2\Phi_x).$$

The Poisson bracket defined by this operator is the global from of the N=2 superconformal algebra. It should be noted that for a=4 there exists another Hamiltonian structure, namely [23]

$$\Phi_t = \partial \frac{\delta}{\delta\Phi} \int [\Phi^4 + (3/2)\ \Phi^2\ D_1 D_2\Phi + \Phi_x{}^2\ ]\ dx d\theta_1 d\theta_2\ .$$

As a result there are twice as many conservation laws for a=4 than for a=-2 or a=1.

The N=2 supersymmetric extension of the Miura map and the Schwarzian derivative are [19]

$$\Phi = a_1\Psi_x + a_2(D_1 D_2\Psi) + (D_2\Psi)(D_1\Psi)\ , \qquad a_1{}^2 + a_2{}^2 = 1,$$

$$\Phi = \frac{1}{2}\left[\frac{(D_1D_2\Omega)}{\Omega} - \frac{3}{2}\frac{(D_1\Omega)(D_2\Omega)}{\Omega^2}\right]\ , \quad \Omega = (D_1\zeta)^2 + (D_2\zeta)^2$$

and again they are canonical maps from, respectively,

$$\mathcal{D}_\Psi = -\tfrac{1}{2} D_1 D_2\ \partial^{-1}$$

$$\mathcal{D}_\zeta = \tfrac{1}{2} [(D_1\zeta)D_1 + (D_2\zeta)D_2]^{-1} \Omega D_1 D_2 \partial^{-1} \Omega [2\zeta_x - (D_1\zeta)D_1 - (D_2\zeta)D_2]^{-1}$$

to $\mathcal{D}_\Phi$. From these operators one could obtain the explicit form of the N=2 super mKdV and the super Schwarzian KdV equations. (N.B. Ψ is an even superfield but ζ is odd).

5. OPEN PROBLEMS

N=1:

. Is the super KdV equation related to super Riemann surfaces?

. Can we understand the origin of the odd nonlocal conservation laws found by Kersten [24]? Are they related to a new odd Hamiltonian structure?

.Are there other integrable super KdV equations, involving negative powers of the fields? Systems of that sort have not been studied so far. Consider for instance an even superfield $R(x,\theta)=u(x) + \theta\sigma(x)$, and the following equation[25]

$$R_t = R_{xxx} + 6RR_x - [(DR)(DR_{xx})/R_x]_x \tag{5}$$

Since the last term contains two odd factors it will vanish when $\sigma=0$ and the KdV equation will be recovered. (Obviously many other such terms can be constructed which are compatible with the basic grading of the KdV equation). It is interesting to note that this equation has an infinite number of conservation laws of the form $\int F(R)dxd\theta$, where F is any function of R! Another interesting property of this equation is that for $R(x,t,\theta) = R(x-vt,\theta)$ it is solved (by direct integration) in terms of the super Weiestrass function [26] $P = \wp(x;\tau) + \theta\delta(\partial\wp(x;\tau)/\partial\tau)$ where $\wp$ is the ordinary Weierstrass function, τ is the modulus and δ is the super modulus. The defining equation for the super Weiestrass function is

$$P_x^{\,2} - 4P^3 + g_2 P + g_3 - 2(DP)(DP_x) = 0 .$$

It remains to be checked whether (5) is integrable.

N=2:

. Prove that (4) with a=1 is integrable.

. What is the relation between (4) and the N=2 super sine-Gordon equation of [27]?

. Is there a non abelian deformation of the N=2 super Miura map? (see [28] for N=0,1). Notice however that N=2 super Kac Moody algebra does not exist [22].

. Do N=2 Gardner transformations exist? Naive deformations (of order ε^2) of the Miura map which can also be deformed to the identity are certainly not integrable [23].

N=3,4:

. Are there integrable N=3,4 super KdV equations? At least for these cases we know that the super second Hamiltonian structure does exist (since there is a corresponding superconformal algebra):

$$\mathcal{D}_\Phi = \tfrac{1}{2}(D_1..D_N\partial^{3-N} + c\delta_{N,4} + (4-N)\Phi\partial + (-)^{N+1}(D_i\Phi)D_i + 2\Phi_x), \qquad i=1,..N\leq 4.$$

For N=3 it was argued that there is no Miura map [29]. However the N=3,4 super Schwarzian derivatives exist and they are well defined [30].

REFERENCES

1. M.B. Green, J.H. Schwarz and E. Witten, "Superstring theory", Cambridge Univ. Press (1987).
2. D. Friedan, Z. Qiu and S. Shenker, Phys, Lett. 151B, 37 (1985).
3. A.A. Belavin, A.M. Polyakov and A.B. Zamolodchikov, Nucl. Phys. B 241,417 (1985).
4. J.L. Gervais, Phys. Lett. B166,277 (1985); B.A. Kupershmidt,Phys. Lett. A 109,417 (1985).
5. A.B. Zamolodchikov, JETP Lett. 46, 160 (1987); Int. J. Mod. Phys. A3; 743 (1988); "Integrable field theory from conformal field theory", preprint.
6. P. Mathieu, "Integrability of perturbed superconformal minimal models", Nucl. Phys. B,to appear.
7. B.A. Kupershmidt and P. Mathieu, Phys. Lett. B 227, 245 (1989); T. Eguchi and S.K. Yang, Phys. Lett B 224, 393 (1989).
8. F. Magri, J. Math. Phys. 19, 1156 (1978).
9. I.M. Gelfand and L.A. Dikii, Funct. Anal. Appl. 10, 259 (1976).
10. R.M. Miura. , J. Math.Phys. 9, 1202 (1968).
11. B.A.Kupershmidt, and G. Wilson, Invent. Math. 62, 403 (1981).
12. G.Wilson, Phys. Lett. A. 132, 445 (1988); J. Weiss, J. Math. Phys. 24, 1405 (1983).
13. P. Mathieu, Phys. A 128, 169 (1988).
14. P. Mathieu, Phys. Lett. B 203, 287 (1988).
15. P. Mathieu, J. Math. Phys. 29, 2499 (1988).
16. Yu. I. Manin and A. O. Radul, Comm. Math. Phys. 98, 65 (1985).

17. B. A. Kupershmidt, Phys. Lett. A 102, 213 (1984).
18. I. Yamanaka and R. Sasaki, Prog. Theor. Phys. 79, 1167 (1988).
19. P. Mathieu, "Super Miura transformations, super Schwarzian derivative and super Hill operators" to appear in "Integrable and Superintegrable Systems " ed. by B. A. Kupershmidt, World Scientific.
20. C.-A. Laberge and P. Mathieu, Phys. Lett. B 215, 718 (1988).
21. P. Labelle and P.Mathieu, work in preparation.
22. M. Chaichian and J. Lukierski, Phys. Lett. B 212, 461 (1988).
23. B.A. Kuperschmidt, private communication.
24. P. H. M. Kersten, Phys. Lett. A 134, 25 (1988)
25. P. Mathieu, unpublished.
26. J. M. Rabin and P.G.O. Freund, Comm. Math. Phys. 114, 131 (1988)
27. E. Napolitano and S. Sciuto, Phys. Lett. B113, 43 (1982)
28. B.A. Kupershmidt, Int. J. Mod. Phys. (to appear); J. Harnad and B.A. Kupershmidt, CRM preprint.
29. T. Khovanova and O. Ovsienko, in "Seminars on Supermanifolds "no 28, p.60, ed. by D. Leites, Univ. of Stockholm, 1988.
30. K. Schoutens, Nucl. Phys. B 292, 150 (1987).

THE CALOGERO–MOSER–KRICHEVER SYSTEM AND ELLIPTIC BOUSSINESQ SOLITONS†

EMMA PREVIATO

Department of Mathematics, Boston University
Boston, MA 02215, USA

Certain special classes of solutions of the KP hierarchy have interesting topological properties: those that come from initial data that are rational in x or in e^{ax} for suitable constants a are dense in the Segal-Wilson Grassmannian [**12**]. The rational solutions to KdV (the t_2–stationary reduction of KP) have been interpreted in [**1**] as stable manifolds of the Calogero-Moser system, which describes the motion of 1–dimensional particles interacting pairwise according to an inverse-square potential.

A generalization of these two facts is still in the making: as for the first aspect, the natural generalization of the solutions rational in x or e^{ax} are those which have a (complete) lattice Λ of periods in the variable x; correspondingly, the inverse-square potential can be modified by giving it one or two periods. These potentials too give rise to completely integrable systems; to generalize the second aspect, one asks whether there be manifolds stable under other flows of the system, parametrizing the t_k–stationary KP reductions, $k \geq 2$.

The basic picture was outlined by Krichever [**8**]; his construction was a solution to the "inverse spectral problem," as the lingo goes. A. Treibich and J.-L. Verdier, working jointly over the past five years, proved the "direct" part, thus giving a complete description of what they called the *elliptic KP solitons*. Notably, they introduced a new tool: an algebraic surface, which hosts the curves that represent the "action variables" of the CMK system, and whose automorphisms give some control over the difficult problem of detecting the stable manifolds. The answer depends on the lattice Λ, a fact that indicates a (still unexplored) significance of the KP equation in number theory. The case of t_2–stationary (KdV) equations was studied by Treibich and Verdier in [**14**]. The case of t_3–stationary (Boussinesq) equations, under an extra restriction, was studied by Verdier and myself: in this report I shall sketch the construction, the details will appear in [**11**].

This paper is organized as follows: the first three sections review the CMK system, pose the problems and sketch the Treibich-Verdier construction and answer. Section 4 is devoted to the Boussinesq case and Section 5 proposes a few remarks to highlight the link of the CMK system with vector bundles and modular varieties, and some questions.

The occasion for this paper was the CRM workshop on Hamiltonian Systems, Transformation Groups and Spectral Transform Methods. I would like to express my warmest thanks to the organizers, John Harnad and Jerry Marsden, for creating such a friendly and stimulating atmosphere; special thanks to Profs. L.A. Dickey and J. Moser for their kind explanations: the former supplied the reference to the beautiful survey [**10**].

1. THE CMK SYSTEMS AND THE KP HIERARCHY

By CMK (for Calogero–Moser–Krichever) I dub essentially three classes of Hamiltonian systems, which give rise to solutions to the KP hierarchy.

†Research partly supported by NSF Grants INT 8600990 and DMS 8802712

Let $(q_1, \dots, q_n;\ p_1, \dots, p_n)$ be position and momenta in a complex phase space and consider the hamiltonian

$$H = \frac{1}{2}\sum_{i=1}^{n} p_i^2 + \sum_{i<j} v(q_i - q_j), \tag{1.1}$$

where v is an even function. The three situations I mentioned occur as consequences of the following ansatz, which provides a Lax pair for the system:

2.1 THEOREM (CF.[**10**],§§1-4). *If L, M are $n \times n$ matrices with entries:*

$$L_{jk} = p_j\delta_{jk} + \sqrt{-1}(1 - \delta_{jk})u(q_j - q_k)$$
$$M_{jk} = \delta_{jk}\sum_{l\neq j} w(q_j - q_l) - (1 - \delta_{jk})z(q_j - q_k)$$

where u is odd, w and z are even, then the equation

$$\sqrt{-1}\dot{L} = [M, L]$$

entails $z = -\dot{u}$, $v = u^2 + const.$, and a functional equation for u and w, which has solutions $u = \frac{1}{q}, \frac{1}{\sin q}, \frac{1}{sn q}$; in each case, the invariants

$$F_k = \frac{1}{k}tr(L^k) \quad k = 1, \dots, n$$

are (generically) functionally independent and in involution, so that the system is completely integrable.

From now on, we shall be concerned only with the two cases $v = \frac{1}{q^2}$, $v = \wp(q)$ ($\wp$ is the Weierstrass $\wp$ function for some normalized lattice $\Lambda =< 1, \tau >$ evaluated at $\frac{\tau}{2} + q$ if you wish to avoid poles at $q = 0$) and we call X the elliptic curve $\mathbf{C}/\Lambda$. The case $v = \frac{1}{\sin^2 q} - \frac{1}{3}$ is intermediate (1 period) and I listed it for completeness: it has received but little attention so far.

Recall that the KP equation

$$u_{yy} = (u_t + u_{xxx} - 6uu_x)_x \tag{1.3}$$

is the first (significant) equation of the hierarchy (cf. [**12**]) of evolution equations for the formal pseudo-differential operator $\mathcal{L} = \partial + u(x)\partial^{-1} + \dots$.

$$\partial_{t_j}\mathcal{L} = [(\mathcal{L}^j)_+, \mathcal{L}]$$

where $x = t_1, y = t_2, t = t_3$ and the KdV

$$u_t = 6uu_x - u_{xxx}$$

and Boussinesq

$$u_{yy} = (u_{xxx} - 6uu_x)_x$$

equations are the reductions of KP that correspond to stationary t_2, t_3 resp. ($(\mathcal{L}^2)_+ = \mathcal{L}^2, (\mathcal{L}^3)_+ = \mathcal{L}^3$, resp.).

The above-defined particle systems were linked to the motion of poles of KP for the first time in [**1**]:

1.4 THEOREM ([**1**]). *The function $f(x,t) = 2\sum_{j=1}^{n}(x-q_j(t))^{-2}$ is a solution of the KdV equation if and only if the set $(q_1(t_0),\dots,q_n(t_0);p_1(t_0),\dots,p_n(t_0))$ is the equilibrium configuration for the hamiltonian $F_2 = \frac{1}{2}tr(L^2)$; moreover, the t dependence of $q_j(t)$ is the flow of the hamiltonian $F_3 = \frac{1}{3}tr(L^3)$. The analogous statement holds for the function $f(x,y) = 2\sum_{j=1}^{n}(x-q_j(y))^{-2}$ and the Boussinesq equation, if we interchange F_2 and F_3 (also a linear change of variables in t and y is needed for the Boussinesq to be normalized as above).*

This naturally leads to two questions: 1. Is there a link with the higher KP reductions and does the link persist for the other types of potentials; 2. Are the equilibrium loci nonempty?

Krichever found a way to answer question 1 by combining the (x–independent) entries of an eigenvector of the matrix L, the jth entry against the simplest function (meromorphic on $\mathbf{P}^1$ or on X) that has a pole at q_j. The combination turns out to be a Baker function for $\mathcal{L}$: $\mathcal{L}\psi = z\psi$, hence the KP hierarchy. The upshot is:

1.5 THEOREM ([7] AND [8]). *The function $u(x,\underline{t}) = \sum_{j=1}^{n} v(x-q_j(\underline{t}))$ gives a solution to the KP hierarchy if $\underline{t} = (t_2,t_3,\dots)$ are the evolutions of the CMK system with potential $v = \frac{1}{q^2}$ or $\wp(q)$.*

1.6 REMARKS: **1.** Krichever's Lax pair for the elliptic case, found as an inverse problem by starting with the eigenvector, introduces a parameter in the matrix L; this parameter can be viewed as a point of the elliptic curve X, and the resulting spectral curve Γ comes equipped with an $n:1$ mapping to X. Γ has generically genus n and the flows linearize on JacΓ. The situation is geometrically very interesting: given an elliptic curve X one produces an n–dimensional family of curves of genus n whose relationship with X is very special: these are the *tangent covers* that will be introduced and classified in the next section.

2. The KP solutions that one gets in Theorem 1.5 are rational (elliptic, resp.) in x. They give all the rational KP solutions that vanish when x is infinite, and the elliptic case is even better but requires nontrivial tools:

1.7 THEOREM ([**16**]). *Let $u(x,y,t)$ be a solution to the KP equation (1.3), defined and meromorphic on $\mathbf{C}\times U$, $U\subset \mathbf{C}\times\mathbf{C}$ an open domain, and doubly-periodic in x with period Λ. Then u comes from an algebraic curve Γ which is an $n:1$ cover of X for some n and up to a linear change of variables $u = 2\sum_{j=1}^{n}\wp(x-q_j(y,t))$; also, the map $U\to X^{(n)}$ $(y,t)\mapsto(q_j(y,t))$ is analytic and can be extended to $\mathbf{C}\times\mathbf{C}$.*

1.8 SIDE REMARKS: **1.** By modifying the form of the Lax pair, it can be proved that the system is still integrable for $u = q^{-2}+q^2$ ([**10**]); this system gives rational KP solutions that do not vanish at infinity (in x), cf. [**17**].

2. A quartic term can be added to H in such a way as to preserve integrability and produce rational solutions to the nonlinear Schrödinger equation, cf. [**18**].

As for the second question, concerning the reduction loci, it has proved much harder but also richer in geometric and number-theoretic implications. Many aspects of it are still open. Let us make a

1.9 DEFINITION. *Let the set of equilibrium points for the hamiltonian F_k be $\mathcal{V}_k(\frac{1}{q^2})$ or $\mathcal{V}_k(\wp(q))$, depending on the potential. The equations defining $\mathcal{V}_2$ are: q_j pairwise distinct;*

$$p_j = 0; \sum_{k \neq j} (q_j - q_k)^{-3} = 0, \quad \sum_{k \neq j} \wp'(q_j - q_k) = 0 \ resp.; \ j = 1, \dots, n$$

Thus $\mathcal{V}_2$ can also be viewed as a sublocus of the symmetric product $\mathbf{C}^{(n)}, X^{(n)}$, resp.; let $\overline{\mathcal{V}}_2$ be the closure of such locus.

1.10 THEOREM ([1]). *Let $v(q) = \frac{1}{q^2}$. The locus $\mathcal{V}_2$ is nonempty if and only if $n = \frac{g(g+1)}{2}$, g any positive integer; in this case $\overline{\mathcal{V}}_2$ is diffeomorphic to $\mathbf{C}^g$. The locus $\mathcal{V}_3$ is nonempty if $n = 2$ even if $(q,p) \in \mathbf{R}^4$ and it is a manifold of dimension 2.*

REMARK: The manifold $\overline{\mathcal{V}}_2$ of Theorem 1.10 turns out to parametrize all x-rational solutions of KdV. It should be noted that while the KdV equilibrium loci have dimension less than n, which is the dimension of the generic orbit, in the Boussinesq example of (1.10) the dimension is the same. The triangular numbers $n = \frac{g(g+1)}{2}$ have a topological significance which is still being unraveled, as shown by P. van Moerbeke's talk at this conference; they also occur in the elliptic case, precisely in the following beautiful result in ordinary differential equations:

1.11 THEOREM ([5]). *Let $R = \frac{d^2}{dx^2} - c\wp(x)$, $c \in \mathbf{R}$. There exists a differential operator of odd order $2g+1$ that commutes with R if and only if $c = g(g+1)$.*

As a consequence of (1.11), it was expected that $\mathcal{V}_2(\wp(q))$ be nonempty at least when $n = \frac{g(g+1)}{2}$, and perhaps only in these cases; however, even the triangular case could not be proved except when $n = 3$, in **[1]** and more recently in **[6]**, where explicit solutions are given. The elliptic potential resisted attack up to the introduction of the algebraic surface S, which we review in the next section. From now on, we consider only the elliptic case.

2. TANGENT COVERS

I will only recall the part of the theory of tangent covers that is needed to classify the curves occurring in the CMK system.

2.1 DEFINITION.

(i) *A finite surjective morphism from a reduced, irreducible closed algebraic curve Γ to the elliptic curve X, which takes a given smooth point $P \in \Gamma$ to a given point $Q \in X$ is said to be a tangent cover if $\iota_\pi(X)$ is tangent to $A_\Gamma(\Gamma^0)$ at the origin of* JacΓ*; here Γ^0 are the smooth points of Γ, A_Γ is the abel map with base point Q into the Jacobian, i.e. the moduli space of isomorphism classes of line bundles of degree zero, and $\iota_\pi = \pi^* \circ A_X$.*

(ii) *A tangent cover $(\Gamma, P) \to (X, Q)$ is said to be minimal if it cannot be factored through another tangent cover $(\Gamma', P') \to (X, Q)$ in such a way that $(\Gamma, P) \to (\Gamma', P')$ is a birational morphism other than an isomorphism.*

2.2 PROPOSITION ([15]). *The following data are equivalent:*

(i) *an X-periodic KP solution in the sense of (1.7)*

(ii) *a solution to the CMK system with potential* $\wp(q)$

(iii) *an* $n:1$ *tangent cover* $\Gamma \to X$.

The tangent covers of a fixed (pointed) elliptic curve X form "half" of the phase space of the CMK system whose potential comes from X. They can be described as follows:

2.3 DEFINITION. *Let* $\mathbf{C}(X)$ *be the field of meromorphic functions of the elliptic curve* X, T *an indeterminate over* $\mathbf{C}(X)$ *and* w *a local parameter near* $Q \in X$. *A monic polynomial of degree* $n \geq 1$, $P(T) \in \mathbf{C}(X)[T]$ *is said to be a tangential polynomial (with respect to* w*) if and only if all the coefficients of* $wP(T-w^{-1})$ *are holomorphic at* Q*: this condition depends only on the tangent vector* $\frac{d}{dw}$*; we denote by* $\Theta(n, X, \frac{d}{dw})$ *the set of tangential polynomials with respect to* w.

2.4 THEOREM ([13]). $\Theta(n, X, \frac{d}{dw})$ *is an affine space of dimension* n. *To every tangent cover* $(\Gamma, P) \to (X, Q)$ *together with a choice of* $\frac{d}{dw}$ *we can associate a unique meromorphic function on* Γ *whose minimal polynomial* $P(T)$ *over* $\mathbf{C}(X)$ *belongs to* $\Theta(n, X, \frac{d}{dw})$*; conversely, any* $P(T) \in \Theta(n, X, \frac{d}{dw})$ *gives rise to a minimal tangent cover.*

A dramatically new way to classify tangent covers requires the introduction of a ruled surface S: essentially, S represents the tangent vector $\frac{d}{dw}$ and the automorphisms of S/X correspond exactly to changing $\frac{d}{dw}$.

2.5 REMARK AND DEFINITIONS. **1.** *The image of the first coboundary in the exact sequence* $0 \to \mathcal{O}_X \to \mathcal{O}_X(Q) \to \mathbf{C}_Q \to 0$ *is the image of the differential of the abel map at* Q.

2. *If* W *is the rank-2 bundle over* X *associated to an element* $\alpha \in H^1(\mathcal{O}_X) \cong Ext^1(\mathcal{O}_X, \mathcal{O}_X)$: $0 \to \mathcal{O}_X \to W \to \mathcal{O}_X \to 0$, *we let* S *be the ruled surface* $=ProjW$ *and* $\pi: S \to X$ *be the natural projection.*

3. *The curves on* S *can be described quite explicitly; the following are standard facts (cf. [3]):* π *has a unique section* C_0 *with self intersection* $C_0 \cdot C_0 = 0$*; if* S_Q *is the fibre over* Q, $C_0 \cdot S_Q = 1$ *and* $S_Q \cdot S_Q = 0$. *The divisors up to linear equivalence form the group* $PicS \cong \mathbf{Z} \oplus \pi^*\ PicX$, *whose identity component is* $Pic^o S \cong X$, *and the divisors up to algebraic equivalence (=numerical equivalence) form the group* $NS(X)$ *free abelian over* C_0 *and* S_Q. *4.* $S \backslash C_0$ *is an additive group* Δ *which can be viewed as the generalized Jacobian of* X *with a cusp* $2Q$: $0 \to \mathbf{G}_a \to \Delta \to X \to 0$, $\mathbf{G}_a$ *(=additive group of* $\mathbf{C}$*) acts as affine transformations on the fibres and gives* $Aut(S/X)$.

2.6 THEOREM ([14]).

(i) *Given a tangent cover* $\pi_n: (\Gamma, P) \to (X, Q)$, *there exists a morphism* $\iota: \Gamma \to S$ *of degree 1 that sends* P *to* $P = C_0 \cap S_Q$ *and* Γ *to a member of the linear system* $|nC_0 + S_Q|$. *Any two such morphisms* ι_1, ι_2 *are related by* $\iota_1 = \sigma \circ \iota_2$, $\sigma \subset Aut(S/X)$.

(ii) (Γ, P, π_n) *is a minimal tangent cover* $\Leftrightarrow$ ι *is an immersion.*

(iii) (Γ, P, π_n) *is a minimal tangent cover* $\Leftrightarrow$ *the arithmetic genus* $g_a(\Gamma) = n$ *(in general,* $g_a(\Gamma) \leq n$*).*

(iv) *Each irreducible divisor in* $|nC_0 + S_Q|$ *is a minimal tangent cover of degree* n *(with* π_n *induced by* π*).*

2.7 THEOREM (**[14]**). *For all* $n \geq 1$ *the linear system* $|nC_0 + S_Q|$ *has dimension* n *and only one fixed point* P*; a divisor in this system is irreducible if and only if it cuts* C_0 *transversely at* P*. The set of irreducible divisors* $V(n, X)$ *in the linear system is an* n*-dimensional affine space and its smooth members form a Zariski-open dense subset.*

In summary, there is a $1:1$ correspondence between isomorphism classes of minimal tangent covers of degree n and points of the affine space of dimension $n-1$: $V(n, X)/Aut(S/X)$.

3. THE STABLE LOCI: KdV

3.1 DEFINITION. *Let* σ *be the natural symmetry of* S *induced by the inversion on* Δ*. We say that a tangent cover* (Γ, P, π) *has a symmetry if* Γ *has an involution that fixes* P *and covers the natural one on* X*. We say that* Γ *is a hyperelliptic tangent cover (HTC) if there is an involution* σ *of* Γ *that fixes* P *and such that* $\Gamma/\sigma \cong \mathbf{P}^1$*; in particular, a HTC is symmetric.*

3.2 THEOREM (**[14]**). *Any symmetric tangent cover admits a unique* $\iota : \Gamma \to S$ *such that* $\iota(\Gamma) \in V(n, X)^\sigma$*, which is a linear space of dimension* $[\frac{n}{2}]$*. For the generic* $C \in V(n, X)^\sigma$*,* C/σ *is smooth of genus* $[\frac{n}{2}]$*.*

3.3 DEFINITIONS.

(i) *The surface* S *has 8 points fixed by* σ*, 4 of them on* $C_0 : s_0, s_1, s_2, s_3$ *and 4 on* $\Delta : r_0, r_1, r_2, r_3$*;* s_i *and* r_i *belong to the same fibre and* s_0, r_0 *are on* S_Q*. We identify a symmetric tangent cover* Γ *with its image* $\iota(\Gamma) \in V(n, X)^\sigma$ *and let* $\mu_i =$ *multiplicities of* Γ *at* r_i*.*

(ii) *Let* $e : S^\perp \to S$ *be the blow up of* S *at the 8 points* s_i, r_i *and* $s_i^\perp, r_i^\perp$ *the corresponding exceptional divisors and let* $\phi : S^\perp \to \tilde{S}$ *be the identification mod* σ*. Then* $\iota : \Gamma \to S$ *can be lifted to* $S^\perp, \tilde{S}$ *and* S/σ *in a Cartesian diagram, in fact* $\iota^\perp : \Gamma \to S^\perp$ *is the strict transform.* $\tilde{S}$ *is a smooth rational surface and* Pic$\tilde{S}$ *has rank 10. We let* $\lambda(n, \mu)$ *be the unique element of Pic* $\tilde{S}$ *such that* $\phi^*(\lambda(n, \mu)) = e^*(nC_0 + S_Q) - s_0^\perp - \sum_{i=0}^3 \mu_i r_i^\perp$*. The* μ_i *necessarily satisfy the congruences:* $\mu_i + \epsilon_i \equiv n \bmod 2, \epsilon_0 = 1, \epsilon_1 = \epsilon_2 = \epsilon_3 = 0$*.*

3.4 THEOREM (**[14]**). *There is a* $1:1$ *correspondence between minimal-hyperelliptic tangent covers and rational (i.e., of geometric genus 0) curves in* $|\lambda(n, \mu)| : \Gamma \mapsto \rho(\Gamma) =$ *the reduced image of* $\iota^\perp(\Gamma)$ *under* ϕ*;*

$$g_a(\Gamma) = (\sum_{i=0}^{3} \mu_i - 1)/2, \; g_a(\rho(\Gamma)) = (2n + 1 - \sum_{i=0}^{3} \mu_i^2)/4$$

3.5 THEOREM ([**14**]). *The following are equivalent:*

(i) $g_a(\rho(\Gamma)) = 0$

(ii) $\rho(\Gamma)$ *is an exceptional divisor*

(iii) $\iota^{\perp}$ *is an immersion.*

If a tangent cover has the properties stated in theorem 3.5 we call it exceptional.

3.6 THEOREM ([**14**]). *For fixed X and n, there is only a finite number of isomorphism classes of HTC of X of degree n and their arithmetic genus is bounded by* $\gamma(n) = \sup\{\ell \in \mathbf{N} | \frac{\ell(\ell+1)}{2} \leq n\}$.

NOTA BENE: No HTC can be minimal as a cover unless $n = 1$ and $\pi =$ identity because for a minimal cover $g_a = n$ (2.6(iii)), thus, it dominates a singular cover which is not hyperelliptic.

3.7 THEOREM ([**14**]). *If we let $\chi(n) =$ the number of the exceptional HTC of degree n and $\psi(n)$ the number of those that in addition have $g_a = \gamma(n)$, then*

(i) $\chi(n)$ *is an increasing function of the order* $O(n \log\log n)$

(ii) $\psi(n) = 1$ *if and only if n is a triangular number;* $\sup\{\psi(\ell) | 0 < \ell \leq n\}$ *goes to infinity with n.*

Very little is known on the non-exceptional HTC, their number and genera but the examples in ([**14**]) show that these will depend on X, suggesting number-theoretic implications of these properties.

At last, the F_2–equilibrium loci of the CMK system hold no more mystery:

3.8 THEOREM ([**14**]). *$\overline{\mathcal{V}}_2$ is nonempty for all n, its connected components are closures of (generalized) Jacobians, and unions of closures of KdV orbits as well; they are in 1:1 correspondence with the (isomorphism classes of) minimal-hyperelliptic tangent covers of degree n; they may have different dimensions but all bounded by $\gamma(n)$.*

4. THE STABLE LOCI: BOUSSINESQ

This section is only meant to illustrate the difference between the $\mathcal{V}_3$ loci and the $\mathcal{V}_2$ of §3; the proofs will appear in [**11**].

RESTRICTION: For the time being, we only know how to treat the Galois-Boussinesq case, which means the following: $P \in \Gamma$ has the property that there exists a function $f \in H^0(3P) \backslash H^0(2P)$ and $\mathbf{C}(\Gamma)$ is Galois over $\mathbf{C}(f)$. Equivalently, Γ has an automorphism of order 3 that fixes P and $\Gamma/\sigma \cong \mathbf{P}^1$. If $(\Gamma, P) \to (X, Q)$ is also a tangent cover, then σ induces an automorphism of order 3 on X (through the tangent vector at Q) and X is unique up to isomorphism.

In the Galois-Boussinesq-tangent-cover case (GBTC), we can extend σ to an automorphism of the surface S, which will have 6 fixed points: s_0, s_1, s_2 on C_0 and r_0, r_1, r_2

on Δ (corresponding to $0 \in \mathbf{G}_a$). As in §3, s_i, r_i belong to the same fibre and s_0, r_0 to S_Q. However, the action of σ on the tangent space at the points on Δ is not a dilation and one needs to blow up twice at r_i in order to find a nonsingular surface $S^{\perp}$. One then needs to blow down three exceptional divisors in order to find a minimal resolution $\tilde{S}'$ of S/σ. The diagram is as follows:

$$\begin{array}{ccc} S^{\perp} & \rightarrow & S^{\perp}/\sigma = \tilde{S} \\ b.d. \downarrow & & \downarrow \\ S' & \xrightarrow{\phi} & S'/\sigma = \tilde{S}' \\ e \downarrow & & \downarrow \\ S & \rightarrow & S/\sigma \\ \downarrow & & \downarrow \\ X & \rightarrow & \mathbf{P}' \end{array}$$

$\tilde{S}'$ is a rational surface, Pic$\tilde{S}'$ is free of rank 11 and we denote a basis by: $\tilde{\ell}$ (generic fibre), $\tilde{C}_0, \tilde{S}_i, \tilde{s}_i, \tilde{v}_i$ $(i = 0, 1, 2)$, where the notation does its best to be suggestive; another triple of divisors that came naturally from the blow up are $\tilde{h}_i$, with $\tilde{\ell} = 3\tilde{S}_i + 2\tilde{v}_i + \tilde{h}_i + \tilde{s}_i$. Each of these curves $\tilde{C}$ has an analog $C^{\perp}$ in S'.

4.1 THEOREM. *There is a 1:1 correspondence between the isomorphism classes of minimal-GB TC of X and the rational curves R on $\tilde{S}'$ such that*
$e_*\phi^*(R) \in |nC_0 + S_Q|$.

4.2 DEFINITIONS.

(i) *Let $\Gamma \in V(n,X)^{\sigma}$, Γ^1 be the strict transform of Γ in S' and $e^*\Gamma = \Gamma^1 + \sum_{i=0}^{2} a_i v_i^{\perp} + \sum_{i=0}^{2} b_i h_i^{\perp} + s_0^{\perp}$. Then necessarily*

$$\begin{cases} 0 \le a_i \le 2b_i \\ 0 \le b_i \le 2a_i \end{cases} \text{and} \begin{cases} b_i + n \equiv \epsilon_i & mod3 \\ a_i + b_i \equiv 0 & mod3 \end{cases} \text{and } a_i \le n - \epsilon_i \tag{4.3}$$

where $\epsilon_0 = 1, \epsilon_1 = \epsilon_2 = 0$.

(ii) *Given $(a;b)$ satisfying (4.3), let $\lambda(n,a,b) \in Pic\tilde{S}'$ be the unique divisor such that $\phi^*\lambda(n,a,b) = e^*(nC_0 + S_0) - s_0^{\perp} - \sum_{i=0}^{2} a_i v_i^{\perp} - \sum_{i=0}^{2} b_i h_i^{\perp}$. Then the minimal GBTC are in $1:1$ correspondence with the rational curves in $|\lambda(n,a,b)|$, $\Gamma \mapsto \rho(\Gamma)$, and*

$$(3(n+1) - \sum_{i=0}^{2}(a_i^2 + b_i^2 - a_i b_i))/9 \;= g_a(\rho(\Gamma)).$$

4.4 THEOREM. *Any solution $(a;b)$ of (4.3) such that $3(n+1)=\sum_{i=0}^{2}(a_i^2+b_i^2-a_ib_i)$ gives an exceptional GBTC of degree n. There exist a finite nonzero number of GBTC of X of degree n for any $n\geq 1$.*

The existence of solutions to the diophantine equation in (4.4) is proved by number-theoretic methods.

To close this section I will give a pleasantly simple example of GBTC for $n=2$, which is smooth of genus 2 and hence gives an equilibrium locus of the same dimension as the generic integral set for the CMK system.

4.5 EXAMPLE: The smooth hyperelliptic curve Γ with affine equation

$$\omega_1^2=\eta_1^6-1$$

has two points at infinity in the coordinate patch $(\frac{\omega_1}{\eta_1^3},\frac{1}{\eta_1})$: let P be the one where $\frac{\omega_1}{\eta_1^3}-1$ has a zero (of order 6). The function $f=\omega_1+\eta_1^3$ has a pole of order 3 at P and fulfills the Galois-Boussinesq condition. Let $\pi\colon\Gamma\to X$ be given by $(\omega_1,\eta_1)\mapsto(\omega=\omega_1,\eta=\eta_1^2)$; this is a 2:1 cover of the elliptic curve: $\omega^2=\eta^3-1$, branched at two points. To show that π gives a tangent cover we use 2.4: the function $\frac{\eta}{\omega}$ gives a local parameter near Q =infinity on X; its pull back on Γ is $\frac{\eta_1^2}{\omega_1}$. The function η_1 on Γ has poles on the fibre over Q, $\eta_1+\frac{\omega_1}{\eta_1^2}$ has a pole at P but not at the other point of the fibre and its minimal polynomial over $\mathbf{C}(X)$, $P(T)=T^2-\eta$, is a tangential polynomial: indeed, all coefficients of $\frac{\eta}{\omega}((T-\frac{\omega}{\eta})^2-\eta)$ are regular at Q.

5. LINKS WITH HITCHIN'S CONSTRUCTION AND NOVIKOV'S PROBLEM

The CMK system offers one setting for two possible applications of the KP hierarchy: to higher-rank vector bundles over a curve and to modular varieties. As for the former direction, I would just like to remark that the CMK system is a situation very similar to Hitchin's integrable system on the phase space $T^*\mathcal{N}$, where $\mathcal{N}$ is the moduli space of stable bundles of any given rank over a Riemann surface of genus ≥ 2. Indeed,

5.1 PROPOSITION. *The points of the CMK phase space correspond to a rank-n bundle E over X, together with a twisted endomorphism $\kappa\colon E\to E\otimes\mathcal{O}_X(Q)$; Krichever's matrix L represents κ in a suitable basis.*

REMARK: The points of the Hitchin phase space are rank-n bundles E over a curve X_1 of genus ≥ 2, together with a section of $\mathrm{Hom}(E,E\otimes K)$, K being the canonical bundle. The system linearizes on $\mathrm{Jac}\,\Gamma$, where Γ is the spectral curve constructed as in the CMK system.

The remark cuts both ways: the CMK system might suggest an answer to Hitchin's open question in [1], "to find explicitly the hamiltonian differential equations which correspond to these systems." On the other hand, Hitchin's spectral-curve construction might be used to generalize the CMK system to a phase space whose points are pairs (E,ϵ), E a bundle over any Riemann surface X_1 and $\epsilon\in H^0(E^*\otimes E\otimes\mathcal{O}_X(Q))$, $Q\in X_1$.

The Novikov problem (cf. [9]) asks for an elliptic analog of the superposition of solitons; a variation of it that has been studied is the question of expressing a KP solution of genus g by lower-genus theta functions. This is of course not disjoint from the question treated in the previous sections, and it would be interesting to find precisely where they intersect. But here I will only state a very pretty fact about Example 4.5 that I get for free out of Belokolos-Enol'skii's investigation in [2] of the Novikov problem in genus 2.

5.2 PROPOSITION. *The curve* Γ *of Example 6.5 has period matrix* $[\tau_{ij}]$ *that belongs to the Humbert surface* H_4*, defined by a modular form of weight 35, or equivalently by the equation* $a\tau_{11} + b\tau_{12} + c\tau_{22} + d(\tau_{12}^2 - \tau_{11}\tau_{22}) + e = 0$*, where* $a, b, c, d, e \in \mathbf{Z}$ *and* $b^2 - 4ac - 4de = 4$*. The KP solution associated to* Γ *can be expressed in terms of elliptic periods.*

REFERENCES

[1] H. Airault, H.P. McKean and J. Moser, Rational and elliptic solutions of the Korteweg-de Vries equation and a related many-body problem, *Commun. Pure Appl. Math.* **30** (1977), 95-148.

[2] E.D. Belokolos and V.Z. Enol'skii, Algebraically integrable nonlinear equations and Humbert surfaces, Plasma theory and nonlinear and turbulent processes in physics, 1-2, *World Sci. Publ.*, Singapore (1988), 20-57.

[3] R. Hartshorne, *Algebraic Geometry*, Springer-Verlag, New York 1977.

[4] N. Hitchin, Stable bundles and integrable systems, *Duke Math. J.* **54** (1987), 91-114.

[5] E.L. Ince, Further investigations into the periodic Lamé function, *Proc. Roy. Soc. Edinburgh*, **60** (1940), 83-99.

[6] A.R. Its and V.Z. Enol'skii, Dynamics of the Calogero-Moser system and the reduction of hyperelliptic integrals to elliptic integrals, *Functional Anal. Appl.* **20** (1986), 62-64.

[7] I.M. Krichever, Rational solutions of the Kadomtsev-Petviashvili equation and integrable systems of n particles on a line, *Functional Anal. Appl.* **12** (1978), 59-61.

[8] I.M. Krichever, Elliptic solutions of the KP equation and integrable systems of particles, *Functional Anal. Appl.* **14** (1980), 282-290.

[9] S.P. Novikov, A periodic problem for the KdV equation, *Functional Anal. Appl.* **8** (1974), 236-246.

[10] M.A. Olshanetsky and A.M. Perelomov, Classical integrable finite-dimensional systems related to Lie algebras, *Physics Reports* **71** (1981), 313-400.

[11] E. Previato and J.-L. Verdier, Boussinesq elliptic solitons: the cyclic case, to appear in *Proceedings Bombay Symp.* 1989.

[12] G. Segal and G. Wilson, Loop groups and equations of KdV type, *Publ. Math. Inst. Hautes Études Sci.* **61** (1985), 5-65.

[13] A. Treibich, Tangential polynomials and elliptic solitons, preprint.

[14] A. Treibich and J.-L. Verdier, Solitons elliptiques, preprint.

[15] J.-L. Verdier, New elliptic solitons, *Algebraic Analysis* **2** (1988), 901-910.

[16] J.-L. Verdier, *Electronic message*, May 1989.

[17] A.P. Veselov, Rational solutions of the KP equation and hamiltonian systems, *Russian Math. Surveys* **35:1** (1980), 239-240.

[18] S. Wojciechowski, Review of recent results on integrability of natural hamiltonian systems, preprint.

MULTIDIMENSIONAL INTEGRABLE TOPS

A. G. REYMAN

Steklov Mathematical Institute Leningrad Branch
Fontanka 27
Leningrad, 191011, USSR

This is a review of recent results on integrable cases of motion of a multidimensional rigid body. All these integrable systems are derived within a group-theoretic construction developed by the author and M. Semenov-Tian-Shansky **[1,2]** which in turn is an outcome of the general r-matrix scheme (or Kostant–Adler–Symes theorem) applied to twisted loop algebras. Hamiltonian systems which arise in this way are defined on cotangent bundles of Lie groups by "natural" Hamiltonians" (which are sums of kinetic and potential energies) and have a Lax representation with a spectral parameter. This allows one to show their complete integrability and produce explicit solutions in terms of theta functions by using the well-known finite-band integration technique. One of the most interesting (and recent) examples here is the Kowalewski top where some improvement is achieved even in the classical 3-dimensional case **[3]**. The reader interested in the physical meaning of various toplike systems is referred to **[4]**.

1. n-DIMENSIONAL RIGID BODY

The position of the n-dimensional rigid body fixed at a point is described by the orthonormal frame $(e_1, \ldots, e_n)$ attached to the body, hence its configuration space is the rotation group $SO(n)$. The velocity space is the tangent bundle $TSO(n)$ which we identify with the product $SO(n) \times so(n)$ via left translations; the phase space is the cotangent bundle $T^*SO(n) = SO(n) \times so(n)$ (the Lie algebra $so(n)$ is identified with its dual by means of the invariant scalar product $(x, y) = -\operatorname{tr} xy$). Let $k \in SO(n)$, denote the top's frame and let $\omega \in so(n)$ and $\ell \in so(n)$ be the angular velocity and angular momentum variables defined in terms of the above splittings. The rotation of the body is determined by prescribing its kinetic energy $E(\omega)$–a quadratic form on $so(n)$ (which in other words is a left-invariant metric on $SO(n)$) and a potential function $U(k)$. It is important that the kinetic energy of a "realistic" body in $\mathbf{R}^n$, i.e. one corresponding to a rotating rigid mass distribution, has a very special form:

$$E(\omega) = -\operatorname{tr} \omega^2 \mathcal{I} \tag{1.1}$$

where $\mathcal{I}$ is the inertia tensor, a symmetric positive definite $n \times n$ matrix. The angular velocity and momentum are linearly related by the Legendre transform:

$$\ell = \mathcal{I}\omega + \omega\mathcal{I} \tag{1.2}$$

which for (1.1) gives

$$E(\omega) = -\frac{1}{2} \operatorname{tr} \ell\omega. \tag{1.3}$$

For instance, if $\mathcal{I} = \operatorname{diag}(b_1, \ldots, b_n)$ then

$$\ell_{ij} = (b_i + b_j)\, \omega_{ij} \tag{1.4}$$

and in terms of angular momentum E takes the form

$$E(\ell) = \sum_{i<j} \frac{1}{b_i + b_j} \ell_{ij}^2. \tag{1.5}$$

Clearly, for $n > 3$ the realistic kinetic energies (1.1) are only a subspace of the space of all quadratic forms on $so(n)$.

The total Hamiltonian on $T^*SO(n)$ is

$$H(k, \ell) = E(\ell) + U(k) \tag{1.6}$$

and the associated equation of motion are

$$\dot{\ell} = [\ell, \omega] - k^{-1} dU(k),$$

$$k^{-1}\dot{k} = \omega, \tag{1.7}$$

where ℓ and ω are related by (1.2). For a free top, when $U(k) = 0$, the equations reduce to the Euler equations $\dot{\ell} = [\ell, \omega]$ on $so(n)$. A set of quadratic integrals of motion for the kinetic energy (1.1) in this case was found by Mischenko in 1970, and was shown to be in involution by Dickey. The natural conjecture was that the free top is completely integrable. The problem was settled by Manakov in 1976 [**5**]. He showed that the Euler equations on $so(n)$ corresponding to the kinetic energy

$$E(\ell) = \sum_{i<j} \frac{b_i - b_j}{a_i - a_j} \ell_{ij}^2 \tag{1.8}$$

have a Lax representation

$$\dot{L}(\lambda) = [L(\lambda), M(\lambda)], \tag{1.9}$$

where

$$L(\lambda) = A\lambda + \ell,$$

$$M(\lambda) = B\lambda + \omega, \tag{1.10}$$

$$A = \operatorname{diag}(a_1, \dots, a_n), \qquad B = \operatorname{diag}(b_1, \dots, b_n),$$

$$\omega_{ij} = \frac{b_i - b_j}{a_i - a_j} \ell_{ij}.$$

The kinetic energy (1.5) is recovered when $A = B^2$. This Lax pair was studied further in [**6**] where it was shown in particular that the spectral invariants of $L(\lambda)$, which are obviously integrals of motion, are in involution and generically form a complete family. We shall now describe a more general Lax-pair construction which gives integrable tops in potential fields.

2. GENERAL CONSTRUCTION

It is natural to generalize our discussion by passing from $SO(n)$ to an arbitrary (compact) Lie group K. Let $\mathfrak{k}$ be the Lie algebra of K, and $\mathfrak{k}^*$ the dual space. We identify the cotangent bundle T^*K with $K \times \mathfrak{k}^*$ by left translation. The construction described below gives a large family of completely integrable Hamiltonian system on T^*K with natural Hamiltonians

$$H(k, \ell) = E(\ell) + U(k), \tag{2.1}$$

where $k \in K$, $\ell \in \mathfrak{k}^*$ and E is a quadratic form on $\mathfrak{k}^*$.

The construction runs as follows. Let G be a Lie group, $\mathfrak{g}$ its Lie algebra, $\mathfrak{g}^*$ the dual space of $\mathfrak{g}$. Let δ be an involution in G and K the fixed subgroup of δ. let $\mathfrak{g} = \mathfrak{k} + \mathfrak{p}$ be the Cartan decomposition of $\mathfrak{g}$ for the induced involution on $\mathfrak{g}$ (i.e. $\delta = 1$ on $\mathfrak{k}$ and $\delta = -1$ on $\mathfrak{p}$), $\mathfrak{g}^* = \mathfrak{k}^* + \mathfrak{p}^*$ the dual decomposition.

Fix two elements A, F in $\mathfrak{p}^*$ and to any point (k, ℓ) of $T^*K = K \times \mathfrak{k}^*$ associate a "Lax matrix" $L(\lambda)$:

$$L(\lambda) = A\lambda + \ell + (Ad^*k^{-1} \cdot F)\lambda^{-1}. \tag{2.2}$$

THEOREM. *Functions of the form*

$$\phi(L(\lambda)), \tag{2.3}$$

where ϕ ranges over coadjoint-invariant functions on $\mathfrak{g}^*$ and λ over the reals, Poisson commute with respect to the canonical Poisson structure on T^*K. For any such function, the corresponding Hamiltonian equations T^*K have a Lax representation

$$\dot{L}(\lambda) = -ad^*M(\lambda) \cdot L(\lambda) \tag{2.4}$$

with $L(\lambda)$ given by (2.2) and $M(\lambda)$ depending on the choice of the Hamiltonian.

From now on, to simplify notation, we shall assume that $\mathfrak{g}$ has a nondegenerate invariant inner product (e.g. $\mathfrak{g}$ is semi-simple) so that all the spaces $\mathfrak{g}$, $\mathfrak{k}$, $\mathfrak{p}$ are identified with their duals and $ad^* = ad$. Then (2.4) takes the usual Lax form

$$\dot{L}(\lambda) = [L(\lambda), M(\lambda)].$$

It is not hard to select among the Hamiltonians (2.3) those which have the natural form (2.1): taking

$$H(k, \ell) = \text{coeff. of } \lambda^{-2} \text{ in } \phi\,(\lambda^{-1}L(\lambda)) \tag{2.5}$$

we have

$$H(k, \ell) = \frac{1}{2}(\ell, \omega(\ell)) + (B, Adk^{-1} \cdot F), \tag{2.6}$$

where $(\ ,\ )$ is the inner product in $\mathfrak{g}$, $B = d\phi(A)$ and $\omega(\ell)$ is the angular velocity determined in terms of ϕ and A. In this case

$$M(\lambda) = B\lambda + \omega. \tag{2.7}$$

There is an explicit formula for $\omega(\ell)$. For instance, when $\mathfrak{g}$ is split semisimple, δ a Cartan involution, and A a regular element,

$$\omega(\ell) = adB\,(adA)^{-1}\ell. \tag{2.8}$$

For example, if $\mathfrak{g} = s\ell(n, \mathbf{R})$ or $g\ell(n, \mathbf{R})$, and $A = \text{diag}\,(a_1, \ldots, a_n)$, the kinetic energy $E = \frac{1}{2}(\ell, \omega(\ell))$ is the same as (1.5). (Notice also that the Lax pair (2.2), (2.7) is an extension of Manakov's pair (1.10)).

The most natural proof of the above theorem makes use of the general r–matrix scheme applied to twisted loop algebras (see e.g. [**1**, **2**]). There is also an elementary proof which proceeds in two steps (see e.g. [**3**]). Consider first the "Lax matrix"

$$L(\lambda) = A\lambda + \ell + s\lambda^{-1} \tag{2.9}$$

where A is fixed and $\ell \in \mathfrak{k}^*$, $s \in \mathfrak{p}^*$ are dynamical variables. Then the invariants of $L(\lambda)$ are in involution with respect to the Lie-Poisson bracket on $(\mathfrak{k}+\mathfrak{p})^*$ for the semi-direct sum structure of $\mathfrak{k}+\mathfrak{p}$. Next, one shows that the mapping from T^*K to $(\mathfrak{k}+\mathfrak{p})^*$ sending $(k+\ell)$ to $\ell \oplus Ad^*k^{-1} \cdot F$ is a Poisson map.

REMARK: Let K_F be the isotropy subgroup of F in K. Clearly, $L(\lambda)$ in (2.2) is left unchanged under the action of K_F on T^*K by left multiplication. Furthermore, the mapping of the quotient space T^*K/K_F into $(\mathfrak{k}+\mathfrak{p})^*$ associated with $L(\lambda)$ is a Poisson embedding. The coadjoint orbits in the semidirect sum dual $(\mathfrak{k}+\mathfrak{p})^*$ are thus the reduced spaces of T^*K with respect to K_F (for various F). The Hamiltonian systems associated with $L(\lambda)$, such as (2.6), can also be interpreted as systems on these orbits. In particular, the orbit in $(\mathfrak{k}+\mathfrak{p})^*$ through $F \in \mathfrak{p}^*$ is simplectically diffeomorphic to the cotangent space $T^*(K/K_F)$.

Also, the Hamiltonian (2.5) (but not $L(\lambda)$) is invariant under the action of the isotropy subgroup K_A by right multiplication. Reduction with respect to K_A gives integrable systems on $T^*(K/K_A)$, with a K–invariant kinetic term (this is not the case for the K_F–reduction). The Lax matrix $L(\lambda)$ does not survive this reduction but there is a Lax pair with the Lax matrix $\tilde{\ell}(\lambda) = Ad^*k \cdot L(\lambda)$.

3. EXAMPLES

The most interesting examples are obtained when (G, K) is a Riemannian symmetric pair of noncompact type, so that G is a semi-simple (noncompact) Lie group, δ a Cartan involution, and K a maximal compact subgroup. Our choice below is motivated by the desire to get systems with configuration space $SO(n)$.

1. $G = SL(n, \mathbf{R})$ or $GL(n, \mathbf{R})$, $K = SO(n)$.

The construction described above gives an n–dimensional top in an arbitrary quadratic potential $f(x) = \sum f_{ij} x_i x_j$ in $\mathbf{R}^n$. The Hamiltonian is

$$H(k, \ell) = \sum_{i<j} \frac{b_i - b_j}{a_i - a_j} \ell_{ij}^2 - \sum_{i,j,s} b_s f_{ij} k_{is} k_{js}. \tag{3.1}$$

Hence any top in a quadratic potential is always completely integrable [**1**, **4**]! This is in striking contrast with what we have for tops in the ordinary homogeneous gravitational field where integrable cases are very scarce. If the symmetric matrix $F = (f_{ij})$ has $(n-1)$ coinciding eigenvalues, so that the isotropy subgroup $K_F \simeq SO(n-1)$, then reduction of the Hamiltonian (3.1) with respect to K_F (at zero momentum) gives the Neumann system on S^{n-1} with Hamiltonian (we have set $A = B$)

$$H = \frac{1}{2} \sum p_i^2 - \sum b_i x_i^2,$$

$$|x| = 1, \qquad (x, p) = 0. \tag{3.2}$$

Note that, for $n = 3$, reducing at a nonzero momentum for the symmetry subgroup $SO(2)$ gives the Neumann system in a magnetic field. Other reductions of (3.1) lead to Neumann like systems on various homogeneous spaces of $SO(n)$.

2. $G = SO(p, q)$, $K = SO(p) \times SO(q)$.

We have a system of two interacting tops of dimensions p and q. Let k, r and ℓ, m be the associated frames and angular momenta. With the obvious (p, q)–block notation the Lax matrix (2.2) becomes

$$L(\lambda) = \begin{pmatrix} 0 & A \\ A^t & 0 \end{pmatrix} \lambda + \begin{pmatrix} \ell & 0 \\ 0 & m \end{pmatrix} + \begin{pmatrix} 0 & k^+ Fr \\ r^t F^t k & 0 \end{pmatrix} \lambda^{-1}. \tag{3.3}$$

For $p = q$ this gives a meaningful system of two interacting p–dimensional tops described by the Hamiltonian

$$\begin{aligned} H = & \frac{1}{2} \sum_{i<j} \frac{a_i b_i - a_j b_j}{a_i^2 - a_j^2} \left(\ell_{ij}^2 + m_{ij}^2 \right) \\ & - \sum_{i<j} \frac{a_i b_j - a_j b_i}{a_i^2 - a_j^2} \ell_{ij} \, m_{ij} - \sum_{i,j,s} b_s \, f_{ij} \, k_{is} \, r_{js}. \end{aligned} \tag{3.4}$$

The tops interact both via their angular momenta and a bilinear potential. In particular, if $A = B$, we get a system of two spherical tops interacting via a bilinear potential. As in the previous example, reduction gives a system of two particles on the sphere S^{n-1} with Hamiltonian

$$\begin{gathered} H = \frac{1}{2} \sum (p_i^2 + q_i^2) - \sum a_1 x_i y_i \\ |x| = 1 = |y|, \qquad (p, x) = 0, \qquad (q, y) = 0. \end{gathered} \tag{3.5}$$

The Lax matrix here becomes

$$L(\lambda) = \begin{pmatrix} 0 & A \\ A^t & 0 \end{pmatrix} \lambda + \begin{pmatrix} p \wedge x & 0 \\ 0 & q \wedge y \end{pmatrix} + \begin{pmatrix} 0 & y \otimes x \\ x \otimes y & 0 \end{pmatrix} \lambda^{-1}. \tag{3.6}$$

Other reductions lead to systems of two interacting points on homogeneous spaces of $SO(n)$. The group $SO(p, 1)$ leads to the p–dimensional Lagrange case of axial symmetry.

Note that since $so(3, 3) = s\ell(4, \mathbf{R})$, two interacting 3–dimensional tops with Hamiltonian (3.4) are the same as the 4–dimensional top with Hamiltonian (3.1) (up to a double covering of the configuration spaces).

3. Let G be the noncompact real form of the exceptional Lie group of type G_2, $K = SO(4)$. This group can be represented by 7×7 matrices and leads to an "exotic" family of integrable 4–dimensional tops, see **[2]** (Adler and van Moerbeke **[7]** show that there are three families of algebraically completely integrable metrics on $so(4)$; for the "exotic" family the additional integral of motion is quartic rather than quadratic.)

4. GENERALIZED KOWALEWSKI TOPS

As is well known, the 3–dimensional heavy top (in a constant gravity field) is integrable only in few cases:

(a) if it is fixed at its center of mass, so that it rotates as a free top (Euler);

(b) if it is axially symmetric (Lagrange) and

(c) if the moments of inertia have a fixed ratio 2:2:1 and the center of mass is located in the equatorial plane perpendicular to the symmetry axis of the inertia tensor (Kowalewski).

It turns out that the Kowalewski top and its various generalizations can be treated within the same construction as above by using a further symmetry reduction [3]. Consider again the symmetric pair $G = SO(p, q)$, $K = SO(p) \times SO(q)$, so that the Lax matrix has the form (3.3):

$$L(\lambda) = \begin{pmatrix} 0 & A \\ A^t & 0 \end{pmatrix} \lambda + \begin{pmatrix} \ell & 0 \\ 0 & m \end{pmatrix} + \begin{pmatrix} 0 & k^t F r \\ r^t F^t k & 0 \end{pmatrix} \lambda^{-1} \tag{4.1}$$

Take the simplest invariant Hamiltonian:

$$H = -\frac{1}{4} \operatorname{Res}_\lambda \lambda^{-1} \operatorname{tr} \left(L(\lambda)^2\right) = $$
$$= -\frac{1}{4} \operatorname{tr} (\ell^2 + m^2) - \operatorname{tr} (k^t F r A^t). \tag{4.2}$$

As we already observed, the isotropy group K_F of F acts as a symmetry group of $L(\lambda)$ and hence of H. We now choose F in such a way that K_F be large: set $F_{ij} = \delta_{ij}$, so that F is the truncated identity matrix. Then K_F consists of matrices of the block form

$$\begin{pmatrix} k & 0 \\ 0 & r \end{pmatrix}, k \in SO(p), r \in SO(q), \text{ where } k = \begin{pmatrix} r & 0 \\ 0 & s \end{pmatrix}, s \in SO(p-q)$$

(we assume that $p \geq q$). Hence $K_F \simeq SO(q) \times SO(p-q)$. Let us now perform Hamiltonian reduction with respect to the $SO(q)$–part of K_F for the zero value of the corresponding momentum; this reduces T^*K to $T^*SO(p)$. Reduction amounts to imposing the constraints

$$P_q \tilde{\ell} P_q + \tilde{m} = 0, \qquad r = 1, \tag{4.3}$$

where

$$\tilde{\ell} = k \ell k^t, \qquad \tilde{m} = r m r^t,$$

and P_q is the orthogonal projection $\mathbf{R}^p \to \mathbf{R}^q$. The reduced phase space is $T^* SO(p)$ with its canonical structure. Inserting the constraints (4.3) into (4.2) gives the Hamiltonian

$$H = \frac{1}{2} \left(\sum_{i,j=1}^{p} \tilde{\ell}_{ij}^2 + \sum_{i,j=1}^{q} \tilde{\ell}_{ij}^2 \right) - \sum_{i=1}^{q} (e_i, g_i), \tag{4.4}$$

where $e_1, \ldots, e_p$ is the orthonormal frame in the top and $g_1, \ldots, g_q$ are the column vectors of the matrix $G = kA$. The Hamiltonian (4.4) involves the right-invariant momentum $\tilde{\ell} = k \ell k^t$ instead of ℓ but this is of no importance since the inversion $k \to$

k^{-1} on $SO(n)$ carries ℓ to $-\tilde{\ell}$. The potential term in (4.4) can be interpreted as a superposition of q homogeneous force fields (g_i are the force vectors viewed from the moving frame) so that the corresponding center of charge vectors lie on the orthogonal axes $e_1, \ldots, e_q$ in $\mathbf{R}^q$. The structure of the kinetic energy in (4.4) precisely reflects the Kowalewski inertia tensor: some of the $\tilde{\ell}_{ij}$ come with coefficient 2 and some with coefficient 1.

The system (4.4) still has a residual $SO(p-q)$ symmetry, so one can add to (4.4) the term $c \sum_{i,j=q+1}^{p} \tilde{\ell}_{ij}^2$ without damaging its integrability. Also, for $q = 2$, one can reduce at a non-zero momentum γ for $SO(2)$ which adds to the Hamiltonian (4.4) a "gyrostatic" term $\gamma \tilde{\ell}_{12}$.

The reduced Lax pair for the generalized p-dimensional Kowalewski top in q fields results from (4.1), (2.7) by inserting the constraints (4.3). After the transformation $L \mapsto \tilde{k} L \tilde{k}^t$ where $\tilde{k} = \begin{pmatrix} k & 0 \\ 0 & 1 \end{pmatrix}$ (see [3]) it can be written more conveniently as follows:

$$L(\lambda) = \begin{pmatrix} 0 & G \\ G^t & 0 \end{pmatrix} \lambda + \begin{pmatrix} \tilde{\ell} & 0 \\ 0 & -P_q \tilde{\ell} P_q \end{pmatrix} + \begin{pmatrix} 0 & F \\ F^t & 0 \end{pmatrix} \lambda^{-1},$$
$$M(\lambda) = \begin{pmatrix} \omega & 0 \\ 0 & 0 \end{pmatrix} - \begin{pmatrix} 0 & F \\ F^t & 0 \end{pmatrix} \lambda^{-1}. \tag{4.5}$$

(Recall that $F_{ij} = \delta_{ij}$ and $G = kA$ is the matrix of Poisson vectors $g_1, \ldots, g_q$).

The classical 3-dimensional case is associated with the group $SO(3, 2)$. This group is (locally) the same as $Sp(2, \mathbf{R})$, hence there is a 4×4 Lax pair in this case. Interestingly, the general scheme automatically gives an extension of the Kowalewski top; namely, the Kowalewski gyrostat in two homogeneous fields (say, gravitational and electric).

Finally, let us emphasize again that the geometric reason for the integrabilty of the Kowalewski top is that it is the reduction of a highly symmetric (and so rather obviously integrable) system with 4 degrees of freedom: the spherical top interacting with a 2-dimensional rotator.

5. FROM LAX EQUATIONS TO ALGEBRAIC GEOMETRY

There exists a standard algebro-geometric technique for solving Lax equations with a spectral parameter (see e.g. [8]) which can be applied to the multidimensional tops described above. We shall briefly describe the situation for the 3-dimensional Kowalewski top, referring the reader to [3, 9] for further details.

The spectral curve Γ defined by the equation

$$\det (L(\lambda) - \mu \cdot 1) = 0 \tag{5.1}$$

is left invariant in the course of Lax evolution. Assuming that generically the spectrum of $L(\lambda)$ is simple (this is the case for the Kowalewski top), one can show that the eigenspaces of $L(\lambda)$ combine into a holomorphic line bundle over the compactified curve Γ. This eigenbundle is therefore described by a point on the Jacobian of Γ. It turns out that if $L(\lambda)$ evolves according to a Lax equation $\dot{L} = [L, M]$ of the r-matrix scheme, then the corresponding evolution of the eigenbundle is linear on $Jac(\Gamma)$.

Consider the 4×4 Lax pair for the 3-dimensional Kowalewski top, related to $sp(2, \mathbf{R})$. This Lax pair has two involutive symmetries, one describing the subalgebra $sp(2, \mathbf{R})$

in $s\ell(4, \mathbf{R})$, the other being $L(\lambda)^t = -L(-\lambda)$ which is common to all Lax matrices of our construction. These give rise to two involutions on the spectral curve Γ : $\tau_1 : (\lambda, \mu) \to (-\lambda, \mu)$ and $\tau_2 : (\lambda, \mu) \to (\lambda, -\mu)$. Also the velocity vector V of the corresponding flow on $Jac(\Gamma)$ satisfies $\tau_1^* V = V$, $\tau_2^* V = -V$. It is natural therefore to consider the quotient curves and the coverings:

$$\Gamma \to \Gamma_1 = \Gamma/\tau_1 \to \Gamma_2 = \Gamma_1/\tau_2. \tag{5.2}$$

The genera of these curves are as follows:

	Γ	Γ_1	Γ_2
general case (two fields, $\gamma \neq 0$)	8	4	1
Kowalewski case	5	3	1

The Kowalewski flow on $Jac(\Gamma)$ is actually confined to $Jac(\Gamma_1)$ and is parallel to the Prym variety of the covering $\Gamma_1 \to \Gamma_2$. Using the machinery of Baker-Akhiezer functions, one can write explicit solutions of the Kowalewski top in terms of Riemann's theta functions of genus 3 associated with Γ_1 or theta functions of genus 2 associated with this Prym variety (see [3]).

References

1. A. G. Reyman, *Integrable systems connected with graded Lie algebras*, J. Sov. Math **19** (1982), 1507–1545.
2. A. G. Reyman and M. A. Semenov-Tian-Shansky, *A new integrable case of the motion of the 4-dimensional rigid body*, Commun. Math. Phys. **105** (1986), 461–472.
3. A. I. Bobenko, A. G. Reyman and M. A. Semenov-Tian-Shansky, *The Kowalewski top 99 years later: a Lax pair, generalizations and explicit solutions*, Commun. Math. Phys. **122** (1989), 321–354.
4. O. I. Bogoyavlensky, *Integrable equations on Lie algebras arising in problems of mathematical physics*, Izv. Akad. Nauk SSSR, Ser. Mat. **48** (1984), 883–938.
5. S. V. Manakov, *Note on the integration of Euler's equations of the dynamics of an n-dimensional rigid body*, Funct. Anal. Appl. **10** (1976), 328–329.
6. A. S. Mishchenko and A. T. Fomenko, *Euler equations on finite-dimensional Lie groups*, Math. USSR Izv. **12** (1978), 371–389.
7. M. Adler and P. van Moerbeke, *Geodesic flow on SO(4) and the intersection of quadrics*, Proc. Nat. Acad. Sci., U.S.A. **81** (1984), 4613–4616.
8. B. A. Dubrovin, V. B. Matveev and S. P. Novikov, *Nonlinear equations of KdV type, finite-zone linear operators and abelian varieties*, Russian Math. Surveys **31** (1976), 59–146.
9. A. G. Reyman and M. A. Semenov-Tian-Shansky, *Group-theoretical methods in the theory of integrable systems*, In: "Modern problems in mathematics", Fundamental developments, vol. **16**, Moscow, VINITI (1987), 119–195. Engl. transl. in: "Encyclopaedia of Mathematical Sciences" vol. **16**, Berlin, Springer (1990).

ON THE ADLER–GEL'FAND–DIKII BRACKET

GEORGE WILSON

Mathematics Department, Imperial College
London, England SW7 2BZ

The Adler-Gel'fand-Dikii (AGD) bracket is a very interesting Poisson bracket first discovered by Adler [**1**] in connexion with the theory of certain integrable systems that generalize the Korteweg-de Vries (KdV) equation. It is also known as the second Hamiltonian structure for scalar Lax equations, or (in the recent Russian literature) simply as the Gel'fand-Dikii bracket. A key property of the AGD bracket is that it makes the composition of (ordinary) differential operators into a canonical transformation in the sense of Hamiltonian mechanics (see [**8**]). In this lecture I want to sketch a way of understanding this result even more thoroughly than was done in the earlier literature (the principal references are [**1, 2, 3, 4, 5, 8**]).

1. THE HAMILTONIAN FORM OF THE KdV EQUATION

We first review quickly the simplest non-trivial example of the phenomena we want to discuss, the KdV equation

$$u_t = u_{xxx} + 6uu_x.$$

It has the Hamiltonian form

$$u_t = \left(\frac{1}{2}\partial^3 + u\partial + \partial u\right)\frac{\delta H}{\delta u} \tag{1.1}$$

where $\partial \equiv \partial/\partial x$ and $H[u] = u^2$. Here $\delta/\delta u$ denotes the Euler-Lagrange operator: if H is any differential polynomial in u (that is, any polynomial in u and its x–derivatives $u^{(j)}$) then

$$\frac{\delta H}{\delta u} = \sum_{j=0}^{\infty} (-\partial)^j\, \partial H/\partial u^{(j)}.$$

The term "Hamiltonian form" deserves elucidation. The idea is that (1.1) is analogous to Hamilton's equations on a (finite-dimensional) manifold. To make that a little more precise, let U be the linear space of (to fix ideas) all smooth periodic functions $u(x)$ with some fixed period in x. Then for any differential polynomial $H[u]$ we have a function on U defined by

$$I_H[u] = \int H[u]dx, \tag{1.2}$$

where the integral is taken over a period. If, as usual, we identify U with a subspace of its dual by identifying $f \in U$ with the linear functional $g \mapsto \int fg dx$, then $dI_H[u]$ is just $\delta H/\delta u$. Thus to complete the analogy with Hamilton's equations we have only to regard the skew operator

$$l = \frac{1}{2}\partial^3 + u\partial + \partial u \tag{1.3}$$

occurring in (1.1) as a skew tensor of type (0, 2) (that is, of the opposite type to a 2–form) on the manifold U. In what sense that is rigorously true is a question that is of no importance in practice. All that matters is that we can use l to define the Poisson bracket of two functions I_H and I_K of the form (1.2): namely, we set

$$\{I_H, I_K\}[u] = \int \frac{\delta K}{\delta u} l \frac{\delta H}{\delta u} dx.$$

The crucial property of the operator l that justifies the term "Hamiltonian" is that this bracket satisfies the Jacobi identity.

We now come to an important philosophical point. If the operator l were invertible, then the inverse operator $\lambda = l^{-1}$ would define a 2–form on U; the fact that the Poisson bracket satisfies the Jacobi identity would then correspond simply to this 2–form being closed, so that U would be a symplectic manifold. However, since l is not quite invertible in any reasonable sense, the standard view on the Hamiltonian nature of the KdV equation (and its generalizations) is that we should simply regard l as defining a Poisson bracket as indicated above, and not seek to interpret it as the inverse of a 2–form on U. In [9] I have argued that this may not be the best viewpoint; in fact I claim that the following situation is typical. Although l is not, strictly speaking, invertible, it is nevertheless obtained by "inverting" a (degenerate) 2–form λ, not on U, but on some slightly larger function space X on which a group G acts in such a way that the quotient space X/G can be identified with U. The (slight) degeneracy of λ is compensated for by its invariance under the group G, and inverting λ really means passing to the quotient by G. This is just the familiar situation of "Hamiltonian reduction", except for the twist that the 2–form λ , as well as the quotient tensor l, is degenerate.

In the KdV example, G is the group PSL(2) and X is the space of all (smooth periodic) functions $\psi(x)$ of one variable x (with non-vanishing derivative). The action of G on X is the usual one by linear fractional transformations $\psi \mapsto (a\psi + b)/(c\psi + d)$, and the projection $X \to X/G$ takes ψ to (half) its Schwarzian derivative:

$$2u = \psi_{xxx}\, \psi_x^{-1} - \frac{3}{2} \psi_{xx}^2\, \psi_x^{-2}.$$

The basic G–invariant 2–form on X corresponds to the skew operator

$$\lambda = -\frac{1}{2} \partial \psi_x^{-1} \partial \psi_x^{-1} \partial;$$

that means (for example) that if we identify the tangent space to X at any point with the space of all smooth periodic functions of x, then the skew form on the tangent space at ψ defined by our 2–form is

$$(f, g) \mapsto \int f(\lambda g) dx.$$

The fact that λ somehow turns into the operator l in (1.3) when we divide by the action of G has the following easily verified consequence: if H is any differential polynomial in u, then the evolution equation[1] $\psi_t = \lambda^{-1} \delta H / \delta \psi$ for ψ is G–invariant and implies the Hamiltonian equation $u_t = l \delta H / \delta u$ for u. In particular, taking $H = u^2$ we obtain the "Ur–KdV" equation

$$\psi_t = \psi_{xxx} - \frac{3}{2} \psi_{xx}^2\, \psi_x^{-1}$$

[1] It is part of the theory that this is always a differential equation, despite the apparent intervention of the inverse of λ.

which implies the KdV equation for u.

This example is considered in detail in [9]: it is sufficiently simple so that all the formulae can be checked by mindless calculations. Here I want to concentrate on the more complicated and interesting example where l defines the AGD bracket. It is in principle the same, but the group PSL(2) is replaced by $GL(n)$.

2. THE HISTORY OF THE AGD BRACKET

As is well known, the KdV equation defines an isospectral deformation of the Schrödinger operator $\partial^2 + u$. A natural generalization is to consider the similar isospectral deformations of a scalar (ordinary) differential operator

$$L = \partial^n + u_{n-1}\,\partial^{n-1} + \cdots + u_1\,\partial + u_0$$

of any order n. These deformations are described by Lax equations

$$L_t = [P,\, L] \equiv PL - LP \tag{2.1}$$

where P is the regular part $L_+^{r/n}$ of a fractional power of L. Thus for each positive integer r, equation (2.1) is a system of evolution equations for the functions $u_i(x,\, t)$, $0 \leq i \leq n-1$. The elegant technology of fractional powers of operators is explained in many places (see, for example [**1, 4, 5**]), and need not concern us here. What interests us is Adler's discovery [**1**] that the systems (2.1) can be written in the alternative Hamiltonian form[2]

$$\boldsymbol{u}_t = l\delta H/\delta \boldsymbol{u}$$

where the bold face symbols stand for column vector, and l is a certain rather complicated skew adjoint matrix of differential operators with coefficients that are differential polynomials in the u_i. The Hamiltonians H have an elegant description as the "residues" (coefficents of ∂^{-1}) of the fractional powers of L, but again that will not concern us here.

In Adler's paper [**1**] the Hamiltonian form of the equations was obtained from the Lax form (2.1) by a straightforward algebraic calculation (using nothing more than the mechanism of fractional powers). However, the calculation provided no reason to expect that the Poisson bracket formed from l (cf. §1 above) would satisfy the Jacobi identity. Nevertheless, Adler conjectured that this would be the case. The first proof of this conjecture was given by Gel'fand and Dikii [**5**]. Their proof, though a fascinating *tour de force*, still amounted to a direct calculation starting from (2.1), and did not really reveal any simple reason why the result should be true. A quite different kind of proof was offered in [**8**]; it went as follows. Consider the "generalized Miura transformation"

$$L = (\partial + v_{n-1})\ldots(\partial + v_1)(\partial + v_0) \tag{2.2}$$

obtained by factorizing the nth order operator L into the first order factors. Thus our original variables u_i are a non-commutative version of the elementary symmetric functions of the new variables v_i; indeed, if in the formulae expressing the u_i in terms of the v_i we set all terms involving derivatives of the v_i equal to zero, then we shall get exactly the elementary symmetric functions (for then factorizing the differential operator $L = f(\partial)$ becomes the same as factorizing an ordinary polynomial $f(X)$).

---

[2] I am referring to the "second" Hamiltonian form of these equations. For brevity (and with apologies to the relevant authors who had studied it earlier) I shall ignore completely the first one.

Now, given any Hamiltonian $H[\boldsymbol{u}]$ that is a differential polynomial in the u_i, we can form the evolutionary system

$$\boldsymbol{v}_t = \partial \delta H / \delta \boldsymbol{v} \tag{2.3}$$

for the v_i; here of course the expressions $\delta H/\delta v_i$ are calculated by regarding H as a differential polynomial in the v_i using (2.2). Let $D \equiv D\boldsymbol{u}/D\boldsymbol{v}$ be the "Frechet Jacobian" of the transformation $\boldsymbol{v} \mapsto \boldsymbol{u}$, that is, the matrix of differential operators (with coefficients depending on the v_i) such that the evolutionary system $\boldsymbol{v}_t = \boldsymbol{a}$ implies $\boldsymbol{u}_t = D\boldsymbol{a}$. Then it is easy to see that (2.3) implies the equations

$$\boldsymbol{u}_t = (D\boldsymbol{\partial}D^\star)\,\delta H/\delta\,\boldsymbol{u},$$

where $D^\star$ is the formal adjoint of D, and the bold face $\boldsymbol{\partial}$ denotes ∂ times the identity matrix. Clearly, $D\boldsymbol{\partial}D^\star$ is a skew adjoint matrix of differential operators with coefficients that are differential polynomials in the v_i.

THEOREM 2.4 [8]. *The skew matrix $D\boldsymbol{\partial}D^\star$ coincides with the operator l found by Adler.*

The merit of this is that the Jacobi identity for the Poisson bracket formed from l now follows by a trivial formality from the corresponding statement for the bracket formed from the very simple skew operator $\boldsymbol{\partial}$, which is easy to prove. Thus theorem 2.4 disposes of the original problem with the operator l.

However, we still have the problem of understanding why (2.4) itself is true. The original proof of (2.4) in [8] was clearly unsatisfactory. The proofs of Drinfel'd and Sokolov [4] and of Dickey [2] are more illuminating and/or simpler; nevertheless, to my taste they fall short of confronting directly what seems to me the basic mystery. What I find most striking in (2.4) is not so much that $D\boldsymbol{\partial}D^\star$ is the operator found by Adler, but that it is any operator at all in the u_i alone: part of the assertion is that the complicated differential polynomials in the variables v_i occuring in the matrix $D\boldsymbol{\partial}D^\star$ are in fact differential polynomials in the "non-commutative elementary symmetric functions" u_i. It is this "cancellation of v's" that we shall try to understand more clearly in the following sections.

EXAMPLE: It is perhaps worth seeing the explicit formulae in the simplest case $n = 2$. In that case the Miura transformation is just

$$\partial^2 + u_1\partial + u_0 = (\partial + v_1)(\partial + v_0),$$

that is,

$$\left.\begin{aligned} u_1 &= v_0 + v_1 \\ u_0 &= v_0 v_1 + v_0' \end{aligned}\right\}$$

The Frechet Jacobian is

$$D = \begin{bmatrix} \partial + v_1 & v_0 \\ 1 & 1 \end{bmatrix}$$

which certainly cannot be expressed in terms of u_0 and u_1; it is only after some surprising computations that the v_i cancel out and we find that

$$D\boldsymbol{\partial}D^* = \begin{bmatrix} -\partial^3 + a\partial + \partial a & \partial^2 + u_1\partial \\ -\partial^2 + \partial u_1 & 2\partial \end{bmatrix}$$

where $a = \frac{1}{2}u_1^2 - u_0 + u_1'$.

3. BEYOND THE MIURA TRANSFORMATION

Let us reformulate the phenomenon of cancellation of v's that we wish to understand. Let $\mu : V \to U$ be the map of function spaces defined by the Miura transformation (2.2); thus V is the space of (to fix ideas) all smooth periodic vector valued functions $\boldsymbol{v}(x)$ and U the corresponding space of functions $\boldsymbol{u}(x)$. On V we have the Poisson bracket defined by the skew operator $\boldsymbol{\partial}$. We can therefore try to define a Poisson bracket on U as follows: given two functions I and J on U (sufficiently smooth, for example, functions obtained by integrating densities as in §1), we can pull them back to get functions $\mu^* I$ and $\mu^* J$ on V, and form the Poisson bracket $\{\mu^* I, \mu^* J\}$ on V. In general there is no reason to expect that this Poisson bracket should be of the form $\mu^* K$ for some function K on U; but the cancellation of v's means that this is indeed the case. There is a kind of explanation for such a phenomenon that seems to me so natural and obvious that we should accept no other. Namely, suppose we had a group G acting on V in such a way that the map μ could be identified with the quotient map $V \to V/G$. Then the functions on V of the form $\mu^* I$ would be simply the G–invariant functions on V. If in addition the Poisson bracket on V were G–invariant, then the bracket of two G–invariant functions would indeed be again a G–invariant function, so we should have a good explanation of our cancellation phenomenon. Thus the problem becomes: find the group G of symmetries of V that has these properties. A good clue as to what G should be is provided by our previous remark that the Miura transformation (2.2) is a differential (non-commutative) version of factorizing a polynomial. In that case the analogue of μ would just be the map $\mathbf{C}^n \to \mathbf{C}^n / \sum_n$ where $\sum_n$ is the symmetric group acting on $\mathbf{C}^n$ by permuting the coordinates. So our problem is to find the "differential symmetric group" G.

Unfortunately, this is a little too naive, and the group G described above does not exist. The reason for this, and the remedy, can be most clearly explained if we take a more algebraic point of view, replacing the spaces U and V by the corresponding fields of rational functions. In the commutative case that means that instead of the map $\mathbf{C}^n \to \mathbf{C}^n / \sum_n$ we consider the field extension

$$\mathbf{C}(u_0, \ldots, u_{n-1}) \quad \subset \quad \mathbf{C}(v_0, \ldots, v_{n-1}) \tag{3.1}$$

where the u_i are the elementary symmetric functions of the v_i. The basic theorem on symmetric functions implies that this is a Galois extension with Galois group $\sum_n$. In the differential case we introduce the <u>differential field</u> $\mathbf{C}(\boldsymbol{v})$ of all rational functions of the symbols v_i and their "derivatives" $v_i^{(j)}$: this field is equipped with the (unique) derivation ∂ such that $\partial v_i^{(j)} = v_i^{(j+1)}$ (and of course we identify v_i with $v_i^{(0)}$). The rational functions of the u_i and their derivatives form a differential (closed under ∂) subfield, and the group G that we seek would be the <u>differential Galois group</u> of the extension $\mathbf{C}(\boldsymbol{u}) \subset \mathbf{C}(\boldsymbol{v})$ (see [7]). The problem now is that, unlike what we have in the commutative case, this is not a Galois extension.

The reason lies in a simple and fundamental difference between factorizing a differential operator, as in (2.2), and factorizing a (commutative) polynomial $f(X)$. Namely, factorizing a polynomial $f(X)$ is equivalent to solving the algebraic equation $f(X) = 0$; but factorizing a differential operator L does not immediately solve the differential equation $L\psi = 0$. The correct statement is that if we can factorize L, then the equation can be solved by quadratures (see [6] ch. 9, and references therein). This assertion can be seen in terms of Galois theory as follows. Let us fix a basis $\{\psi_0, \ldots, \psi_{n-1}\}$ of solutions of $L\psi = 0$ compatible with the factorization (2.2): by that we mean that for any r with $0 \leq r \leq n-1$, $\{\psi_0, \ldots, \psi_{r-1}\}$ is a basis of solutions of the rth order equation

$$(\partial + v_{r-1}) \ldots (\partial + v_0)\psi = 0.$$

Explicitly, if ω_{r-1} denotes the Wronskian determinant of $\{\psi_o, \ldots, \psi_{r-1}\}$, and if we set $\rho_i = \omega_{i-1}/\omega_i$ (and $\omega_{-1} \equiv 1$), then we have

$$v_i = \partial \log \rho_i. \tag{3.2}$$

Thus instead of (3.1) we now have a tower of three differential fields

$$\mathbf{C}(\boldsymbol{u}) \quad \subset \quad \mathbf{C}(\boldsymbol{v}) \quad \subset \quad \mathbf{C}(\boldsymbol{\psi}). \tag{3.3}$$

Instead of the symmetric group $\sum_n$ we now have the group $G = GL(n, \mathbf{C})$ acting on the top field $\mathbf{C}(\boldsymbol{\psi})$ in an obvious way (a matrix g acts on the column vector $\boldsymbol{\psi}$ as usual, and this extends uniquely to a differential ($\equiv$ commuting with ∂) automorphism of the field). The fixed field of G is exactly $\mathbf{C}(\boldsymbol{u})$, that is, G is the (differential) Galois group of the extension $\mathbf{C}(\boldsymbol{u}) \subset \mathbf{C}(\boldsymbol{\psi})$ in (3.3). The Galois group of the extension $\mathbf{C}(\boldsymbol{v}) \subset \mathbf{C}(\boldsymbol{\psi})$ in (3.3) is the subgroup $B \subset G$ of upper triangular matrices; since this is solvable group, the main theorem of Galois theory gives the fact mentioned above that the equation can be solved by quadratures once we have factorized the operator L. More relevant for us is the fact that B is not normal in G, so that the field extension $\mathbf{C}(\boldsymbol{u}) \subset \mathbf{C}(\boldsymbol{v})$ that we were originally interested in is not Galois: the symmetry group that we were seeking earlier would have to be the flag manifold G/B, which is not a group.

The way out of this difficulty is clear: instead of just the Miura transformation $\mu : V \to U$ we have to consider the tower of maps $X \to V \to U$, where X is the space (with G–action) of vector valued functions $\boldsymbol{\psi}(x)$ with invertible Wronskian, and the map[3] $X \to V$ is defined by (3.2). Our "obvious" explanation of the cancellation of v's has to be modified to the following: on X there must be a G–invariant Poisson bracket that induces the bracket defined by the skew operator $\boldsymbol{\partial}$ on $V = X/B$ and the AGD bracket $\boldsymbol{l}$ on $U = X/G$.

It is easy to calculate what this bracket must be, if it exists; for if l_ψ denotes the skew matrix of operators defining it, then we must have

$$\frac{D\boldsymbol{v}}{D\boldsymbol{\psi}}\, l_\psi \left(\frac{D\boldsymbol{v}}{D\boldsymbol{\psi}}\right)^* = \boldsymbol{\partial}. \tag{3.4}$$

On the other hand, we have

$$\frac{D\boldsymbol{v}}{D\boldsymbol{\psi}} = \frac{D\boldsymbol{v}}{D\boldsymbol{\rho}}\,\frac{D\boldsymbol{\rho}}{D\boldsymbol{\psi}}$$

where the variables ρ_i are defined just before (3.2), and

$$\frac{D\boldsymbol{v}}{D\boldsymbol{\rho}} = \partial \text{ diag } \left(\rho_0^{-1}, \ldots, \rho_{n-1}^{-1}\right).$$

Hence $D\boldsymbol{v}/D\boldsymbol{\psi}$ has the form $\boldsymbol{\partial}A$ (for some matrix of differential operators A); inserting that into (3.4) and calculating formally, we find that

$$l_\psi^{-1} = A^* \boldsymbol{\partial} A.$$

We have again encountered the situation of §1 : l_ψ is the <u>inverse</u> of a differential operator λ, so that we have a <u>2–form</u> (rather than a Poisson bracket) on X.

[3]Stricly speaking, this map is defined only on the open subset of X where all the Wronskians ω_i are invertible.

At this stage we still have several problems. One of them is to clarify the process by which the 2-form on X turns into a tensor l of the opposite type on $U = X/G$. I shall not discuss that here (it is treated in detail in [**10**]). Another problem is that X/G is actually not the whole of U: indeed, X/G is clearly the space of nth order differential operators with trivial monodromy, which is in fact just one leaf in the symplectic foliation of U defined by the Poisson structure l. That too I shall overlook here; the reason is that the question we are trying to discuss is really purely local in x, and has nothing to do with monodromy. I am using the intuitive language of function spaces merely as a device for avoiding the proper tedious algebraic formulation of the theory. The most serious problem with our discussion so far, however, would remain equally acute if we reformulated everything in algebraic language: namely, although the calculation above has given us a perfectly well defined 2-form λ on X, it is completely unclear that it is G-invariant, which was the reason we wanted it in the first place. In the next section we shall give a different description of λ that will make its G-invariance obvious.

4. A FORMULA FOR λ

We first recall the classical notion of the <u>bilinear concomitant</u> (see [**6**]). Given an (ordinary) differential operator L and two functions ψ and ϕ, the bilinear concomitant $\langle\psi,\, \phi\rangle_L$ is the unique expression bilinear in the derivatives of ψ and ϕ such that

$$\frac{\partial}{\partial x}\langle\psi,\, \phi\rangle_L = (L\psi)\phi - \psi(L^*\phi).$$

If now $L\psi = 0$ and $L^*\phi = 0$, then this shows that $\langle\psi,\, \phi\rangle_L$ is constant. Thus we have a pairing

$$\ker L \times \ker L^* \to \mathbf{C},$$

which is easily seen to be non-singular, so that $\ker L^*$ gets canonically identified with the dual space to $\ker L$. In fact if $\boldsymbol{\psi} = \{\psi_0, \ldots, \psi_{n-1}\}$ is a basis for $\ker L$, then the dual basis $\boldsymbol{\phi} = \{\phi_0, \ldots, \phi_{n-1}\}$ for $\ker L^*$ is obtained by taking the last column in the inverse of the Wronskian matrix of $\boldsymbol{\psi}$.

Now, as in §3, consider the space X of all smooth periodic vector valued functions $\boldsymbol{\psi}(x)$ with invertible Wronskian. For each i with $0 \leq i \leq n-1$ we have the function $\bar{\psi}_i : X \to \mathbf{C}$ whose value at $\boldsymbol{\psi}$ is $\int \psi_i(x)dx$. Similarly, we denote by $\bar{\phi}_i$ the function $\boldsymbol{\psi} \mapsto \int \phi_i(x)dx$, where ϕ_i is calculated from $\boldsymbol{\psi}$ as indicated above. Let $L_{\boldsymbol{\psi}}$ denote the (unique) n^{th} order differential operator with leading term ∂^n such that $\psi_i \in \ker L_{\boldsymbol{\psi}}$ for $0 \leq i \leq n-1$. Then on X we have the 2-form

$$\sum_{i=0}^{n-1} d\bar{\phi}_i \wedge L_{\boldsymbol{\psi}}\, d\bar{\psi}_i. \tag{4.1}$$

In down-to-earth terms, this means that the skew matrix of differential operators defining the 2-form is

$$\lambda = (D\boldsymbol{\phi}/D\boldsymbol{\psi})^* \boldsymbol{L},$$

where the bold face $\boldsymbol{L}$ denotes L times the identity matrix.

THEOREM 4.2. *This 2-form on X coincides with the one constructed in §3.*

This solves the main problem we were left with at the end of §3, since the formula (4.1) makes it obvious that the form is $GL(n)$-invariant.

There are two ways of proving Theorem 4.2. The first is by direct calculation starting from the formula (4.1); the essential point is to study how the 2–form (4.1) behaves with respect to factorization of the operator L. The second, ultimately more enlightening, way is to understand the group theoretical origin of the two versions of the 2–form that we have encountered.

5. GROUP THEORETICAL DESCRIPTION OF λ

We first describe the space X in a more group-theoretical way. Let $G = GL(n, \mathbf{C})$, and let $LG = C^\infty(S^1, G)$ be the loop group of all smooth maps from the unit circle to G (equivalently, 2π–periodic G–valued functions of one variable x). Let $\mathfrak{g} = \mathfrak{gl}(n, \mathbf{C})$ be the Lie algebra of G; we write

$$\mathfrak{g} = \mathfrak{g}_- \oplus \mathfrak{g}_0 \oplus \mathfrak{g}_+$$

for its decomposition as direct sum of strictly lower triangular, diagonal and stricly upper triangular matrices, and we write $\alpha = \alpha_- + \alpha_0 + \alpha_+$ for the corresponding decomposition of a matrix $\alpha \in \mathfrak{g}$. Set

$$Y = \{\Psi \in LG : (\Psi_x \Psi^{-1})_+ = I\},$$

where I denotes the matrix with 1's on the diagonal above the main one and all other entries zero (that is, the sum of the simple root vectors for the Lie algebra $\mathfrak{g}$). Let G_-, G_0, and G_+ be the subgroups of G corresponding to the Lie algebras $\mathfrak{g}_-, \mathfrak{g}_0, \mathfrak{g}_+$: thus G_0, for example, is the group of all invertible diagonal matrices. Clearly, Y is preserved under left multiplication by LG_-, and also under right multiplication by G (thought of as the group of constant loops in LG).

LEMMA 5.1. *The map $\Psi \mapsto$ (first row of Ψ) induces a bijection from the quotient space $LG_- \backslash Y$ to the space X of §3.*

From now on we shall identify X with $LG_- \backslash Y$. This description of X has the virtue of having an obvious generalization when we replace $GL(n)$ by any simple Lie group G; more important here is that it enables us to understand the origin of the 2–form(s) on X that we have met in the preceding sections.

The loop group LG comes equipped with a natural 2–form defined as follows. On the loop algebra $L\mathfrak{g}$ we have the skew form

$$\omega(f, g) = \int tr[f(x)g'(x)]dx.$$

Thinking of $L\mathfrak{g}$ as the tangent space to LG at the identity, we obtain a unique left invariant 2–form ω on LG that agrees with the above skew form at the identity. Because the trace is invariant under conjugation, ω is also invariant under the action of G on LG by <u>right</u> multiplication. Further, the restriction of ω to $Y \subset LG$ satisfies the condition needed for it to be the pull-back of a 2–form λ_1 on the quotient space X.

PROPOSITION 5.2. *This 2–form on X coincides with the one given by (4.1).*

For the rest of the article we shall pretend that every element $h \in G$ has a (necessarily unique) factorization of the form $h = h_- h_0 h_+$ with $h_- \in G_-$, $h_0 \in G_0$, $h_+ \in G_+$.

Striclty speaking, we should replace G by the dense open subset of elements for which that is true, and Y and X by the corresponding dense open subsets. We write $\pi : LG \to LG_0$ for the projection $\pi(\Psi) = \Psi_0$ where $\Psi = \Psi_-\Psi_0\Psi_+$ denotes the factorization of Ψ as a product of loops in G_-, G_0 and G_+. Since π commutes with left multiplication by elements of LG_-, it induces a projection $p : X \to LG_0$. Explicitly, p is given by

$$p\psi = \text{diag}\ (\rho_0, \ldots, \rho_{n-1}),$$

where the ρ_i are the quotients of Wronskians introduced in §3.

Now let ω_0 denote the restriction to LG_0 of the 2-form ω on LG. Thus we get another 2-form $\lambda_2 = p^*\omega_0$ on X.

PROPOSITION 5.3. *This 2-form on X coincides with the one found in* §3.

The main theorem is now that the 2-forms λ_1 and λ_2 on X coincide. An equivalent formulation is as follows.

PROPOSITION 5.4. *The restrictions to Y of the two 2-forms ω and $\pi^*(\omega_0)$ on LG coincide.*

The material in this section has been heavily influenced by the "group-theoretical" description of the AGD bracket given by Drinfel'd and Sokolov [4]; indeed, if we now pass to the quotient space X/G, we get exactly their description of the bracket. The most important difference between the present version of the theory and that of [4] is that in [4] everything is formulated (as far as possible) in terms of Lie algebras, rather than the corresponding groups, so that, for example, the space X, which from our point of view seems fundamental, does not appear.

Details of this work will be published elsewhere.

REFERENCES

1. M. Adler, *On a trace functional for formal pseudo-differential operators and the symplectic structure of the Korteweg-de Vries equations*, Invent. Math **50** (1979), 219–248.
2. L. A. Dickey, *A short proof of a Kupershmidt-Wilson theorem*, Commun. Math. Phys. **87** (1982), 127–129.
3. V. G. Drinfel'd and V. V. Sokolov, *Equations of Korteweg-de Vries type and simple Lie algebras*, Dokl. Akad. Nauk SSSR **258:1** (1981), 11–16; Sov. Math. Doklady **23** (1981), 457–462.
4. V. G. Drinfel'd and V. V. Sokolov, *Lie algebras and equations of Korteweg-de Vries type*, Itogi Nauki i Tekhniki, ser. Sovremennye Problemy Matematiki **24** (1984), 81–180; J. Sov. Math. **30** (1985), 1975–2036.
5. I. M. Gel'fand and L. A. Dikii, "A family of Hamiltonian structures connected with integrable nonlinear differential equations," Preprint Inst. Appl. Math., USSR, 1978, no. 136. English translation in: I. M. Gel'fand, Collected papers, volume 1 (S. G. Gindikin et al, eds) Springer-Verlag, 1987, pp. 625–646.
6. E. L. Ince, "Ordinary differential equations," Longmans Green and Co., London, 1926.
7. I. Kaplansky, "An introduction to differential algebra," Hermann, Paris, 1957.
8. B. A. Kupershmidt and G. Wilson, *Modifying Lax equations and the second Hamiltonian structure*, Invent. Math **62** (1981), 403–436.
9. G. Wilson, *On the quasi-Hamiltonian formalism of the KdV equation*, Phys. Lett. **132A** (1988), 445-450.
10. G. Wilson, *On antiplectic pairs in the Hamiltonian formalism of evolution equations*, Preprint (1989).

KAC-MOODY-VIRASORO SYMMETRIES OF INTEGRABLE NONLINEAR EVOLUTION EQUATIONS

PAVEL WINTERNITZ

Centre de recherches mathématiques, Université de Montréal
CP 6128 – A, Montréal, Québec H3C 3J7, Canada

ABSTRACT. Local point symmetries of integrable evolution equations in more than two dimensions are shown to be realized by infinite dimensional Lie algebras with a Kac-Moody-Virasoro structure. Two examples are presented: the three wave resonant interaction equations and the second equation in the Kadomtsev-Petviashvili hierarchy of Jimbo and Miwa.

1. INTRODUCTION

Lie groups of local point transformations, leaving systems of differential equations invariant, provide powerful tools for obtaining exact analytic solutions of nonlinear differential equations [1], [2], [3]. An interesting phenomenon occurs when symmetry group methods are applied to integrable nonlinear evolution equations in more than two space-time dimensions. Namely, it turns out that the corresponding symmetry groups are infinite dimensional. Moreover, their Lie algebras have a very specific algebraic structure. Indeed, they are all infinite dimensional subalgebras of Kac-Moody-Virasoro (KMV) algebras [4], [5].

The purpose of this contribution is to illustrate how the KMV algebras occur and the use they can be put to. To do this we shall discuss two particular systems, namely the three-wave resonant interaction equations (3WRI) in three dimensions [6], [7], and the first two equations in the Kadomtsev-Petviashvili series, as presented by Jimbo and Miwa [8].

The symmetries to be considered here are Lie point ones. That is, the group transformations have the form

$$\tilde{x} = \Lambda_g(x,u), \qquad \tilde{u} = \Omega_g(x,u), \tag{1}$$

where x and u are the independent and dependent variables respectively, g represents the group parameters and Λ and Ω are some locally defined functions. The solution set of the considered equations is invariant, that is if $u = f(x)$ is a solution, and $\tilde{u}, \tilde{x}$ are defined, then $\tilde{u} = \tilde{f}(\tilde{x})$ is also a solution [1]. The Lie algebra of the symmetry group is realized by vector fields of the form

$$v = \sum_{i=1}^{p} \eta_i(x,u)\,\partial_{x_i} + \sum_{\alpha=1}^{q} \phi_\alpha(x,u)\,\partial_{u^\alpha}, \quad \begin{matrix} x \in \mathbb{R}^p \\ u \in \mathbb{R}^q. \end{matrix} \tag{2}$$

The functions η_i and ϕ_α are obtained by solving a system of linear partial differential equations, the "determining equations" of the symmetry. They are in turn obtained from the requirement that the transformations (1) leave the considered equations invariant. The algorithm for determining the vector fields (2) is standard [1], [2], [3] and has been implemented in various computer programs (in MACSYMA [9], REDUCE [10], and other languages [11]). Solving the determining equations usually introduces a finite number of integration constants. The number of independent significant constants is the dimension of the underlying Lie algebra. For some equations, in particular multidimensional integrable evolution equations, the solution of the determining equations gives rise to arbitrary functions of the independent and dependent variables. These can be expanded into Taylor, or Laurent series, with arbitrary coefficients. This in turn gives rise to infinite dimensional Kac-Moody-Virasoro algebras as symmetry algebras.

Equations that have been recently analyzed from this point of view and shown to have KMV symmetry algebras, include the Kadomtsev-Petviashvili equation **[12]**, the modified KP equation [9], [11], the Davey-Stewartson equation **[13]**, the dispersive long wave equations **[14]**, the 3WRI equations **[15]**, and others **[16]**, **[17]**.

The commutation relations for an untwisted KMV algebra can be summed up as:

$$\begin{aligned}
[\, L_m,\, L_n \,] &= (m-n)\, L_{m+n} + \frac{1}{12} m(m^2-1)\, \delta_{m,-n}\, C \\
[\, T^a_m,\, T^b_n \,] &= f^{abc}\, T^c_{m+n} + m\delta^{ab}\, \delta_{m,-n}\, K \\
[\, L_m,\, T^a_n \,] &= -nT^a_{m+n} \\
[\, C,\, L_m \,] &= [\, C, T^a_m \,] = [\, K, L_m \,] = [\, K,\, T^a_n \,] = [\, K,\, C \,] = 0 \\
& 1 \le a,\, b,\, c \le N, \qquad m,\, n \in \mathbb{Z}
\end{aligned} \tag{3}$$

The coefficients f^{abc} are the structure constants of some finite dimensional Lie algebra L_0, usually assumed to be simple. The symmetry algebras of integrable nonlinear multidimensional evolution equations all have the following features:

(1) They allow a Levi decomposition

$$L = S \triangleright R \tag{4}$$

where S is semisimple and R is a solvable ideal. The semisimple part is a Virasoro algebra or the direct sum of several Virasoro algebras. The radical R is contained in a Kac-Moody algebra and is based on a solvable finite dimensional algebra L_0.

(2) The loop parameter is the evolution variable of the considered equation, usually time.

(3) The central elements are represented trivially $K = C = 0$.

2. SYMMETRIES OF THE THREE WAVE RESONANT INTERACTION EQUATIONS

The 3WRI equations can be written in characteristic coordinates as

$$\begin{aligned}
u_{1x_1} &= i\,\varepsilon_1\, u_2^*\, u_3^*, \qquad & u_{2x_2} &= i\,\varepsilon_2\, u_3^*\, u_1^*, \qquad & u_{3x_3} &= i\,\varepsilon_3\, u_1^*\, u_2^*, \\
\varepsilon_j^2 &= 1, & u_{jx_j} &= [\, \frac{\partial}{\partial t} + (\vec{v}_k\, \nabla)\,]\, u_j,
\end{aligned} \tag{5}$$

where u_i are three complex wave amplitudes, $\vec{v}_k$ are 3 velocity vectors, assumed to be linearly independent, ε_j are real normalized coupling constants and the stars denote complex conjugation.

Equations (5) have been integrated by inverse scattering techniques [6], [7], but we shall not dwell upon this aspect. Instead, we concentrate on the symmetry group [15]. In any case, symmetry techniques in general provide different types of solutions than inverse scattering and related methods.

Using the standard symmetry group algorithm [1] and the MACSYMA program [9], we find that the general element of the symmetry algebra can be written as

$$V = \sum_{i=1}^{3} [\, V_i\,(f_i) \,+\, W_i\,(s_i)\,], \tag{6}$$

where

$$\begin{aligned} V_1\,(f_1) &= f_1\,(x_1)\,\frac{\partial}{\partial x_1} - \frac{1}{2}\,f_1'\,(x_1)\,(u_2\,\partial_{u_2} + u_3\,\partial_{u_3} + u_2^*\,\partial_{u_2^*} + u_3^*\,\partial_{u_3^*} \\ W_1\,(s_1) &= i\,s_1\,(x_1)\,(u_3\,\partial_{u_3} - u_2\,\partial_{u_2} - u_3^*\,\partial_{u_3^*} + u_2^*\,\partial_{u_2^*}), \end{aligned} \tag{7}$$

and V_2, V_3, W_2, W_3 are obtained by cyclic permutations. The functions $f_i(x_i)$ and $s_i(x_i)$ are all real and C^∞ on some open set $I \subset \mathbb{R}$; the prime denotes differentiation with respect to the argument x_1. The commutation relations for the Lie algebra can be written as

$$\begin{aligned} [\, V_i\,(f_i),\, V_j\,(f_j)\,] &= \delta_{ij}\,V_i\,(f_i\,f_j' - f_i'\,f_j), \\ [\, V_i\,(f_i),\, W_j\,(s_j)\,] &= \delta_{ij}\,W_j\,(f_i\,s_j'), \\ [\, W_i\,(s_i),\, W_j\,(s_j)\,] &= 0. \end{aligned} \tag{8}$$

We see that the symmetry algebra of the 3WRI equations is the direct sum of 3 isomorphic Lie algebras, each of which is the semidirect sum of a centerless Virasoro algebra and an infinite dimensional Abelian algebra:

$$L = L_1 \oplus L_2 \oplus L_3, \qquad L_i = \{V_i(f_i)\} \triangleright \{W_i(s_i)\} \tag{9}$$

The KMV structure of this algebra is thus extremely simple and is best seen by choosing a Laurent polynomial basis for the functions f_i and s_i. We then have, e.g. for L_1:

$$\begin{aligned} &V_{1n} \equiv V_1\,(-x_1^{n+1}), \quad W_1 = W_1(x_1^n), \quad n \in \mathbb{Z} \\ &[V_{1m},\, V_{1n}\,] = (m-n)\,V_{1m+n}, \quad [\,V_{1m},\, W_{1n}] = -nW_{1n+m}, \; [\,W_{1m},\, W_{1n}\,] = 0 \end{aligned} \tag{10}$$

These are precisely the commutation relations for a $\hat{u}(1)$ KMV algebra, when the underlying finite dimensional algebra is one-dimensional and hence we have $f^{abc} = 0$ in (3).

The Lie algebra (7) can be straightforwardly integrated to provide the local point transformations of the symmetry group. Restricting ourselves to one parameter subgroups, corresponding to chosen functions $f_i(x_i)$ and $s_i(x_i)$, we find that $V_1(f_1)$ generates the transformation

$$\begin{aligned}
\tilde{x}_1 &= F_1^{-1}(\lambda + F_1(x_1)), \qquad \tilde{x}_2 = x_2, \qquad \tilde{x}_3 = x_3, \\
\tilde{u}_1 &= u_1, \quad \tilde{u}_2 = [\, f_1(x_1)/f_1(\tilde{x}_1)\,]^{1/2} u_2, \quad \tilde{u}_3 = [\, f_1(x_1)/f_1(\tilde{x}_1)\,]^{1/2} u_3 \\
F_1(x_1) &= \int^{x_1} \frac{ds}{f_1(s)}.
\end{aligned} \tag{11}$$

These are quite general smooth transformations of x_1 accompanied by an appropriate gauge transformation of u_2 and u_3. In particular, for $f = 1$, we have translations, $f = x_1$ dilations and $f = (x_1)^2$ local projective transformations.

The vector field $W_1(s_1)$ generates gauge transformations, namely:

$$\tilde{x}_i = x_i, \quad \tilde{u}_1 = u_1, \quad \tilde{u}_2 = e^{-i\lambda s_1(x_1)} u_2, \quad \tilde{u}_3 = e^{i\lambda s_1(x_1)} u_3. \tag{12}$$

Infinite dimensional symmetry groups can be used in exactly the same way as finite dimensional ones to perform symmetry reduction. Thus, the 3WRI system (5) can be reduced to a system of ordinary differential equations, for solutions that are invariant under a suitable two-dimensional subgroup of the symmetry group. To do this in a systematic manner, we need to first classify the low dimensional subalgebras of the symmetry algebra into conjugacy classes. This can be done using methods originally developped for finite dimensional Lie algebras, **[18]**, **[19]**, and generalized to subalgebras of KMV algebras **[12]**, **[15]**, **[17]**. We refer to the original article **[15]**, for a subalgebra classification for the 3WRI symmetry algebra. Here we give just one example of solutions obtained by symmetry reduction.

Consider the subalgabra generated by

$$\begin{aligned}
A &= V_1(1) + \kappa_1 V_1(1) + \kappa_2 V_3(1), \\
B &= V_1(x_1) + V_2(x_2) + V_3(x_3) + aW_1(1) + bW_2(1) \\
\kappa_1^2 &= \kappa_2^2 = 1, \qquad a,\ b \in \mathbf{R}.
\end{aligned} \tag{13}$$

We calculate the invariants of the corresponding subgroup and express the wave amplitudes in terms of them to obtain

$$\begin{aligned}
u_1 &= (-\kappa_1 x_2 + \kappa_2 x_3)^{-1} \; exp\; ib\; ln\,(-\kappa_1 x_2 + \kappa_2 x_3)\, F_1(\xi) \\
u_2 &= (x_1 - \kappa_2 x_3)^{-1} \exp\,[-ia\; ln\,(x_1 - \kappa_2 x_3)\, F_2(\xi) \\
u_3 &= (x_1 - \kappa_1 x_2)^{-1} \exp\,[i\,(a-b) ln\,(x_1 - \kappa_1 x_2)\, F_3(\xi) \\
\xi &= \frac{x_1 - \kappa_1 x_2}{x_1 - \kappa_2 x_3}, \qquad F_i = \rho_i(\xi)\, e^{i\phi_i(\xi)}.
\end{aligned} \tag{14}$$

Substituting into the 3WRI equations (5) and separating the real and imaginary parts, we obtain a system of 6 ordinary differential equations for ρ_i and ϕ_i. Introducing a new variable

$$\begin{aligned} u(\eta) &= \left[\int \rho_2^2(\xi)\, d\xi - \mu(\xi)\right] \lambda^{-1}(\xi) \\ \lambda(\xi) &= (\kappa_2 \varepsilon_1 \varepsilon_3)^{1/3} \frac{1}{\sqrt{3}} (\xi - 1)^{1/3} \xi^{1/3} \\ \eta(\xi) &= \frac{1}{\sqrt{3}} (\kappa_1 \varepsilon_1 \varepsilon_3)^{-2/3} \int \xi^{-2/3} (\xi - 1)^{-2/3} d\xi \\ \mu(\xi) &= \mu_1 \xi + \mu_0, \end{aligned} \tag{15}$$

where μ_1 and μ_0 are appropriately chosen constants, we find that $u(\eta)$ satisfies a prototype [**20**] equation

$$\begin{aligned} [u'' - 2\mathcal{P}(\eta, 0, 1)\, u]^2 &= -4u'^3 + K\mathcal{P}u'^2 \\ &+ (12\mathcal{P}u^2 - K\mathcal{P}'u + H)u' - 4\mathcal{P}'u^3 + K\mathcal{P}^2u^2 - H'u + g_0. \end{aligned} \tag{16}$$

Here $\mathcal{P}$ is a Weierstrass elliptic function, $H(\eta)$ satisfies the Lamé equation

$$H'' - 2\,\mathcal{P}\,H = 0, \tag{17}$$

K and g_0 are constants.

Eq. (16) was integrated by Bureau [**20**], [**21**], and collaborators as part of a systematic study of equations of the form

$$y''^2 + A(t, y, y')\, y'' + B(t, y, y') = 0, \tag{18}$$

having the Painlevé property (no movable critical points). The solution is expressed in terms of the Painlevé transcendent P VI.

Once $u(\eta)$ is known, the wave moduli ρ_i and phases ϕ_i are simply expressed in terms of this function.

3. KMV ALGEBRAS AS CONDITIONAL SYMMETRIES OF CONDITIONALLY INTEGRABLE EVOLUTION EQUATIONS

Jimbo and Miwa [**8**] have introduced families of evolution equations, involving an increasing number of independent variables. They gave τ-function solutions of these families of equations, representing N solitons, for N arbitrary. One of the families is the Kadomtsev-Petviashvili hierarchy. The first two equations of this hierarchy are the potential KP equations itself, involving 3 independent variables, and an equation involving four independent variables, that has been called the Jimbo–Miwa equation [**22**]. In standard notations the PKP and JM equations can be written as

$$w_{xxxx} + 6w_x w_{xx} + 3w_{yy} - 4w_{xt} = 0, \tag{19}$$

$$w_{xxxy} + 3w_x w_{xy} + 3w_y w_{xx} + 2w_{yt} - 3w_{xz} = 0, \tag{20}$$

respectively. It has been shown [22] that the JM equation on its own does not pass the usual integrabality indicators. Thus, eq. (20) does not pass the Painlevé test and its solitary wave solutions can, in general, not be combined into multisolitons [22]. If, on the other hand, one restricts to a subclass of solutions, namely those that for a fixed value of $z = z_0$ satisfy the PKP equation (19), then those simultaneous solutions of both equations do pass the integrability tests. This type of situation has been call "conditionally integrability" [22] and its meaning is that the linear techniques of the inverse spectral transform and generalizations thereof can be used to obtain a subset of solutions of the considered equation, satisfying additional conditions (i.e. all the preceeding equations in a hierarchy).

The symmetry algebra of the PKP equation alone does have a KMV structure [23]. That of the J.M. equation alone does not, though it is infinite dimensional. Indeed, this algebra is generated by the vector fields

$$\begin{aligned}
P_t &= \partial_t, \qquad P_z = \partial_z, \qquad D_1 = z\partial_z + y\partial_y \\
D_2 &= 3t\partial_t + x\partial_x - 2y\partial_y - w\partial_w \\
R(g) &= g(t)\,\partial_x + \frac{2}{3}g'(t)\,x\partial_w \\
Y(h) &= h(z)\,\partial_y + \frac{3}{4}\,th'(z)\,\partial_x - \frac{1}{4}[2xh' + 3tyh'']\,\partial w \\
X(t) &= k(z)\,\partial_x - yk'(z)\,\partial_w \\
W(G) &= G(z,t)\,\partial_w .
\end{aligned} \tag{21}$$

There is no Virasoro subalgebra, since that would require the presence of elements of the type $f(z)\partial_z + \ldots$, or $g(t)\,\partial_t + \ldots$.

The KMV structure of the symmetry algebra is recovered [24], if we take the conditional integrability of the J.M. equation into account. Thus, if we look for local point transformations that leave the simultaneous solution set of the two equations (19) and (20) invariant, we find that the Lie algebra of this group has a basis consisting of the vector fields:

$$\begin{aligned}
&Y(h), \qquad X(k), \qquad W(G), \\
&Z(f) = f(z)\,\partial_z + \frac{1}{4}[xf' + \frac{2}{3}ytf'' + \frac{9}{32}t^3\,f'''\,]\,\partial_x \\
&\quad + \frac{1}{2}\,[yf' + \frac{9}{16}t^2 f''\,]\,\partial_y + \frac{3}{4}tf'\,\partial_t - \frac{1}{4}[wf' \\
&\quad + xyf'' + \frac{3}{4}t\,(y^2 + \frac{3}{4}tx)\,f''' + \frac{9}{32}yt^3\,f^{IV}\,]\,\partial_w . \\
&T(g) = g\partial_t + \frac{1}{32}\,[16yg' + 9t^2\,g''\,]\,\partial_x + \frac{3}{4}tg'\,\partial_y - \frac{1}{32}\,[\,4(3tx + 2y^2)\,g'' \\
&\quad + 9yt^2\,g'''\,]\,\partial_w .
\end{aligned} \tag{22}$$

The commutation relations include the relation

$$[Z(f_1),\ Z(f_2)] = Z(f_1\, f_2' - f_1' f_2)$$

so that $Z(f)$ is a centerless Virasoro algebra. The entire algabra (22) can be interpreted as a KMV subalgebra of $\hat{sl}\ (14, \mathbb{R})$ [**24**].

4. CONCLUSION

Subalgebras of KMV algebras keep occuring as point symmetries of integrable evolution equations in more than two dimensions. They can even be used to pick out integrable cases amongst generically nonintegrable ones, [**13**] or to establish that an equation is only conditionally integrable [**24**]. The corresponding infinite dimensional symmetry groups are a very useful tool for obtaining large classes of solutions. On the other hand, it is quite possible to construct equations that have Kac-Moody-Virasoro symmetries and are not integrable [**25**]. What is missing at this stage is a theorem, linking integrability, in some well defined sense, and KMV point symmetries, in a rigorous manner.

Acknowledgements. The research reported here was partly supported by research grants from NSERC of Canada and the FCAR du Québec.

References

1. Olver P.J., "Applications of Lie Groups to Differential Equations," Springer, New York, 1986.
2. Bluman G. and Kumei S., "Symmetries and Differential Equations," Springer, New York, 1989.
3. Ovsiannikov L.V., "Group Analysis of Differential Equations," Academic Press, New York, 1982.
4. Kac V.G., "Infinite Dimensional Lie Algebras," Birkhauser, Boston, 1983.
5. Goddard P. and Olive D., Int. J. Mod. Phys. **A1** (1986), 303.
6. Cornille H., J. Math. Phys. **20** (1979), 1653.
7. Kaup D.J., Physica **1D** (1980), 5.
8. Jimbo M. and Miwa T., Publ. Res. Inst. Math. Sci. Kyoto Univ. **19** (1983), 943.
9. Champagne B. and Winternitz P., "Preprint CRM 1278," Montréal, 1985.
10. Schwarz F., Computing **34** (1985), 91.
11. Schwarz F., SIAM Review **30** (1988), 450.
12. David D., Kamran N., Levi D. and Winternitz P., Phys. Rev. Lett. **55** (1985), 2111; J. Math. Phys. **27** (1980), 1225.
13. Champagne B. and Winternitz P., J. Math. Phys. **29** (1988), 1.
14. Paquin G. and Winternitz P., "CRM 1625," Montréal, 1989.
15. Martina L. and Winternitz P., Ann. Phys. (N Y) **96** (1989), 231.
16. Winternitz P., "Symmetries and Nonlinear Phenomena," World Scientific, 358-375, 1988.
17. Winternitz P., "Preprint CRM 1619 Montréal 1989," (to appear in Partially Integrable Nonlinear Evolution Equations and their Physical Applications, Kluwer, Dordrecht), Montréal, 1989.
18. Patera J, Winternitz P. and Zassenhaus H., J. Math. Phys. **16** (1975), 1597; Ibid. **16** (1975), 1615; Ibid. **17** (1976), 717.
19. Patera J., Sharp R.T., Winternitz P. and Zassenhaus H., J. Math. Phys. **18** (1977), 2259.
20. Bureau F.J., Annals di Mat. Pura Appl. (IV) **91** (1972), 163.
21. Bureau F.J., Garcet A. and Goffar J., Ann. di Mat. Pura Appl. (IV) **92** (1972), 177.
22. Dorizzi B., Grammaticos B., Ramani A. and Winternitz P., J. Math. Phys. **27** (1986), 2848.
23. David D., Levi D. and Winternitz P., Phys. Lett. **A118** (1986), 390.
24. Rubin J. and Winternitz P., J. Math. Phys. **31** (1990 to appear), Montréal.
25. David D., Levi D. and Winternitz P., Phys. Lett. **A129** (1988), 161.

II. TODA SYSTEMS AND CONTROL THEORY

A.M. BLOCH, R.W. BROCKETT, Y. KODAMA, and T. RATIU

R.W. BROCKETT and Anthony BLOCH

Yoshimasa NAKAMURA

SPECTRAL EQUATIONS FOR THE LONG WAVE LIMIT OF THE TODA LATTICE EQUATIONS

A.M. BLOCH[1]*, R.W. BROCKETT[2]**, Y. KODAMA[1]*** AND T. RATIU[3]****

[1] Department of Mathematics, The Ohio State University
231 W 18th Ave, Columbus, OH 43210, USA
[2]Division of Applied Sciences, Harvard University
Cambridge, MA 02138, USA
[3] Department of Mathematics, University of California at Santa Cruz
Santa Cruz, CA 95064, USA

ABSTRACT. In this paper we discuss various aspects of the Hamiltonian structure and integrability of a continuous long wave limit of the Toda lattice on the infinite line.

INTRODUCTION

In recent work (Bloch, Brockett and Ratiu (1989), Bloch, Flaschka and Ratiu (1989)) we examined various aspects of the non-periodic Toda lattice. In particular, we analyzed the gradient structure of the Toda systems, using a double bracket formalism apparently first introduced in this context by Brockett (1988).

Brockett first began analyzing differential equations of the double bracket type, $\dot{L} = [L, [L, N]]$, as systems for matching and sorting numbers. In Bloch, Brockett and Ratiu (1989) (see also Bloch, Flaschka and Ratiu (1989)), we showed how a special case of these equations gives rise to the (generalized) Toda lattice equations. In an accompanying paper, Brockett and Bloch (1990) show that the continuous long wave approximation of the Toda lattice equations may also be written in the double bracket form and can act (in the absence of shocks) as a continuous sorter.

In a certain sense these equations go back to Euler in that a special case of the equations for the propagation of plane sound waves of finite amplitude can be recast in this form (see Gümral and Nutku (1989)). In the Toda lattice context the equations can be found in the work of Kupershmidt (1985). In the work of Kupershmidt and Gümral and Nutku it is shown also that these equations are tri-Hamiltonian or even quadri-Hamiltonian.

Several different sets of variables with corresponding Hamiltonian structures can be used to study these equations. In the first part of this paper, we write down the bi-Hamiltonian structure in two useful sets of variables that enable one to obtain (from a different point of view) the infinite set of integrals discussed in Brockett and Bloch (1990). In the second part of the paper we show how these equations fit into the general formalism of Kodama (1988) and Kodama and Gibbons (1989 a,b) for analyzing equations of hydrodynamic type. This gives us another method of determining the equations and conserved integrals. It also lays the ground work for explicit integration of the equations, a topic which we hope to discuss in detail in a future publication.

---

*Supported in part by NSF Grant DMS-87-01576 and AFOSR Grant AFOSR-ISSA-87-0077
**Supported in part by NSF Grant CDR-85-00108, AFGOSR Grant AFOSR-96-0197 and ARO Grant DA-AL03-86-K-0171
***Supported in part by NSF Grant DMS-88-05521
****Supported in part by NSF Grant DMS-87-031801

1. HAMILTONIAN STRUCTURES FOR THE CONTINUOUS TODA EQUATIONS

Recall that in Flaschka's form (Flaschka (1976), Moser (1974)), the Toda equations can be written

$$\begin{aligned} \dot{a}_k &= a_k(b_{k+1} - b_k) \\ \dot{b}_k &= 2(a_k^2 - a_{k-1}^2) . \end{aligned} \tag{1.1}$$

Taking the simple continuum limit described in Brockett and Bloch (1990), obtained by replacing $b_k(t)$ by $x(t,z)$ and $a_k(t)$ by $y(t,z)$, we obtain the system

$$\begin{aligned} \frac{\partial x}{\partial t} &= 2(y^2)_z \\ \frac{\partial y}{\partial t} &= (x)_z y . \end{aligned} \tag{1.2}$$

In Brockett and Bloch (1990), the initial-boundary problem is considered for t on $[0,\infty)$, z on $[0,1]$. Here we work with densities and consider the system *locally* in space, so that spatial boundary conditions play no role. There are a number of different forms of this equation which are useful for different analytical purposes.

These equations have a natural bi-Hamiltonian structure (see e.g. Kupershmidt (1985)).

PROPOSITION 1.1. *The system (1.2) has the bi-Hamiltonian structure*

$$J_0 = \begin{pmatrix} 0 & \partial_z y \\ y\partial_z & 0 \end{pmatrix} \tag{1.3}$$

with corresponding Hamiltonian

$$H_2 = \frac{1}{2}(x^2 + 2y^2) , \tag{1.4}$$

and

$$J_1 = \begin{pmatrix} 4y\partial_z y & x\partial_z y \\ y\partial_z x & y\partial_z y \end{pmatrix} \tag{1.5}$$

with corresponding Hamiltonian

$$H_1 = x , \tag{1.6}$$

where $(\dot{x},\dot{y})^T = J\nabla H$.

Actually, there is a third compatible Hamiltonian structure, but we omit this here.

Suppose we make the transformation $x = x$, $w = \ln y$. The transformed system is then

$$\begin{aligned} w_t &= x_z \\ x_t &= 2(e^{2w})_z . \end{aligned} \tag{1.7}$$

This (modulo constants) is the system given in Gümral and Nutku (1989). We have

PROPOSITION 1.2. *The system (1.7) has the bi-Hamiltonian structure*

$$J_0 = \begin{pmatrix} 0 & \partial_z \\ \partial_z & 0 \end{pmatrix} \tag{1.8}$$

with corresponding Hamiltonian

$$H_2 = \frac{1}{2}(x^2 + 2e^{2w}) \tag{1.9}$$

and

$$J_1 = \begin{pmatrix} 2e^{2w}\partial_z + 2\partial_z e^{2w} & x\partial_z \\ \partial_z x & 2\partial_z \end{pmatrix} \tag{1.10}$$

with Hamiltonian

$$H_1 = x\,. \tag{1.11}$$

Using the recursion operator $R = J_1 J_0^{-1}$ we can generate infinitely many conserved quantities (as in Nutku (1989) and Brockett and Bloch (1990)).

The first few, in the x, y variables, are

$$H_1 = x$$

$$H_2 = \frac{x^2}{2} + y^2$$

$$H_3 = \frac{x^3}{3} + 2xy^2$$

$$H_4 = \frac{x^4}{4} + 3x^2y^2 + \frac{3}{2}y^4\,. \tag{1.12}$$

Note that the H_n's are related to the I_n's of Brockett and Bloch (1990) by $I_n = n\int_0^1 H_n dz$ We shall rederive these integrals from a different point of view in the next section. Further, we indicate how one can derive a second sequence of integrals, both sequences being necessary for obtaining a complete set of solutions to the problem.

Note also that by using the recursion operator R, we can show that the integrals above are in involution.

2. THE SPECTRAL FORM OF THE CONTINUOUS TODA EQUATIONS

We can also derive the continuous Toda equations discussed above by taking the classical limit (see Kodama (1988)) of the spectral form of the discrete Toda equations. In spectral form we have the equations

$$L\psi = \lambda\psi \tag{2.1}$$

$$\psi_t = B\psi \tag{2.2}$$

whose compatibility relation is

$$L_t = [B, L]. \tag{2.3}$$

Here

$$L = a_n\Delta + b_n + \Delta^{-1}a_n \tag{2.4}$$

$$B = a_n\Delta - \Delta^{-1}a_n \tag{2.5}$$

where Δ is the unit shift operator.

We can now take the classical limit as follows (see Kodama (1989)).

Let $\Delta = e^{\varepsilon\frac{\partial}{\partial z}}$ where ε is the lattice spacing and set $\psi = \exp\left\{\frac{1}{\varepsilon}S\right\}$. Now let $\varepsilon \to 0$,taking therefore the stationary phase approximation. Setting $p = \frac{\partial S}{\partial z}$, we find

PROPOSITION 2.1. *The dispersionless (classical) limit of the spectral equations (2.1) and (2.2) with L and B given by (2.4) and (2.5) respectively, is the set of equations*

$$2y\cosh\, p + x = \lambda \tag{2.6}$$

$$\frac{\partial p}{\partial t} = \frac{\partial}{\partial z}(2y\sinh\, p) \tag{2.7}$$

(with $a_n = y$ and $b_n = x$).

Note that p here gives the conserved density.

We remark that it is shown in Brockett and Bloch (1990) that (2.3) can be written in double bracket form. Also in that paper a quite different perspective is given on the formula (2.6). Note also that if boundary conditions are included, the relationship between L and the problem of interest here becomes a bit tenuous due to the difficulty of defining L at the boundary.

We also wish to make some remarks here on the connection of the above equations with Kupershmidt's (1985) work. We note first that Kupershmidt writes down a set of generalized Toda systems of which the one discussed here is the first. Also, in his formalism, the operator L in the Lax equation $L_t = [B, L]$ is a formal asymptotic series in Δ, and the operators B_n in the higher symmetries are taken to be the polynomial part in Δ of $\frac{L^n}{n}$. In our case, to get the higher systems we take B_n to equal the anti-symmetrization of $\frac{L^n}{n}$.

It is possible to show that the continuous limit of the generalized Toda systems of Kuperschmidt is equivalent to the dispersionless KP hierarchy (see Kodama and Gibbons (1989a)) to within a change of variables. Thus the continuous Toda equations discussed here are part of the dispersionless KP hierarchy and may thus be solved explicitly using the methods of Kodama and Gibbons (1989b). These matters will be analysed in detail in a future publication.

Now we can recover the dispersionless Toda equations (1.2) from (2.6) and (2.7) as follows. We have

$$\frac{\partial}{\partial t}(2y\cosh p + x) = 2\frac{\partial y}{\partial t}\cosh p + 2y\sinh p\frac{\partial p}{\partial t} + \frac{\partial x}{\partial t}$$

$$= 2\frac{\partial y}{\partial t}\cosh p + 2x\sinh p\left(2\frac{\partial y}{\partial z}\sinh p + 2y\cosh p\frac{\partial p}{\partial z}\right) + \frac{\partial x}{\partial t} \tag{2.8}$$

by (2.7).

Now $\frac{\partial}{\partial z}(2y\cosh p + x) = 2\frac{\partial y}{\partial z}\cosh p + 2y\sinh p\frac{\partial p}{\partial z} + \frac{\partial x}{\partial z} = 0$. Hence $2y\sinh p\frac{\partial p}{\partial z} = -2\frac{\partial y}{\partial z}\cosh p - \frac{\partial x}{\partial z}$. Thus from (2.8) we have

$$\left(2\frac{\partial y}{\partial z} - 2y\frac{\partial x}{\partial z}\right)\cosh p - 4y\frac{\partial y}{\partial z} + \frac{\partial x}{\partial t} = 0.$$

This gives precisely the continuous Toda equations (1.2).

Now we show how one obtains the integrals (1.12) by inversion of the series in p given by (2.6), i.e. from

$$p = \cosh^{-1}\frac{\lambda - x}{2y} \tag{2.9}$$

If we expand (2.9) in an asymptotic series about $\lambda = \infty$,we obtain the integrals as the coefficients of the inverse powers of λ, i.e.

$$p = \ln\lambda - \ln y - \frac{H_1}{\lambda} - \frac{H_2}{\lambda^2} - \cdots . \tag{2.10}$$

Note that we also obtain the integral $H_0 = \ln y$ in this expression.

Further, $\frac{\partial p}{\partial \lambda}$ also gives a series whose coefficients are the integrals of motion and $\int \frac{\partial p}{\partial \lambda}dz = \xi(\lambda)$ of Brockett and Bloch (1990).

Note that in fact

$$\frac{\partial p}{\partial \lambda} = \frac{1}{2y\sinh p} = \frac{1}{\sqrt{(\lambda - x)^2 - 4y^2}} . \tag{2.11}$$

As remarked in Section 1, one also obtains a second sequence of integrals by expanding (2.9) about $\lambda = 0$, i.e.

$$p = \cosh^{-1}\frac{x}{2y} + \frac{1}{\sqrt{x^2 - 4y^2}}\lambda + \frac{x}{2(x^2 - 4y^2)^{\frac{3}{2}}}\lambda^2 + \cdots . \tag{2.12}$$

Thus we obtain the non-polynomial integrals

$$\hat{H}_0 = \cosh^{-1}\frac{x}{2y},$$

$$H_{-1} = \frac{1}{\sqrt{x^2 - 4y^2}},$$

$$H_{-2} = \frac{x}{(x^2 - 4y^2)^{\frac{3}{2}}},$$

$$\vdots \tag{2.13}$$

(This sequence plays no role in the work of Brockett and Bloch (1990).) Using the methods of Kodama and Gibbons (1989b), we hope to use the two sequences of integrals to give a complete set of solutions to the equations.

We remark finally that the equations discussed here, being of hydrodynamic type, are subject to shock formation. In Brockett and Bloch (1990) this matter is analyzed for the boundary value problem in some detail. Techniques for analyzing the problem on the infinite line may be found in Lax (1964) (see also Smoller (1983) and references therein).

References

A. M. Bloch, R. W. Brockett and T. Ratiu (1989), A New Formulation of the Generalized Toda Lattice Equations and their Fixed Point Analysis via the Momentum Map, Bull etin of the AMS, to appear.

A. M. Bloch, H. Flaschka and T. Ratiu (1989), A Convexity Theorem for Isospectral Manifolds of Jacobi Matrices in a Compact Lie Algebra, Duke Mathematical Journal, to appear.

R. W. Brockett (1988), Dynamical Systems that Sort Lists, Solve Linear Programming Problems and Diagonalize Symmetric Matrices, Proceedings of the 1988 IEEE Conference on Decision and Control, IEEE, New York.

R. W. Brockett and A. M. Bloch (1990), Sorting with the Dispersionless Limit of the Toda Lattice, to appear.

H. Flaschka (1976), The Toda Lattice, Phys. Rev. 39, 1924-1925.

H. Gümral and Y. Nutku (1989), Multi-Hamiltonian Structure of Equations of Hydrodynamic Type, preprint.

Y. Kodama (1988), A Solution Method for the Dispersionless KP Equations, Prog. in Theoretical Physics, Supp. 94, 184-194.

Y. Kodama and J. Gibbons (1989a), A method for Solving the Dispersionless KP Hierarchy and its Exact Solutions II, Phys. Lett. A. 135, No. 3, 167-170.

Y. Kodama and J. Gibbons (1989b), Integrability of the Dispersionless KP Hierarchy, Proc. of the Workshop on Nonlinear and Turbulent Processes in Physics, Kiev, to appear.

B. A. Kupershmidt (1985), Discrete Lax Equations and Differential- Difference Calculus, (Astérisque 23), Soc. Math. France.

P. D. Lax (1964), Development of Singularities of Solutions of Nonlinear Hyperbolic Partial Differential Equations, J. Math. Phys. 5 No. 5, 611-613.

J. Moser (1974), Finitely Many Mass Points on the Line under the Influence of an Exponential Potential, Batelles Recontres, Springer Notes in Physics, 417-497.

J. Smoller (1983), Shock Waves and Reaction-Diffusion Equations, Springer-Verlag, New York.

SORTING WITH THE DISPERSIONLESS LIMIT OF THE TODA LATTICE

R.W. BROCKETT* [1] and ANTHONY BLOCH** [2]

* Division of Applied Sciences, Harvard University
Cambridge, MA 02138, USA

** Department of Mathematics, Ohio State University
Columbus, OH 43210, USA

ABSTRACT. This investigation arose out of a study of finite dimensional systems of differential equations which sort lists of real numbers. We consider the system of nonlinear partial differential equations $\partial x/\partial t = 2\partial y^2/\partial z$; $\partial y/\partial t = (\partial x/\partial z)y$ $y(t,0) = y(t,1) = 0$ as an initial value problem. This system, which can be regarded as a long wave continuum limit of the Toda lattice, has solutions which, insofar as they avoid shocks, possess an infinite set of conserved quantities and behave as a sorter.

1. EQUIMEASURABLE REARRANGEMENTS

Following Hardy, Littlewood and Polya [1] we say that two measurable functions x and $\tilde{x}$ mapping [0,1] into $\mathbf{R}^1$ are *equimeasurable* if for any measurable set $A \subset \mathbf{R}^1$

$$\mu(x^{-1}(A)) = \mu(\tilde{x}^{-1}(A))$$

In this paper we work with Borel measure. Equimeasurability defines an equivalence relation on measurable functions; we write $[x]$ to denote the equivalence class containing x. In a certain sense, the elements of $[x]$ are rearrangements of each other. If we define $\phi(\omega) = \mu\{z|x(z) \geq \omega\}$, then $\phi : R^1 \to [0,1]$ is monotone decreasing from 1 to 0 and $\phi^{-1} \in [x]$. It is the unique monotone decreasing function in $[x]$ and can also be characterized as the element of $[x]$ which minimizes

$$c = \int_0^1 zx(z)\, dz$$

We can associate with $[x]$ a convex set, a function space analog of the Schur-Horn polytope which plays such a central role in the momentum map theories. Let $CH[x]$ denote the set of all measurable convex combinations of elements of $[x]$. Notice that for any real α the shifted function $x_\alpha(z) = x((z+\alpha) \bmod 1) \in [x]$. The representation

$$\bar{x} = \int_0^1 x((z+\alpha) \bmod 1)\, d\alpha$$

shows that $CH[x]$ contains the constant function whose value is the average of an element of $[x]$.

Obviously the p-means

$$m_p = \left(\int_0^1 x^p(z)\, dz\right)^{1/p} \qquad ; \qquad p = 0, 1, 2, 3, \ldots$$

[1] Supported in part by the the U.S. Army Research Office under grant DAAL03-86-K-0171 the National Science Foundation under grant CDR-85-00108 and DARPA grant AFOSR-89-0506.

[2] Supported in part by the National Science Foundation under grant DMS-87-01576 and the Air Force Office of Scientific Research under grant AFOSR-ISSA-87-0077.

are the same for any $x \in [x]$ and, as we will see, the set $\{m_p\}_{p=1}^{\infty}$ characterizes $[x]$. If elements of $[x]$ satisfy $|x(z)| \leq \alpha$, then for $|s| > \alpha$ the sum

$$\hat{x}(s) = \sum_{p=0}^{\infty} \tfrac{1}{s}(m_p/s)^p$$

and the corresponding integral

$$\hat{x}(s) = \int_0^1 \frac{dz}{s - x(z)}$$

converge. Later we will give a formula for the monotone increasing representative of $[x]$ in terms of $\hat{x}$. Notice that as a function of the complex variable, $\hat{x}$ has an analytic continuation to the interior of the disk $|s| \leq \alpha$ except possibly for the segment of the real axis which lies between the minimum and maximum value of x.

We give three examples which shed light on subsequent developments. If x is piecewise constant, and takes on the values $\lambda_1, \lambda_2, \ldots, \lambda_k$ over intervals of length $\mu_1, \mu_2, \ldots, \mu_k$, respectively, then $\hat{x}$ is a proper rational function

$$\hat{x}(s) = \sum_{i=1}^{k} \frac{\mu_i}{s - \lambda_i}$$

If x is piecewise linear with slopes $1/\alpha_1, 1/\alpha_2, \ldots, 1/\alpha_k$, then $\hat{x}$ takes the form

$$\hat{x}(s) = \ln \prod_{i=1}^{k} \left(\frac{s - \beta_i}{s - \beta_{i-1}} \right)^{\alpha_i}$$

Finally, if n is a positive integer and if $x(z) = a \cos \pi n z$, then $\hat{x}(s)$ is

$$\hat{x}(s) = \frac{a}{\sqrt{s^2 - a^2}}$$

Notice that $\hat{x}(s)$ is independent of $n \in \mathbf{Z}$ because on [0,1] $\cos \pi z$ and $\cos \pi n z$ are just rearrangements of each other.

In general $\hat{x}(s)$ has a representation in terms of a nonnegative Borel measure μ (c.f. Bernstein's representation of completely monotonic functions)

$$\hat{x}(s) = \int_{-\alpha}^{\alpha} \frac{d\mu(\lambda)}{(s - \lambda)}$$

and the problem of determining x from $\hat{x}$ is that of finding μ since, formally speaking, $(dx/dz)^{-1} dz = d\mu$.

Appealing to the definition of $\hat{x}$ we see that if x is bounded and if Γ is a circle centered at 0 having radius larger than any value of $|x(z)|$, then

$$\frac{1}{2\pi i} \int_{\Gamma} s\hat{x}(s)\, ds = \int_0^1 x(z)\, dz$$

Moreover, if $x = x_1 + x_2$ with $x_1 x_2 = 0$, then $\hat{x} = \hat{x}_1 + \hat{x}_2$. Define x_q^- to be the function which is equal to $x(z)$ if $x(z) < q$ and equal to zero if $x(z) \geq q$, then

$$\int_0^1 x_q^-(z)\, dz = \frac{1}{2\pi i} \int_{\Gamma} s\hat{x}_q^-(s)\, ds$$

Thus if any singularity of $\hat{x}$ at q is removable, then for $r = q + \alpha$

$$F(q) \stackrel{\text{def}}{=} \frac{1}{2\pi i} \int_{s=re^{i\theta}-\alpha} s\hat{x}(s)\, ds$$

Because $\hat{x}$ is analytic off the interval of the real axis which lies between $\alpha = \inf x(z)$ and $\beta = \sup x(z)$, we see that if $\hat{x}(s)$ does not have a pole at r, the two contours shown in figure 1 can be used to define contour integrals which are equal. Assuming that we take any branch cuts for $\hat{x}$ to lie on the real axis between α and β,

$$F(q) = \frac{1}{2\pi i} \lim_{\epsilon \to 0} \int_{-\infty}^{q} \eta[\hat{x}(\eta + i\epsilon) - \hat{x}(\eta - i\epsilon)]\, d\eta$$

This integral is monotone increasing as a function of q and from the definition of F,

$$f(q) = (dF/dq) = q/(dx/dz)$$

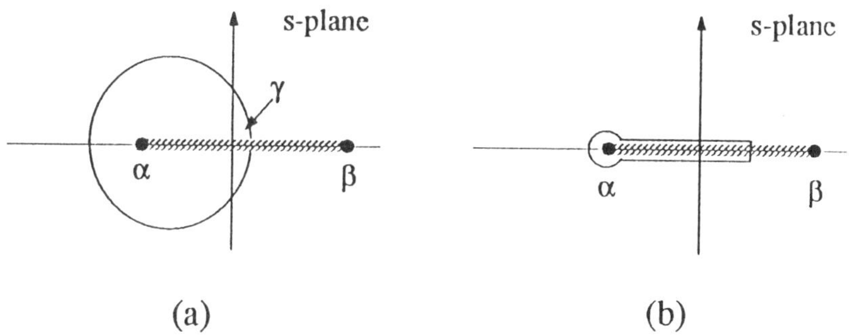

Figure 1: The evaluation of $F(q)$.

LEMMA 1. Let $[x]$ be such that its monotone increasing representative $\tilde{x}$ is bounded, differentiable, strictly monotone increasing on [0,1] and let $\hat{x}$ be defined as above. Then the inverse of $\tilde{x}$ is given by

$$z(\tilde{x}) = \lim_{\epsilon \to 0} \frac{1}{2\pi i} \int_{\alpha}^{\tilde{x}} (\hat{x}(\sigma + i\epsilon) - \hat{x}(\sigma - i\epsilon))\, d\sigma$$

This lemma can be said to describe a sorting procedure in that $\hat{x}$ depends only on $[x]$ and yet the final expression gives a monotone function. It may be compared with the well known inverse function theorems of complex function theory based on Rouché's theorem. See, for example, Chapter 9 of [2].

2. SORTING AND INVARIANTS

In some recent work on analog computing [3] the question of sorting lists with ordinary differential equations was investigated. Here we investigate partial differential equations of the evolution type with the goal of identifying a system whose solutions evolve in time in such a way as to converge to the monotone rearrangement of its initial data. To this end we will make use of a second function y which is initially small and finally zero but serves as a "swap space" for the sorting process.

In [3] we explored the role of a differential equation involving a double commutator of real symmetric matrices, $\dot{H} = [H,[H,N]]$, as a sorter. In [4] it is shown that by restricting H and specializing N this equation reduces to various forms of the Toda lattice, thus suggesting that the Toda lattice can be thought of as an analog system for sorting lists of real numbers. Here we look at a version of this equation in which H is a bounded self-adjoint operator whose domain is a space of functions defined on a subinterval of the real line. Deift *et. al.* investigate infinite dimensional Toda models [5] and Kupershmidt [6] develops some related ideas.

One way to write the equations for the Toda lattice is in the Flaschka form

$$\begin{aligned} \dot{b}_k &= 2(a_k^2 - a_{k-1}^2) \\ \dot{a}_k &= (b_{k+1} - b_k)a_k \end{aligned}$$

these equations have a variety of continuum limits. We focus on a particularly simple one in which $b_k(t)$ is replaced by $x(t,z)$ and $a_k(t)$ is replaced by $y(t,z)$ both being defined on $[0,\infty) \times [0,1]$ and satisfying

$$\begin{aligned} \frac{\partial x}{\partial t} &= 2\frac{\partial y^2}{\partial z} \\ \frac{\partial y}{\partial t} &= \left(\frac{\partial x}{\partial z}\right) \cdot y \end{aligned}$$

The original equations suggest the boundary conditions $y(t,0) = y(t,1) = 0$ and in this form the limit has a number of interesting properties.

LEMMA 2. If x and y are classical solutions of $x_t = 4yy_z$; $y_t = yx_z$ for $(t,z) \in [0,\infty) \times [0,1]$ with $y(t,0) = y(t,1) = 0$, then (a) the equilibrium points correspond to $y \equiv 0$. (b) the first moment of x decreases monotonically according to

$$\frac{d}{dt}\int_0^1 zx(t,z)\,dz = -4\int_0^1 y^2(t,z)\,dz$$

and (c) for each positive integer p

$$\begin{aligned} I_p &= \int_0^1 \sum_{\substack{k=0 \\ k \text{ even}}}^{p} x^{p-k} y^k \binom{p}{k}\binom{k}{k/2}\,dz \\ &= \frac{1}{2\pi}\int_0^1\int_0^{2\pi} (x + 2y\cos\theta)^p\,d\theta\,dz \end{aligned}$$

is constant along trajectories.

PROOF: Part (a) is obvious. The statement (b) follows from integrating $zx_t = 4zy_zy$, applying integration-by-parts to the right-hand side. Notice that this equality implies that x is becoming more nearly sorted as time evolves.

To prove (c) we compute the derivative of I_p,

$$\frac{d}{dt}I_p = \int_0^1 \sum_{\substack{k=0 \\ k \text{ even}}}^{p} c_k(4(p-k)x^{p-k-1}y^{k+1}y_z + kx^{p-k}y^k x_z)\,dz$$

where

$$c_k = \binom{p}{k}\binom{k}{k/2}$$

We integrate the second term in parentheses by parts (making use of $y(t,0) = y(t,1) = 0$) to get

$$\frac{d}{dt}I_p = \sum_{\substack{k=0\\k\text{ even}}}^{p} \frac{c_k k x^{p-k+1}}{p-k+1} y^k + \int_0^1 \sum_{\substack{k=0\\k\text{ even}}}^{p} c_k \left(4(p-k)x^{p-k-1}y^{k+1}y_z - \frac{k^2}{p-k+1}x^{p-k+1}y^{k-1}y_z\right) dz$$

The coefficient of $x^{p-k-1}y^{k+1}$ in this integral is

$$\begin{aligned} f_k &= c_k 4(p-k) - c_{k+2}\frac{(k+2)^2}{(p-k-1)} \\ &= \binom{p}{k}\frac{k!4(p-k)}{(k/2)!(k/2)!} - \binom{p}{k}\frac{(p-k)(p-k-1)(k+2)!(k+2)^2}{(k+1)(k+2)(k/2)!(k/2)!(k/2+1)^2(p-k-1)} \\ &= 0 \end{aligned}$$

which establishes the desired result. For k an even integer the Cauchy integral formula lets one relate the central binomial coefficient to the integral of the power of the cosine

$$\int_0^{2\pi} 2^k \cos^k\theta \, d\theta = 2\pi \binom{k}{k/2}$$

For k odd this integral vanishes. Thus we can also express I_p as the double integral indicated at the end of the lemma statement.

By Lemma 2, the following are all constant along classical solutions.

$$\begin{aligned} I_1 &= \int_0^1 x \, dz \\ I_2 &= \int_0^1 x^2 + 2y^2 \, dz \\ I_3 &= \int_0^1 x^3 + 6xy^2 \, dz \end{aligned}$$

With respect to part (a) we note that even more is true, z can be replaced by any function m with a positive derivative. This gives

$$\begin{aligned} \frac{\partial}{\partial t}\int_0^1 m(z)x(t,z)\, dz &= \int_0^1 4m(z)y_z(t,z)y(t,z)\, dz \\ &= -\int_0^1 m_z(z)y^2(t,z)\, dz \end{aligned}$$

Thus if $m_z(z)$ is monotone increasing, the integral $\int_0^1 m(z)x(t,z)\, dz$ is monotone decreasing. Secondly, by combining this with the fact that I_2 is constant we get

$$\frac{\partial}{\partial t}\int_0^1 (x(t,z) - m(z))^2 + 2y^2(t,z)\, dt = \int_0^1 2m_z(z)y^2(t,z)\, dz$$

Thus if $m_z(z) \leq \varepsilon < 0$, any solution which exists on $[0, \infty)$ must be such that

$$\phi = \int_0^1 (x(t,z) - m(z))^2 + 2y^2(t,z)\, dz$$

approaches a limit. In view of the previous expression for $d\phi/dt$ this can only happen if y^2 goes to zero. We summarize with a lemma.

LEMMA 3. For any initial data with $y(0,z) \geq 0$ and $y(t,0) = y(t,1) = 0$ such that the initial value problem has a smooth solution on $[0,\infty) \times [0,1]$, y goes to zero as t goes to infinity and x approaches an extreme point of $CH[\tilde{x}]$ with $[\tilde{x}]$ being determined by the invariants.

3. THE INVERSION LEMMA

As noted above, for a given set of invariants, $I_0, I_1, I_2, \ldots$, there is at most one monotone decreasing $\tilde{x}$ which satisfies

$$I_p = \int_0^1 \tilde{x}^p\, dz \qquad ; \qquad p = 1, 2, \ldots$$

We need to show that in the situation where the I's are derived from a pair of bounded functions x and y via

$$I_p = \tfrac{1}{2\pi} \int_0^1 \int_0^\pi (x(z) + 2y(z)\cos\theta)^p\, d\theta dz$$

there always exists such an $\tilde{x}$. This may be thought of as a kind of spectral theorem, analogous to establishing the existence of a diagonal matrix which is similar to a symmetric tridiagonal matrix.

Before stating the lemma, we point out that by the Riemann-Lebesgue lemma one can assert that I_p, as given by the integral involving z and θ, can be written as

$$I_p = \lim_{n\to\infty} \int_0^1 (x(z) + 2y(z)\cos nz)^p\, dz$$

This suggests that, in some sense, the $\tilde{x}$ we seek is defined by the high frequency limit of $x(z) + 2y(z)\cos nz$. However, it seems simpler not to pursue this line of reasoning directly but to eliminate θ in the following way. Define $\xi(s)$ by

$$\xi(s) = \frac{1}{2\pi} \int_0^1 \int_0^{2\pi} \frac{d\theta dz}{s - x(z) - 2y(z)\cos\theta}$$

An integration with respect to θ yields

$$\xi(s) = \int_0^1 \frac{dz}{\sqrt{(s - x(z))^2 - 4y^2(z)}} \qquad ; \qquad (s - x(z))^2 > y^2(z)$$

This can be used to generate an explicit inversion formula.

LEMMA 4. If x and y are real valued continuous functions defined on [0,1] with $y(z) > 0$ on (0,1), then there exists $\tilde{x}$ defined on [0,1] such that for $p = 0, 1, 2, \ldots$

$$\frac{1}{2\pi} \int_0^1 \int_0^{2\pi} (x(z) + 2y(z)\cos\theta)^p\, d\theta dz = \int_0^1 \tilde{x}^p(z)\, dz$$

PROOF: The representation given above makes it clear that ξ can be taken to be analytic off the line segment $[\alpha, \beta] = [\min(x - |y|), \max(x + |y|)]$ and since $y^2(s)$ is positive, $\xi(s)$ has no poles. Thus ξ can be written in the form required by Lemma 1. The use of the contour integration formula in the slit form allows us to express the inverse of $\tilde{x}$ as

$$z(\tilde{x}) = 2\int_{\alpha}^{\tilde{x}} \int_{0}^{1} \left(\frac{dz}{\sqrt{(q - x(z))^2 - 4y^2(z)}} \right) dq$$

As an example, we note that if $x = a$ and $y = b$, then $\tilde{x} = a + b\cos \pi z$. More generally, if (x, y) is piecewise constant, then $\tilde{x}$ can be taken to be "piecewise sinusoidal."

4. HYPERBOLIC FORM AND SHOCKS

Examples as elementary as $\dot{x} = x^2$, having solution $x(t) = \frac{-1}{t - 1/x_0}$ indicate that one may expect finite escape times for differential equations with quadratic right-hand sides. The ordinary finite dimensional Toda lattice does not have a finite escape time for real initial data because the integrals of the motion keep Σa_i^2 and Σb_i^2 bounded. The situation for the limiting form under investigation here is different in that even though $\int (x^2 + 2y^2)dz$ is constant, this does not preclude the generation of shocks.

If in place of $x_t = 4yy_z;\ y_t = x_z y$ we introduce $r = (x + 2y)$ and $s = x - 2y$, then the equations take the form

$$\begin{aligned} r_t &= 2yr_z \quad &;\quad r(t,0) &= s(t,0) \\ s_t &= -2ys_z \quad &;\quad r(t,1) &= s(t,1) \end{aligned}$$

with the boundary conditions indicated. Since y does not change sign, these equations place in evidence a wave picture of the solution. The r wave travels to the left with a z-dependent speed y that goes to zero near $z = 0$ and $z = 1$ whereas the s wave travels to the right at the same speed. Because of the tendency of nonlinear hyperbolic systems to generate solutions with shocks, it is comforting to see that at least for some initial data solutions exist on $[0, \infty)$.

EXAMPLE: Consider an x and y pair of the form

$$\begin{aligned} x(t,z) &= a(t)z + b(t) \\ y^2(t,z) &= c(t)(z - z^2) \end{aligned}$$

substituting into $x_t = 4yy_z;\ y_t = yx_z$ gives

$$\begin{aligned} \dot{a}(t) &= -4c(t) \\ \dot{b}(t) &= 2c(t) \\ \dot{c}(t) &= 2a(t)c(t) \end{aligned}$$

The first and third of these equations are equivalent to $\ddot{a} = -8ac = 2a\dot{a}$. This integrates to give an equation of the Riccati type

$$\dot{a}(t) = a^2(t) - 4c(0) - a^2(0)$$

which has the solution

$$a(t) = -\sqrt{a^2(0) + 4c(0)}\tanh\left(\sqrt{a^2(0) + 4c(0)}t\right)$$

Obviously $c(0)$ is nonnegative. If $c(0)$ is positive initially, it remains so and a is monotone decreasing. It approaches a limit as t goes to infinity

$$\lim_{t\to\infty} a(t) = -\sqrt{a^2(0)+4c(0)}$$

Thus for this particular initial data, the equations admit a smooth solution for all time (although y_z is infinite at the boundaries).

This may be an appropriate place to remark that if one introduces $v = y^2$, then the equations become

$$\begin{aligned} x_t &= 2v_z \\ v_t &= 2x_z v \end{aligned}$$

This transformation regularizes the above solution in that v_z is bounded on [0,1].

The general situation is explained by Lemma 5.

LEMMA 5. Let $x(0,z)$ and $y(0,z)$ be defined of [0,1] and differentiable there. The solution of $x_t = 4yy_z; y_t = yx_z$ with $y(t,z) = 0$ and $y(t,0) = y(t,z) = 0$ develops shocks if either $x(t,z)+2y(0,z)$ has local minima on (0,1) or $x(0,z)-2y(0,z)$ has local maxima on (0,1).

PROOF: Consider the sketch in figure 2. The case where $x(0,0) > x(0,1)$ and $x(0,z)+2y(0,z)$ has a local minimum is illustrated. Because

$$\frac{\partial}{\partial t}(x(t,z)+2y(t,z)) = y(t,z)\frac{\partial}{\partial z}(x(t,z)+2y(t,z))$$

we see that as time evolves the local maxima marked a and c move to the left, always tangent to the horizontal lines indicated and the local minimum marked b moves to the left along the horizontal line indicated.

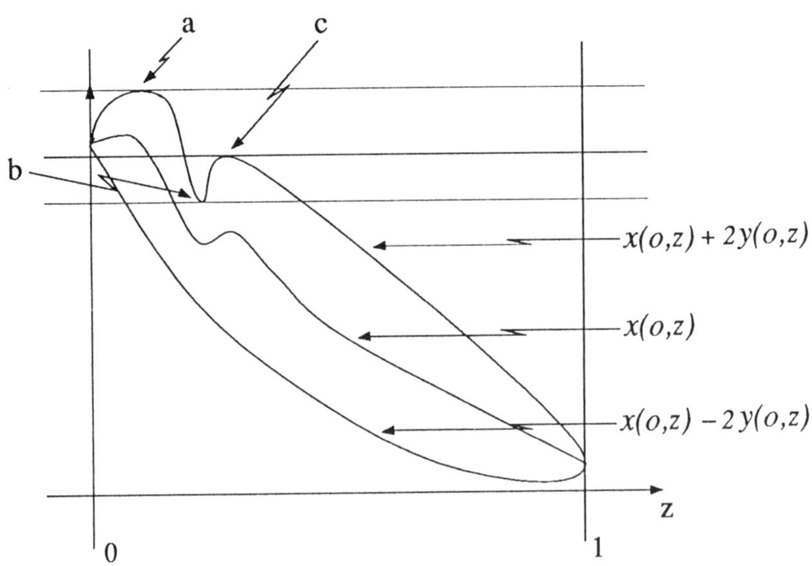

Figure 2: The generation of shocks.

Assuming that the solutions evolve smoothly leads to a contradiction because on one hand $x + 2y$ must go to something monotone (Lemma 2) and yet $x + 2y$ can never be monotone because of the above constraints. The other cases involving $x(0,0) < x(0,1)$, etc., can be treated similarly. We do not give further details.

The nondiffusive version of Burgers' equation arises as a "restriction to the diagonal." To see this notice that the change of scale $t \to t/2$ and $2y = u$ converts the (x, y)-pair to the form

$$\begin{aligned} x_t &= uu_z \\ u_t &= ux_z \end{aligned}$$

If the initial data satisfies $x(t,0) = \pm u(t,0)$ and if m satisfies the nondiffusive form of Burgers' equation $m_t = mm_z$ with $m(t,0) = x(t,0) = \pm u(t,0)$, then setting $x(t,z) = m(t,z)$ and $\pm u(t,z) = m(t,z)$ gives a solution to the x, u equations. Thus we may say that we obtain the nondiffusive form of Burgers' equation by changing the time scale and restricting the dispersionless limit of the Toda lattice to the case of $x = \pm u = \pm 2y$. Because the solutions of Burgers' equation with the present boundary data inevitably develop shocks, this observation also makes it clear that we must avoid certain initial data if we are to make effective use of this system to do equimeasurable rearrangements.

5. THE DOUBLE BRACKET EQUATION

In view of the way in which the finite dimensional theory proceeds it is natural to try to express the generating function $\hat{x}(s)$ as tr $(Is - L)^{-1}$ for some operator L. We now discuss this.

If L is an element of a Lie algebra $\mathcal{L}$, there is then for each $N \in \mathcal{L}$ a differential equation in $\mathcal{L}$, namely

$$\dot{L}(t) = [L(t), [L(t), N]]$$

In an earlier paper [3] it was pointed out that along solutions of this equation evolving in the space of symmetric matrices (thought of as the imaginary part of the skew hermetians) satisfies the identity

$$\frac{d}{dt}\mathrm{tr}(LN) = \mathrm{tr}(([L,N])([L,N]))$$

and thus tr LN is a monotone function of time. It is easy to see that because $[L, N]$ is skew symmetric, along solutions of this equation we have

$$\frac{d}{dt}\mathrm{tr}(L^k(t)) = 0$$

This implies, among other things, that solutions of this equation exist for all positive (and negative) time since tr $L^2(t) = \mathrm{tr}(L^T(t)L(t))$ is constant along solutions.

The equations of this paper fit into the pattern of associating the Toda lattice with an equation of the form $\dot{L} = [L, [L, N]]$ in the following way. Let E denote the shift operator acting on $L_2(-\infty, \infty)$ by the formula

$$E\psi(z) = \psi(z+1)$$

Consider the above equation in L with L taken to be the bounded, self-adjoint operator on $L_2(-\infty, \infty)$

$$L = x(t,z) + y(t,z)E + E^{-1}y(t,z)$$

and N being multiplication by z. Evaluating $[L, z]$ we get

$$\begin{aligned} [L,z] &= [x(t,z) + y(t,z)E + E^{-1}y(t,z), z] \\ &= y(t,z)E - E^{-1}y(t,z) \end{aligned}$$

and

$$\begin{aligned}[L,[L,z]] &= [x(t,z)+y(t,z)E+E^{-1}y(t,z), y(t,z)E-E^{-1}y(t,z)] \\ &= y(t,z)(x(t,z)-x(t,z+1))E - y(t,z-1)(x(t,z)-x(t,z-1))E^{-1} \\ &\quad + 2y^2(t,z-1) - 2y^2(t,z)\end{aligned}$$

From this we obtain the differential-difference equations (c.f. Kupershmidt [4])

$$\begin{aligned}\frac{\partial x(t,z)}{\partial t} &= 2(y^2(t,z-1)-y^2(t,z)) \\ \frac{\partial y(t,z)}{\partial t} &= y(t,z)(x(t,z)-x(t,z+1))\end{aligned}$$

Appealing to Taylor series approximation, these equations give rise to a hierarchy of partial differential equations characterized by the accuracy of the expansion for $x(t,z+1)$ and $y(t,z-1)$. The simplest member of this family is the pair investigated above. The next in the series is

$$\begin{aligned}x_t &= -4y_z y + 2y_z^2 + 2y_{zz}y \\ y_t &= -yx_z - \tfrac{1}{2}yx_{zz}\end{aligned}$$

In this case $r = x + 2y$ and $s = x - 2y$ satisfies

$$\begin{aligned}r_t &= yr_z + yr_{zz} + 2y_z^2 \\ s_t &= -ys_z - ys_{zz} + 2y_z^2\end{aligned}$$

References

[1] G. H. Hardy, J. E. Littlewood and G. Polya, *Inequalities*, 2nd edition, Cambridge University Press, Cambridge, England, 1952.

[2] E. Hille, *Analytic Function Theory*, Ginn, Boston, 1959 (Chapter 9).

[3] R. W. Brockett, "Dynamical Systems that Sort Lists, Solve Linear Programming Problems and Diagonalize Symmetric Matrices," *Proceedings of 1988 IEEE Conference on Decision and Control*, IEEE, New York, 1988.

[4] A. Bloch, R. W. Brockett and T. Ratiu, "A New Formulation of the Generalized Toda Lattice Equations and their fixed Point Analysis via the Momentum Map," *Bulletin of the AMS*, (to appear).

[5] P. Deift, L. C. Li and C. Tomei, "Toda Flows with Infinitely Many Variables," *J. of Functional Analysis*, Vol. 64, No. 3, 1985, pp. 358-402.

[6] B. A. Kupershmidt, *Discrete Lax Equations and Differential–Difference Calculus*, (Astèrisque, Vol. 23) Soc. Math. France, 1985. (page 121)

[7] J. Smoller, *Shock Waves and Reaction-Diffusion Equations*, Springer-Verlag, New York, 1983.

[8] A. C. Newell, *Solitons in Mathematics and Physics*, SIAM, Philadelphia, 1985.

MODULI SPACES OF CONTROLLABLE LINEAR DYNAMICAL SYSTEMS AND NONLINEAR INTEGRABLE SYSTEMS OF TODA TYPE

YOSHIMASA NAKAMURA

Department of Mathematics, Gifu University
Gifu 501-11, Japan

ABSTRACT. Nonlinear integrable flows of Toda type on moduli spaces of controllable linear differential and delay-differential dynamical systems with multi-input are found. They are generalizations of the cyclic-Toda flow which completely parametrizes spaces of certain rational functions.

1. INTRODUCTION

As was reported in [1], one of the basic problems in linear control theory is to investigate the geometry of moduli spaces of linear systems, namely, parameter spaces of families of linear dynamical systems. In a typical case, the moduli space is a quotient space of triples of real matrices (A, B, C), where $A \in \mathbf{R}^{k\times k}, B \in \mathbf{R}^{k\times m}, C \in \mathbf{R}^{n\times k}$, by the $GL(k,\mathbf{R})$ action $g : (A, B, C) \to (gAg^{-1}, gB, Cg^{-1})$, $g \in GL(k,\mathbf{R})$. These matrices represent the autonomous linear dynamical system $\frac{dx}{ds}(s) = Ax(s) + Bu(s), y(s) = Cx(s)$, $s \in \mathbf{R}$, where $x(s), u(s)$ and $y(s)$ are state, control action (input) and observed (output) variables, respectively. The identification under the $GL(k,\mathbf{R})$ action is due to the fact that change of coordinates in state variables do not have any effect on input/output behavior. Linear algebra of course provides the natural tools used in linear control theory. For instance, if the matrices A and B satisfy the generic rank condition $rank(B\ AB\ \ldots\ A^{k-1}B) = k$, then the system is called to be *controllable* and has a critical importance. However, even if complex systems, the geometry of the associated moduli spaces is still imperfectly understood.

In classical mechanics, the phase spaces of completely integrable nonlinear systems have been investigated fully through the invariant tori theorem. Therefore if there exists a completely integrable flow on a moduli space, it can give an information about the geometry of moduli space. Indeed, for a scalar input/output case, where $m = n = 1$, the author [2] proved that each flow of a new nonlinear completely integrable system is diffeomorphic to one of the connected components of (level manifolds of) the space $\mathcal{R}(k)$ of rational functions $f(z) = C(zI - A)^{-1}B$ of degree k. This nonlinear system is called the *cyclic-Toda hierarchy*. The function $f(z)$ is the transfer function of the linear system which uniquely describes the effect of controls on observations. The invariant tori/cylinder theorem then says that each connected component is diffeomorphic to the cylinder $T^l \times \mathbf{R}^{k-l-1}$ for some $l \in \{0, 1, \ldots, k-2\}$. This is an improved version of the early works by Moser [3] and Krishnaprasad [4]. They showed that the usual finite nonperiodic Toda equation induces a flow on the simplest connected component diffeomorphic to $\mathbf{R}^{k-1}$.

It is to be noted that the space $\mathcal{R}^{\mathbf{C}}(k)$ of rational functions over $\mathbf{C}$, where (A, B, C) are complex and $m = n = 1$, also emerges in Yang-Mills theory. Donaldson [5] proved that $\mathcal{R}^{\mathbf{C}}(k)$ is homeomorphic to the extented moduli space $\tilde{\mathcal{M}}(k)$ of $SU(2)$ $k-$monopoles. In this case the complex cyclic-Toda hierarchy parametrizes $\mathcal{R}^{\mathbf{C}}(k)$ [6].

In this paper, we consider the moduli space of slightly different controllable linear systems which have *multi-input* ($m \geq 1$) and *no-output*. It will be shown that there exists a nonlinear integrable flow of Toda type on the moduli space. A moduli space of delay-differential systems is also discussed which is related to the moduli space of $SU(m)$ Yang-Mills $k-$instantons.

2. MODULI SPACE OF CONTROLLABLE SYSTEMS

Consider linear dynamical systems of the form

$$\frac{dx}{ds}(s) = Ax(s) + Bu(s), \tag{1}$$

where $x(s)$ is a $k-$vector of state variables, $u(s)$ is an $m-$vector of control variables, A and B are $k \times k$, $k \times m$ constant matrices over $\mathbf{R}$, respectively. It is well-known that the system is controllable if and only if

$$rank\left(B\ AB\ \ldots\ A^{k-1}B\right) = k. \tag{2}$$

This condition guarantees that for any states x_0, x_1 in $\mathbf{R}^k$ and times s_0, s_1 s.t. $s_0 < s_1$ there exist a control action $u(s)$ and a solution $x(s)$ of (1) satisfying $x(s_0) = x_0$ and $x(s_1) = x_1$. Let us define the quotient space $\Sigma_{k,m}$ of matrices (A, B) by the $GL(k, \mathbf{R})$ action

$$g : (A, B) \rightarrow (gAg^{-1}, gB), \tag{3}$$

where $g \in GL(k, \mathbf{R})$. Namely,

$$\Sigma_{k,m} = \left\{(A, B)\,\middle|\, rank(B\ AB\ \ldots\ A^{k-1}B) = k\right\}/\,GL(k, \mathbf{R}).$$

Since any linear change of coordinates $x \rightarrow gx$ in the state space $\mathbf{R}^k$ transforms (1) into the equivalent system $\frac{dx}{ds}(s) = gAg^{-1}x(s) + gBu(s)$, the space $\Sigma_{k,m}$ can be viewed as the moduli space of all controllable linear systems of the form (1). The topology of $\Sigma_{k,m}$ has been studied by many authors (see [7] and references therein). It is known that $\Sigma_{k,m}$ is a connected non-compact algebraic manifold of dimension km and $\Sigma_{k,1} \cong \mathbf{R}^k$ and $\Sigma_{1,m} \cong \mathbf{R} \times \mathbf{R}P^{m-1}$.

Let $(A_0, B_0) \in \Sigma_{k,m}$. By using the $GL(k, \mathbf{R})$ action (3), we can always assume that A_0 is upper-Hessenberg, that is,

$$A_0 = \begin{pmatrix} * & \cdots & \cdots & * \\ * & \ddots & & \vdots \\ & \ddots & \ddots & \vdots \\ 0 & & * & * \end{pmatrix}.$$

Here g takes the form of sequence of Householder matrices. Let t be a set of $k-$real parameters

$$t = (t_0, t_1, \ldots, t_{k-1}) \in \mathbf{R}^k. \tag{4}$$

We consider the QR factorization of exponential of $\sum_{j=0}^{k-1} t_j {A_0}^j$;

$$exp\left(\sum_{j=0}^{k-1} t_j {A_0}^j\right) = Q^{-1}(t)R(t), \tag{5}$$

where Q is an orthogonal matrix and R is an upper-triangular matrix with positive diagonal entries. Since the l.h.s. is nonsingular, there always exist the unique factors Q

and R. It is easy to see $Q(\mathbf{0}) = I$ and $R(\mathbf{0}) = I$, where $\mathbf{0} = (0, \ldots, 0)$ in $\mathbf{R}^k$. Taking derivatives of (5), we have

$$QA_0{}^jQ^{-1} = -\frac{\partial Q}{\partial t_j} \cdot Q^{-1} + \frac{\partial R}{\partial t_j} \cdot R^{-1}.$$

Noting that the first and the second terms of the r.h.s. are skew-symmetric and upper-triangular, respectively, we obtain

$$\frac{\partial Q}{\partial t_j} \cdot Q^{-1} = ((QA_0{}^jQ^{-1})_L)^\top - (QA_0{}^jQ^{-1})_L, \tag{6}$$

$$\frac{\partial R}{\partial t_j} \cdot R^{-1} = QA_0{}^jQ^{-1} + ((QA_0{}^jQ^{-1})_L)^\top - (QA_0{}^jQ^{-1})_L,$$

where M_L indicates the strictly lower-triangular part of M and $M^\top$ denotes the transpose of M. For any $(A_0, B_0) \in \Sigma_{k,m}$ let us define a nontrivial change of matrices

$$A(t) = Q(t)A_0Q^{-1}(t), \quad B(t) = R(t)B_0. \tag{7}$$

Taking derivatives of (7) and using (6), we derive the system $(0 \leq j \leq k-1)$ of nonlinear and linear equations

$$\frac{\partial A}{\partial t_j}(t) = \left[(A^j(t)_L)^\top - A^j(t)_L,\ A(t)\right], \tag{8}$$

$$\frac{\partial B}{\partial t_j}(t) = \left(A^j(t) + (A^j(t)_L)^\top - A^j(t)_L\right) B(t) \tag{9}$$

with the initial value $A(0) = A_0$ and $B(0) = B_0$.

Next we consider the compatibility of the system (8) and (9). For simplicity, we set

$$F_j = F_j(A) = (A^j(t)_L)^\top - A^j(t)_L.$$

Since $A(t) = R(t)A_0R^{-1}(t)$, A(t) is upper-Hessenberg. The integrability condition for (8) leads to $\left[\frac{\partial F_j}{\partial t_i} - \frac{\partial F_i}{\partial t_j} + [F_j, F_i], A\right] = 0$. Thus if F_j solve the system $(0 \leq i, j \leq k-1)$ of zero-curvature equations

$$\frac{\partial F_j(A)}{\partial t_i} - \frac{\partial F_i(A)}{\partial t_j} + [F_j(A),\ F_i(A)] = 0, \tag{10}$$

then (8) is integrable. Converse is not trivial but can be proved along the same line as in [5], consequently, (8) is equivalent to (10). Whereas the system (8) and (10) guarantees the integrability of (9). We see that the system (8) and (9) is compatible.

Now we shall prove the system (8) and (9) induces a nonlinear integrable flow on the moduli space $\Sigma_{k,m}$ of linear dynamical systems (1) with (2).

THEOREM 1. *If (A_0, B_0) is an element of the moduli space $\Sigma_{k,m}$, then $(A(t), B(t))$ is so for any $t \in \mathbf{R}^k$.*

PROOF: Using the expression of solutions (7), we see

$$\begin{aligned}&(B(t)\ A(t)B(t)\ \ldots\ A^{k-1}(t)B(t))\\ &= \left(R(t)B_0\ Q(t)A_0Q^{-1}(t)R(t)B_0 \ldots Q(t)A_0{}^{k-1}Q^{-1}(t)R(t)B_0\right)\\ &= Q(t)exp\left(\sum_{j=0}^{k-1} t_jA_0{}^j\right)\left(B_0\ A_0B_0\ \ldots\ A_0{}^{k-1}B_0\right).\end{aligned}$$

Here we use the fact that $Q^{-1}(t)R(t) = exp\left(\sum_{j=0}^{k-1} t_j A_0{}^j\right)$ commutes with A_0. It is proved that $rank\left(B\ AB \ldots A^{k-1}B\right) = k$ for any $t \in \mathbf{R}^k$. ∎

The system (8) and (9) is a modification of the cyclic-Toda hierarchy introduced in [2] to multi-input case without output. When A_0 is similar to a Jacobi matrix, the system (8) is reduced to the usual finite nonperiodic Toda equations introduced in [3]. We shall call the system the cyclic-Toda hierarchy over the moduli space $\Sigma_{k,m}$. The hierarchy induces $(A_0, B_0) \to (QA_0Q^{-1}, RB_0)$. On the other hand, under the $GL(k, \mathbf{R})$ action (3) by $g = Q^{-1}$, the matrices (A_0, B_0) change as $(A_0, B_0) \to (Q^{-1}A_0Q, Q^{-1}B_0)$. Combining these two actions, we have an alternative form of the cyclic-Toda flow

$$(A_0, B_0) \to \left(A_0,\ exp\left(\sum_{j=0}^{k-1} t_j A_0{}^j\right) B_0\right). \tag{11}$$

This implies that the cyclic-Toda hierarchy over $\Sigma_{k,m}$ can be linearized on the space

$$\mathcal{B}_{A_0} = \left\{(A_0, B)\,\middle|\, rank(B\ A_0B\ \ldots\ A_0{}^{k-1}B) = k\right\}, \tag{12}$$

where A_0 is a fixed cyclic matrix. For any fixed cyclic matrix A_0, we define a vector field $\{X_l\}_{0 \le l \le k-1}$ on $\Sigma_{k,m}$ by

$$X_l = \sum_{n=1}^{m} \sum_{i,j=1}^{k} a_{ij}^l b_{jn} \frac{\partial}{\partial b_{in}}, \tag{13}$$

where a_{ij}^l and b_{ij} are the (i,j) elements of $A_0{}^l$ and B, respectively. The action (11) implies that $\{X_l\}_{0 \le l \le k-1}$ is a complete vector field. Furthermore, since A_0 is cyclic, X_l are in involution. Theorem 1 asserts that X_l span a k–dimensional subspace of the tanjent space of $\Sigma_{k,m}$ at any point. Let us define

$$\mathcal{L}_{A_0} = \left\{(A_0, B)\,\middle|\, B = exp\left(\sum_{j=0}^{k-1} t_j A_0{}^j\right) B_0\ ,\ (A_0, B_0) \in \Sigma_{k,m}\right\}. \tag{14}$$

Clearly $\mathcal{L}_{A_0}$ is a connected component of $\mathcal{B}_{A_0}$ and X_l act transitively on $\mathcal{L}_{A_0}$.

3. MODULI SPACE OF CONTROLLABLE DELAY-DIFFERENTIAL SYSTEMS

In this section we consider linear dynamical systems of the form

$$\frac{dx}{ds}(s) = A_1x(s) + A_2x(s-h) + Bu(s), \tag{15}$$

where $A_1 \in \mathbf{R}^{k\times k}, A_2 \in \mathbf{R}^{k\times k}, B \in \mathbf{R}^{k\times m}$ and h is a positive constant describing a time delay. It is important to know whether (15) is pointwise complete or not. If A_1 and A_2 commute, then (15) is pointwise complete. Moreover, it is known [8] that the condition

$$rank\left(B\ A_1B\ A_2B\ A_1{}^2B\ A_1A_2B\ A_2A_1B\ A_2{}^2B \ldots A_2{}^{k-1}B\right) = k$$

is a necessary and sufficient condition for a controllability for any pointwise complete system. In this section we consider the moduli space of delay-differential controllable systems (15) satisfying

$$[A_1,\ A_2] = 0. \tag{16}$$

We note that under (16) the controllability condition can be expressed as

$$rank\,(\lambda I - A_1 \quad \mu I - A_2 \quad B) = k \tag{17}$$

for all $(\lambda, \mu) \in \mathbf{C}^2$. Define the moduli space $\Omega_{k,m}$ by the quotient space of matrices (A_1, A_2, B) satisfying (16) and (17) by the $GL(k, \mathbf{R})$ action

$$g : (A_1, B_2, B) \to (gA_1g^{-1}, gA_2g^{-1}, gB), \tag{18}$$

where $g \in GL(k, \mathbf{R})$. In the complex case, where (A_1, A_2, B) are complex matrices and $g \in GL(k, \mathbf{C})$, the resulting moduli space $\Omega^{\mathbf{C}}_{k,m}$ is naturally related to the framed moduli space of $SU(m)$ Yang-Mills k−instantons. A nonlinear integrable flow on the framed moduli space is found in [9].

Let $(A_{10}, A_{20}, B_0) \in \Omega_{k,m}$. We consider a deformation on $\Omega_{k,m}$ parametrized by a set of real parameters

$$t = (t_{00}, t_{10}, t_{01}, t_{20}, t_{11}, t_{02}, \dots, t_{0k}) \in \mathbf{R}^{(k+1)(k+2)/2}. \tag{19}$$

Using the factors Q and R of the QR factorization

$$exp\left(\sum_{i,j} t_{ij} A_{10}{}^{i} A_{20}{}^{j}\right) = Q^{-1}(t)R(t), \tag{20}$$

we introduce the matrices

$$A_1(t) = QA_{10}Q^{-1},\ A_2(t) = QA_{20}Q^{-1},\ B(t) = RB_0. \tag{21}$$

Along a line similar to that in §2, we derive the system of nonlinear and linear equations

$$\begin{aligned} \frac{\partial A_1}{\partial t_{ij}}(t) &= \left[((A_1{}^iA_2{}^j)_L)^{\mathsf{T}} - (A_1{}^iA_2{}^j)_L,\ A_1\right], \\ \frac{\partial A_2}{\partial t_{ij}}(t) &= \left[((A_1{}^iA_2{}^j)_L)^{\mathsf{T}} - (A_1{}^iA_2{}^j)_L,\ A_2\right], \end{aligned} \tag{22}$$

$$\frac{\partial B}{\partial t_{ij}}(t) = \left(A_1{}^iA_2{}^j + ((A_1{}^iA_2{}^j)_L)^{\mathsf{T}} - (A_1{}^iA_2{}^j)_L\right) B(t), \tag{23}$$

with $A_1(0) = A_{10}, A_2(0) = A_{20}, B(0) = B_0$. The compatibility of (22) and (23) can be proved by using $[A_{10}, A_{20}] = 0$. We have

THEOREM 2. *If (A_{10}, A_{20}, B_0) is an element of the moduli space $\Omega_{k,m}$, then $(A_1(t), A_2(t), B(t))$ is so for any $t \in \mathbf{R}^{(k+1)(k+2)/2}$.*

PROOF: We consider the action (18) by $g = R^{-1}$ to get

$$(\lambda I - A_1(t) \quad \mu I - A_2(t) \quad B(t)) = \left(\lambda I - R^{-1}QA_{10}Q^{-1}R \quad \mu I - R^{-1}QA_{20}Q^{-1}R \quad B_0\right).$$

Since $[A_{10}, A_{20}] = 0$, $Q^{-1}R$ commutes with A_{10} and A_{20}. Thus we see

$$rank\,(\lambda I - A_1(t) \;\; \mu I - A_2(t) \;\; B(t)) = rank\,(\lambda I - A_{10} \;\; \mu I - A_{20} \;\; B_0)$$

for any $(\lambda, \mu) \in \mathbf{C}^2$ and $t \in \mathbf{R}^{(k+1)(k+2)/2}$. ∎

It is proved that for any initial value (A_{10}, A_{20}, B_0) of $\Omega_{k,m}$ the integrable system (22) and (23) induces a flow on $\Omega_{k,m}$, the moduli space of the delay-differential controllable systems.

4. CONCLUSIONS

In this paper we have found a class of nonlinear integrable systems on the moduli spaces $\Sigma_{k,m}$ and $\Omega_{k,m}$ of controllable linear dynamical systems. Theorems 1 and 2 enable us to investigate the topology of these spaces in terms of the invariant tori theorem in classical mechanics. We can see that $\Sigma_{k,m}$ and $\Omega_{k,m}$ admit foliations whose leaves are diffeomorphic to certain cylinders. This leads to a new problem in the geometry of foliations and dynamical systems on moduli spaces.

Acknowledgement. The author is grateful for the support of the Centre de Recherches Mathématiques at the Université de Montréal.

REFERENCES

1. W. H. Fleming, Chirman, "Future Directions in Control Theory: A Mathematical Perspective," Soc. Indust. Appl. Math., Philadelphia, 1988.
2. Y. Nakamura, *Geometry of rational functions and nonlinear integrable systems*, preprint, 1989.
3. J. Moser, *Finitely many points on the line under the influence of an exponential potential–An integrable system*, Lec. Notes in Phys., Springer-Verlag, Berlin **38** (1975), 467–497.
4. P. S. Krishnaprasad, *Symplectic mechanics and rational functions*, Richerche Automat. **10** (1979), 107–135.
5. S. K. Donaldson, *Nahm's equations and the classification of monopoles*, Commun. Math. Phys. **96** (1984), 387–407.
6. Y. Nakamura, *Moduli space of SU(2) monopoles and the complex cyclic-Toda hierarchy*, Commun. Math. Phys. (to appear).
7. U. Helmke, *The topology of a moduli space for linear dynamical systems*, Comment. Math. Helvetici **60** (1985), 630-655.
8. M. G. Frost and C. Storey, *A note on the controllability of linear constant delay-differential systems*, Int. J. Control **28** (1978), 673–679.
9. Y. Nakamura, *Nonlinear integrable flow on the framed moduli space of instantons*, preprint, 1989.

III. PERTURBATIONS OF INTEGRABLE SYSTEMS

Mark J. ABLOWITZ and B.M. HERBST

Daniel DAVID

ON HOMOCLINIC BOUNDARIES IN THE NONLINEAR SCHRÖDINGER EQUATION

MARK J. ABLOWITZ and B.M. HERBST

Program in Applied Mathematics, University of Colorado
Boulder, CO 80309, USA

ABSTRACT. Chaos in nonlinear finite dimensional dynamical systems is often associated with presence of homoclinic orbits, in which case the chaos is triggered by crossing through the homoclinic orbit. In infinite dimensional systems even the concept of inside and outside a homoclinic orbit is not well-defined. In this paper we present a straightforward analysis of the nonlinear Schrödinger equation both demonstrating and illustrating with a concrete example how one can identify 'inside' and 'outside' of the homoclinic orbit. With this information one can formulate conditions for temporal chaos to occur as a result of the perturbations induced by the standard numerical schemes.

1. INTRODUCTION

Homoclinic orbits have for a long time been identified as a possible source of chaos in nonlinear finite dimensional dynamical systems, see for example, [1,2]. For example, when an integrable equation, containing a homoclinic orbit, is perturbed, generally speaking, the homoclinic orbit is broken. This allows frequent crossings of the homoclinic orbit of the unperturbed system which, in turn, may lead to horseshoes and chaos. In two dimensional systems a homoclinic orbit naturally divides phase space into two regions, inside and outside the homoclinic orbit. A crossing of the homoclinic orbit is identified as an orbit passing from inside the homoclinic orbit to its outside, or vice versa. Melnikov's method allows one to measure the distance between the stable and unstable manifolds, broken under the perturbation and gives precise information regarding homoclinic crossings.

In higher dimensions, especially infinite dimensional phase space, the situation is considerably more complicated, since phase space is no longer divided into two regions by a homoclinic orbit. Still, for nonlinear dynamical systems of arbitrary but finite dimensions, Melnikov type methods are available for detecting homoclinic crossings and associated horseshoes, and chaos [2]. In infinite dimensions, temporal chaos has been associated with the integrable infinite dimensional equations of soliton theory. In particular, it is found that small perturbations of the NLS and sine-Gordon equations may lead to temporal chaos. Detailed investigations have revealed that again the presence of homoclinic orbits act as the source of the chaos, see for example, [3,4,5,6] and the references therein. Moreover, we have found the (somewhat) surprising situation that the numerical solutions obtained from the standard discretizations (finite differences, spectral-and pseudo- spectral) of these equations, display instabilities and frequently even weak chaos whenever initial conditions are chosen in the vicinity of homoclinic orbits, see, [7,8,9].

The homoclinic orbits asscociated with the sine Gordon and NLS equations have been obtained from inverse spectral theory. Relatively complicated formulae describing a broad class of periodic solutions in terms of Riemann Theta functions of N phases can be computed (see, for example [3,4,10,11]) and via limits, Bäcklund transformations, etc., elementary formulae describing homoclinic orbits can be found [3,4]. Actually we have found that by employing the N-soliton solution of Hirota [12], the homoclinic orbits of the NLS equation are related, through a simple transformation, to the dark-hole soliton solutions of the defocusing NLS equation. This gives a concrete demonstration of the existence of a homoclinic structure consisting of an arbitrary (denumerable) number of homoclinic orbits to a single 'fixed point'; the number being related to the number of dark-hole solitons in the defocusing equation, see [9]. Moreover, the same ideas hold for other nonlinear wave equations [13].

Having demonstrated the existence of a homoclinic structure one can use inverse scattering theory for periodic problems in order to explain how homoclinic crossings, in infinite dimensional phase space, are again responsible for the instabilities and chaos, similar to the finite dimensional dynamical systems. It turns out that one can still identify an inside and outside to each, single homoclinic orbit, in a relatively simple manner [3,4], since homoclinic orbits are connected to multiple eigenvalues of the periodic spectrum of the associated linear problem. All the relevant information concerning the homoclinic structures of the integrable equations of soliton theory, can of course be obtained from the inverse scattering theory. Here we attempt to explain the basic concepts in a simple setting. In fact, as we show here, one can understand the concepts of 'inside' and 'outside' the homoclinic orbits of the NLS equation, in terms of perturbation theory. Although we focus our attention on the NLS equation, the ideas also apply to related equations.

2. A LINEARIZED ANALYSIS

It is convenient to study the NLS equation in the form given by,

$$iq_t + q_{xx} + 2q^2q^* = 0, \tag{1}$$

where $i^2 = -1$, and q^* denotes the complex conjugate of q. Periodic boundary conditions are specified, $q(x+L,t) = q(x,t)$ and the initial condition is given by $q(x,0) = f(x)$.

We perturb (see for example [14]) the plane-wave solution (with zero wave number),

$$\tilde{q}(x,t) = a\exp(i\eta), \tag{2}$$

where $\eta = 2a^2t + \gamma$, a is a real constant and $\gamma \in [0, 2\pi)$ an arbitrary phase. Considering small perturbations of the form,

$$q(x,t) = \tilde{q}(x,t)(1 + \epsilon(x,t)) \tag{3}$$

where $|\epsilon| << 1$, we substitute (3) into (1), keep linear terms in ϵ and find that,

$$\epsilon_t = i\epsilon_{xx} + i2|a|^2(\epsilon + \epsilon^*).$$

Assuming

$$\epsilon(x,t) = \hat{\epsilon}_{-n}(t)\exp(-i\mu_n x) + \hat{\epsilon}_n(t)\exp(i\mu_n x) \tag{4}$$

where $\mu_n = 2\pi n/L$, it follows easily that the growth rate, σ_n, of the n-th mode, $\hat{\epsilon}_n$, is given by

$$\sigma_{n\pm} = \pm\mu_n\sqrt{4|a|^2 - \mu_n^2}. \tag{5}$$

Hence, the plane-wave solution, (2), is unstable with respect to small perturbations, provided,

$$0 < \mu_n^2 < 4|a|^2. \tag{6}$$

Moreover, it follows that the 'direction' of exponential growth, in Fourier space, is given by

$$\begin{pmatrix} \hat{\epsilon}_n \\ \hat{\epsilon}_{-n} \end{pmatrix} = (\mu_n^2 - i\sigma_{n\pm}) \begin{pmatrix} -\exp(i\phi) \\ \exp(-i\phi) \end{pmatrix}, \tag{7}$$

where $\phi \in [0, 2\pi)$. This translates into an initial condition,

$$q(x,0) = a + \epsilon_0(\mu_n^2 + i\sigma_{n\pm})\cos(\mu_n x + \phi), \tag{8}$$

in the limit as $\epsilon_0 \to 0$. Note that the plus sign corresponds to the unstable manifold, and the minus sign to the stable manifold. We have shown [9] explicitly, by choosing a suitable phase, ϕ, how this initial condition approximates the initial condition for an orbit, homoclinic to the plane-wave solution. In fact, since the 'direction' of the homoclinic orbit is known, one can investigate what happens if initial conditions are chosen on 'opposite sides' of the homoclinic orbit. For $a = \frac{1}{2}$, and $L = 2\sqrt{2}\pi$, we solve (1) numerically using the following inital conditions,

$$q(x,0) = a + 0.01\cos(\mu x), \tag{9}$$

$$q(x,0) = a + 10^{-4}(1+i)\cos(\mu x), \tag{10}$$

and

$$q(x,0) = a + 0.01i\cos(\mu x),. \tag{11}$$

The results are shown in Figures 1a-c. Since the period of the solution shown in Figure 1a is about half that of the one shown in Figure 1c, the former is naturally and conceptually thought of as 'inside' and the latter as 'outside' the homoclinic orbit. But how can one determine this boundary in an analytical way? To explain how, we need a minimum of ideas from scattering theory which we outline below.

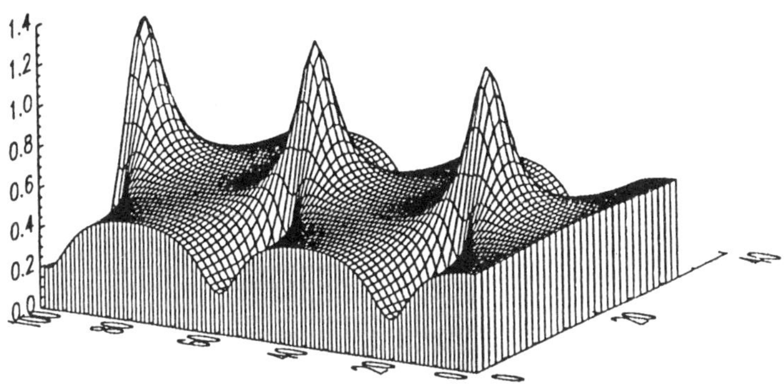

Figure 1(a): Inside the homoclinic boundary.

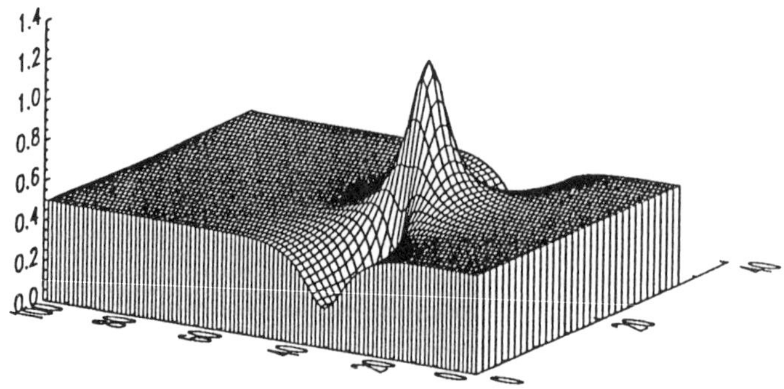

Fiqure 1(b): The homoclinic boundary.

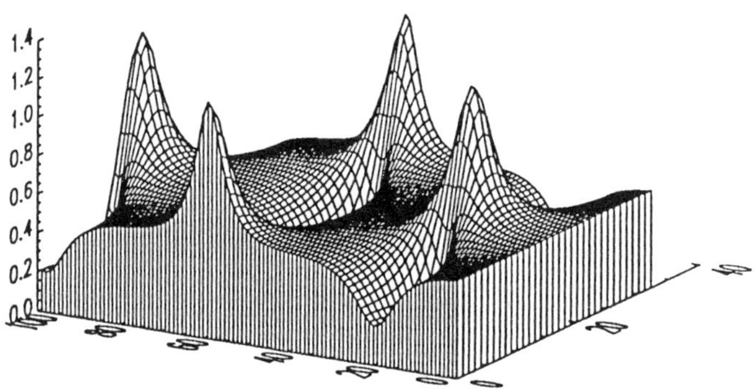

Figure 1(c): Outside the homoclinic boundary.

3. SCATTERING THEORY

It is well-known that the NLS equation, (1), arises as the compatibility condition between two linear operators [15],

$$\begin{aligned} v_{1x} + i\zeta v_1 &= q v_2, \\ v_{2x} - i\zeta v_2 &= -q^* v_1 \end{aligned} \tag{12}$$

and

$$\begin{aligned} v_{1t} &= (-2i\zeta^2 + iqq^*)v_1 + (2q\zeta + iq_x)v_2, \\ v_{2t} &= (-2q^*\zeta + iq_x^*)v_1 + (2i\zeta^2 - iqq^*)v_2. \end{aligned} \tag{13}$$

As an immediate consequence of these operators, we note that the squared eigenfunctions satisfy,

$$\pm\, i(v_j^2)_t + (v_j^2)_{xx} + 4qq^*(v_j^2) - 2q^2(v_{\hat{\jmath}}^2) = 0,\ j = 1,2,\ \hat{\jmath} = 2,1. \tag{14}$$

On the other hand, one observes that a perturbation, $\hat{q}$, around a general solution, $\tilde{q}$, of the NLS equation, i.e.,

$$q = \tilde{q} + \hat{q}, \tag{15}$$

where $|\hat{q}| << 1$, satisfies the linearized equation,

$$i\hat{q} + \hat{q}_{xx} + 4qq^*\hat{q} + 2q^2\hat{q} = 0. \tag{16}$$

(Note that the perturbation is of a slightly different form than the one we used in the previous section.)

A comparison of (16) and (14) shows that

$$\begin{pmatrix} v_1^2 \\ v_2^2 \end{pmatrix} = \begin{pmatrix} \hat{q} \\ -\hat{q}^* \end{pmatrix}. \tag{17}$$

This shows that the time evolution of the squared eigenfunctions describe the behavior of a linearization around the potential, q. The general solution of the linear systems, (12) and (13), is given by,

$$v(x,t,\zeta) = c_1 v^+(x,t,\zeta) + c_2 v^-(x,t,\zeta), \tag{18}$$

where c_1 and c_2 are arbitrary constants and

$$\begin{aligned} v^+(x,t,\zeta) &= \exp[ik(x+2\zeta t)] \begin{pmatrix} \exp(\frac{1}{2}i\eta) \\ \frac{i}{a}(k+\zeta)\exp(-\frac{1}{2}i\eta) \end{pmatrix}, \\ v^-(x,t,\zeta) &= \exp[-ik(x+2\zeta t)] \begin{pmatrix} \exp(\frac{1}{2}i\eta) \\ \frac{i}{a}(-k+\zeta)\exp(-\frac{1}{2}i\eta) \end{pmatrix}, \end{aligned} \tag{19}$$

with $k^2 = a^2 + \zeta^2$. According to our earlier remarks, it is clear that the plane-wave solution is unstable with respect to small perturbations only if an eigenvalue, ζ, has nonzero imaginary part. Accordingly, we must determine the eigenvalues, ζ.

4. THE SPECTRUM OF THE PLANE-WAVE SOLUTION

In order to determine the spectrum associated with the plane-wave solution, we first fix two independent eigenfunction bases for (12),

$$\phi(x,\zeta) = \begin{pmatrix} \phi_1(x,\zeta) \\ \phi_2(x,\zeta) \end{pmatrix}, \; \bar{\phi}(x,\zeta) = \begin{pmatrix} \phi_2^*(x,\zeta^*) \\ \phi_1^*(x,\zeta^*) \end{pmatrix}, \tag{20}$$

and impose the boundary condition,

$$\phi(x_0,\zeta) = \begin{pmatrix} 1 \\ 0 \end{pmatrix}. \tag{21}$$

where x_0 is arbitrary and generally depends on time, see for example, [10]. However, it is easily verified that the general solution, (18), does not satisfy these boundary conditions for all times. Instead, the bases, (20), satisfying the boundary conditions, are related to the general solution, (18), through

$$\phi(x,\zeta) = v\exp(-\tfrac{1}{2}i\eta).$$

This implies that ϕ and $\bar{\phi}$ are solutions of the spatial operator, (12), but satisfy a modified temporal system from (13). However, the compatibility condition remains the NLS equation, (1), as it should.

Henceforth we only deal with the spatial operator, (12), and from now on we suppress the explicit time dependence of the eigenfunctions in our notation.

The spectrum of (12), with potential given by the plane-wave solution, (2), is obtained through a straightforward application of Floquet theory. Defining the Bloch eigenfunction,

$$\psi(x,\zeta) = \begin{pmatrix} \psi_1(x,\zeta) \\ \psi_2(x,\zeta) \end{pmatrix}, \tag{22}$$

as

$$\psi(x+L,\zeta) = \lambda\psi(x,\zeta), \tag{23}$$

it follows in the usual way that (see, for instance, [16]),

$$\lambda^2 - 2a_R\lambda + 1 = 0, \tag{24}$$

where, for the simple potential, (2), the Floquet discriminant, a_R, can be explicitly calculated as,

$$a_R = \cos(\sqrt{a^2+\zeta^2}L). \tag{25}$$

The stable ($|\lambda| = 1$) spectrum is now defined as

$$\{\zeta \mid a_R \text{ real and } -1 \leq a_R \leq 1\},$$

and double points as,

$$\{\zeta \mid a_R = \pm 1 \text{ and } \frac{da_R}{d\zeta} = 0\}.$$

According to (25) the double points are given by,

$$\zeta^2 = \left(\frac{n\pi}{L}\right)^2 - a^2, \tag{26}$$

for any integer n different from zero.

It is necessary to distinguish between multiple points on the real axis, i.e., $0 \leq a^2 \leq (n\pi/L)^2$, and multiple points on the imaginary axis, i.e., $0 < (n\pi/L)^2 < a^2$. The latter condition is exactly the linearized instability condition, (6), obtained earlier. Thus, the instability condition, (6), is the condition for the discrete double eigenvalues to lie on the imaginary axis. This is also the condition required for the squared eigenvalues to grow exponentially in time, i.e., the condition for the plane-wave to be unstable under small perturbations.

There is no instability associated with the real double points.

Armed with this background material, it is now possible to investigate the fate of the double complex eigenvalues under small perturbations in the potential, in more detail.

5. A PERTURBATION ANALYSIS

Henceforth we are only concerned with the behavior of the discrete spectrum of the spatial operator, (12), under small perturbations. Since the discrete spectrum is invariant under the time-flow, we may consider our potential, q, at $t = 0$ and continue to suppress all time dependence in our notation. The perturbed potential is of the form,

$$q = a + \epsilon q^{(1)} \tag{27}$$
$$= a + \epsilon \exp(i\phi)\cos(\mu_n x), \tag{28}$$

where a is a real constant, $\mu_n = 2\pi n/L$ (for integer n), and we investigate the influence of the $O(\epsilon)$ perturbation. Assuming the perturbation expansion,

$$\zeta = \zeta^{(0)} + \epsilon^{\frac{1}{2}}\zeta^{(1)} + \epsilon\zeta^{(2)} + \cdots \tag{29}$$

and

$$\boldsymbol{v} = \boldsymbol{v}^{(0)} + \epsilon^{\frac{1}{2}}\boldsymbol{v}^{(1)} + \epsilon\ \boldsymbol{v}^{(2)} + \cdots \tag{30}$$

we substitute into the spatial operator, (12), gather terms at the various orders of ϵ, to obtain
at $O(\epsilon^0)$:

$$\mathcal{L}\boldsymbol{v}^{(0)} = 0, \tag{31}$$

where

$$\mathcal{L} := \begin{pmatrix} 1 & 0 \\ 0 & -1 \end{pmatrix}\frac{\partial}{\partial x} - \begin{pmatrix} 0 & a \\ a & 0 \end{pmatrix} + i\zeta\begin{pmatrix} 1 & 0 \\ 0 & 1 \end{pmatrix}. \tag{32}$$

at $O(\epsilon^{\frac{1}{2}})$:

$$\mathcal{L}\boldsymbol{v}^{(1)} = \begin{cases} -i\zeta^{(1)}v_1^{(0)} \\ -i\zeta^{(1)}v_2^{(0)} \end{cases}, \tag{33}$$

and at $O(\epsilon)$:

$$\mathcal{L}\boldsymbol{v}^{(2)} = \begin{cases} -i\zeta^{(2)}v_1^{(0)} + q^{(1)}v_2^{(0)} - i\zeta^{(1)}v_1^{(1)} \\ -i\zeta^{(2)}v_2^{(0)} + q^{(1)*}v_1^{(0)} - i\zeta^{(1)}v_2^{(1)} \end{cases}. \tag{34}$$

The general solution of (31) is obtained form (18) and is given by,

$$\boldsymbol{v}^{(0)} = A^+\boldsymbol{\phi}^+ + A^-\boldsymbol{\phi}^-, \tag{35}$$

where

$$\boldsymbol{\phi}^{\pm} = \exp(\pm ikx)\begin{pmatrix} 1 \\ (i/a)(\pm k + \zeta) \end{pmatrix}.$$

(Here and henceforth, we use ζ instead of $\zeta^{(0)}$.)

Proceeding to $O(\epsilon^{\frac{1}{2}})$, we note that the solvability condition, assuming L-periodic or anti-periodic eigenfunctions, for the system

$$\mathcal{L}\boldsymbol{v} = \boldsymbol{F}, \tag{36}$$

with

$$\boldsymbol{F} = \begin{pmatrix} F_1 \\ F_2 \end{pmatrix}$$

is given by the orthogonality condition

$$\int_0^L (F_1 w_1^* + F_2 w_2^*)dx = 0$$

for all $\mathbf{w}$ in the nullspace of the Hermitian operator, $\mathcal{L}^H$, where,

$$\mathcal{L}^H = -\begin{pmatrix} 1 & 0 \\ 0 & -1 \end{pmatrix}\frac{\partial}{\partial x} - \begin{pmatrix} 0 & a \\ a & 0 \end{pmatrix} - i\zeta^*\begin{pmatrix} 1 & 0 \\ 0 & 1 \end{pmatrix}. \tag{37}$$

It follows that the nullspace of $\mathcal{L}^H$ is spanned by

$$\begin{pmatrix} (\phi_2^{\pm})^* \\ (\phi_1^{\pm})^* \end{pmatrix}, \tag{38}$$

and the general solvability condition assumes the form

$$\int_0^L (F_1\phi_2^{\pm} + F_2\phi_1^{\pm})dx = 0. \tag{39}$$

Applied to (33), the solvability condition yields $\zeta^{(1)} = 0$. This knocks out all terms containing $v^{(1)}$ in (34) and we don't need to solve for it (if one proceeds to higher order terms, a nontrivial solution for $\boldsymbol{v}^{(1)}$ may be required, as in the usual Sturm-Louisville case). The solvability condition applied to (34) yields the system of equations,

$$\begin{pmatrix} -\frac{1}{2}\frac{(k+\zeta)^2}{a^2}\exp(i\phi) + \frac{1}{2}\exp(-i\phi) & \frac{2\zeta}{a}\zeta^{(2)} \\ \frac{2\zeta}{a}\zeta^{(2)} & -\frac{1}{2}\frac{(-k+\zeta)^2}{a^2}\exp(i\phi) + \frac{1}{2}\exp(-i\phi) \end{pmatrix}\begin{pmatrix} A^+ \\ A^- \end{pmatrix} = 0, \tag{40}$$

provided $k = \frac{1}{2}\mu_n$. This condition on k implies that a specific double point is selected, in sympathy with the perturbation in the eigenfunction.

In the case of the imaginary double points, nontrivial solutions, $A^{\pm}$, exist if,

$$(\zeta^{(2)})^2 = -\frac{a^2}{4\zeta^2}\sin(\phi+\theta)\sin(\phi-\theta), \tag{41}$$

where $\tan\theta = \mathrm{Im}(\zeta)/k$. In the case of a real double point, the condition becomes,

$$(\zeta^{(2)})^2 = \frac{a^2}{8\zeta^2}\left[\cos(2\phi) + 1 - 2\left(\frac{k^2}{a^2}\right)\right]. \tag{42}$$

In the case of imaginary double points, it is clear that the perturbation, $\zeta^{(2)}$, can be real, zero or purely imaginary, depending on the choice of ϕ in the perturbed potential. Clearly $\phi = \pm\theta$ is special – *it corresponds to a perturbation in the direction of the homoclinic orbit.* This choice is also consistent with the homoclinic perturbation, (8), obtained from linearized analysis. The values for which $\zeta^{(2)}$ is imaginary or real defines the 'inside' and 'outside' of the homoclinic orbit. We shall illustrate this with a concrete example, shortly.

In the case of real double points it follows from (42) that the perturbation can only be imaginary (recall that $a^2 < k^2$ for real double points). The situation with real double points is very different from that of imaginary double points. We have already observed that they do not spawn any instability and now it becomes clear that there is also no homoclinic boundary associated with them.

Let us now illustrate the existence of a homoclinic boundary by a concrete example.

Example: Choose the initial condition

$$q(x,0) = \tfrac{1}{2} + \epsilon \exp(i\phi)\cos(\mu x),$$

where $\mu = 2\pi/L$ and $L = 2\sqrt{2}\pi$, i.e., $\zeta = i/(2\sqrt{2})$. For this initial condition, (41) becomes,

$$(\zeta^{(1)})^2 = \tfrac{1}{2}\sin(\phi + \pi/4)\sin(\phi - \pi/4). \tag{43}$$

Thus, for

(i) $\phi = 0$, $\zeta^{(2)}$ is imaginary. According to Figure 1, this corresponds to an 'inside' orbit.

(ii) $\phi = \pi/4$, $\zeta^{(2)}$ is zero. This corresponds to the homoclinic orbit in Figure 1.

(iii) $\phi = \pi/2$, $\zeta^{(2)}$ is real and this corresponds to the 'outside' orbit.

6. REMARKS ON NUMERICALLY INDUCED CHAOS

The main results described here, are well-suited for a description for the chaos resulting from adding forcing and damping terms to integrable equations, such as the NLS or sine-Gordon equations. Numerical studies show that double eigenvalues split under the perturbations and by monitoring the evolution of the eigenvalues in the complex plane, one can determine when homoclinic crossings occur, see for example, [17], which contains a description of a numerical code for calculating the spectrum of the sine-Gordon equation.

As alluded to in the introduction, we have observed that the standard discretizations of the NLS and related equations, may lead to instabilities and even chaos in the numerical solutions. We have also established that the integrable discretization of Ablowitz and Ladik [18], which has its own homoclinic structure, displays none of the instabilites. Consequently, there is little doubt that the instability is deeply connected with presence of a homoclinic orbit. For this reason we refer to the instability as Numerical Homoclinic Instability (NHI). A more precise description of the instability is undoubtedly possible using the results described in this article as a framework.

All the standard schemes we have investigated are nonintegrable and, of course, by virtue of their nonintegrability, do not respect the time invariance of the discrete spectrum of the scattering problem, (12). This means that the discrete spectrum is allowed to wander around in the complex plane as time evolves. In addition, one can view the nonintegrable schemes as perturbations of the integrable equation, implying that double points will split, as our analysis indicates. This happens to both real and imaginary double points. The former is relatively harmless (and not a homoclinic boundary – as indicated earlier), introducing additional degrees of freedom to the system which show up as small higher frequency components in the spatial strucure (see, for example, [11]). However, the time dependence of the imaginary points may have serious consequences. This allows for the possibility of homoclinic crossings where the imaginary double points split, regroup and then split in the opposite direction (i.e., $(\zeta^{(2)})^2$ either positive or negative). The next step is to investigate this scenario numerically. Since we know the locations of the perturbed double points initially, one can trace their time evolution rather easily, by searching (iteratively) in the regions where they were located at the previous time step. This amounts to a somewhat modified version of the procedure described by Overman, McLaughlin and Bishop [17] for the sine-Gordon equation – but the concept of homoclinic orbits corresponding to certain multiple eigenvalues is the same.

Finally we note that the integrable scheme of Ablowitz and Ladik [18] has its own homoclinic structure, i.e., it does respect the time invariance of the discrete spectrum, which precludes the splitting of double points and the corresponding irregular temporal behavior.

ACKNOWLEDGEMENTS. We appreciate many valuable comments, and the preprint of [3] given to us by G.M. Forest.

This work (MJA) is partially supported by the NSF, Grants No. DMS-8916182, the Office of Naval Research, Grant No. N00014-88- K-0447 and the Air Force Office of Scientific Research, Grant No. AFOSR-90-0039. One of us (BMH) would like to express his appreciation for the hospitality of the University of Colorado and the support of the University of the Orange Free State and colleagues in the department of Applied Mathematics.

†Permanent address: Department of Applied Mathematics, University of the Orange Free State, Bloemfontein 9300, South Africa.

References

[1] J. Guckenheimer and P. Holmes. Nonlinear oscillations, dynamical systems, and bifurcations of vector fields. Springer, New York (1983).

[2] S. Wiggins. Global bifurcations and chaos. Springer, New York (1988).

[3] N. Ercolani, M.G. Forest and D.W. McLaughlin. Geometry of the modulational instability Part III: Homoclinic orbits for the periodic sine-Gordon equation. To appear Physica D (1989).

[4] N. Ercolani, M.G. Forest and D.W. McLaughlin. Notes on Melnikov integrals for models of the driven pendulum chain. Preprint (1989).

[5] A.R. Bishop, D.W. McLaughlin, M.G. Forest and E.A. Overman II. Phys. Lett., **127**A, p335 (1988).

[6] G. Terrones, D.W. McLaughlin E.A. Overman II and A.J. Pearlstein. Stability and bifurcation of spatially coherent solutions of the damped-driven NLS equation. Preprint (1989).

[7] B.M. Herbst and M.J. Ablowitz. Phys. Rev. Lett., **62**, p2065 (1989).

[8] B.M. Herbst and M.J. Ablowitz. On numerical chaos in the nonlinear Schrödinger equation. Report INS # 120, Clarkson University (1989). To be published, Proceedings of the workshop on integrability, Oléron, France. Springer (1989).

[9] M.J. Ablowitz and B.M. Herbst. On homoclinic structure and numerically induced chaos for the nonlinear Schrödinger equation. Program in Applied Mathematics Report 1, University of Colorado, Boulder (1989). To appear in SIAM J. Appl. Math., April (1990).

[10] Y.-C. Ma and M.J. Ablowitz. Stud. Appl. Math., **65**, p113 (1981).

[11] E.R. Tracy. Topics in nonlinear wave theory with applications. Ph.D. Thesis, University of Maryland, College Park (1984).

[12] R. Hirota. Direct methods of finding exact solutions of nonlinear evolution equations, in Bäcklund Transformations, R.M. Muira, ed., Lecture Notes in Mathematics 515, Springer -Verlag, New York (1976).

[13] E.E. Ryan. Ph.D. thesis (in preparation).

[14] J.T. Stuart and R.C. DiPrima. Proc. R. Soc. London, **A362**, p27 (1978).

[15] M.J. Ablowitz and H. Segur. Solitons and the Inverse Scattering Transform. SIAM, Philadelphia, 1981.

[16] W. Magnus and S. Winkler. Hill's equation. Dover, New York (1979).

[17] E.A. Overman II, D.W. McLaughlin and A.R. Bishop. Physica **19D**, p1 (1986).

[18] M.J. Ablowitz and J.F. Ladik. Stud. Appl. Math., **55**, p213(1976).

INTEGRABILITY AND CHAOS IN NONLINEARLY COUPLED OPTICAL BEAMS

DANIEL DAVID

Center for Nonlinear Studies, MS B258, Los Alamos National Laboratory
Los Alamos, NM 87545, USA

ABSTRACT. This paper presents a study, using dynamical systems methods, of the equations describing the polarization behaviour of two nonlinearly coupled optical beams counterpropagating in a nonlinear medium. In the travelling-wave regime assumption, this system possesses a Lie-Poisson structure on the manifold $C^2 \times C^2$. In the case where the medium is assumed to be isotropic, this system exhibits invariance under the Hamiltonian action of two copies of the rotation group, S^1, and actually reduces to a lower-dimensional system on the two-sphere, S^2. We study the dynamics on the reduced space and examine the structure of the phase portrait by determining the fixed points and infinite-period homoclinic and heteroclinic orbits; we concentrate on presenting some exotic behaviour that occurs when some parameters are varied, and we also show special solutions associated with some of the above-mentioned orbits. Last, we demonstrate the existence of complex dynamics when the system is subject to certain classes of Hamiltonian perturbations. To this end, we make use of the Melnikov method to analytically show the occurence of either horseshoe chaos, or Arnold diffusion.

1. INTRODUCTION

Integrable Hamiltonian systems often exhibit a transition to chaotic, or complex, behaviour when they are subject to small periodic perturbations. In low-dimensional cases (e.g., a one-degree-of freedom system), chaoticity usually shows its presence in the form of Smale horseshoes in the structure of the associated Poincaré map. This structure is inherently characterized by combined stretching and folding in the above map, as well as an invariant Cantor-like set as the infinite-time iterate of this map. A method, first developed by Melnikov [1963] and Arnold [1964], and later developed further for extended types of systems by Holmes and Marsden [1982a] and Wiggins [1988], is the tool that is used here to establish in an analytical way that, for the perturbed systems considered in this paper, iterating the Poincaré map indeed implies the existence of a horseshoe structure, by virtue of the Melnikov and Poincaré–Birkhoff–Smale theorems. If the dimensionality of the system is high enough, an additional feature may take place, Arnold diffusion, as a signature that the KAM torii are insufficient to confine orbits within a single region of chaoticity. This can occur through resonance overlaps, thus giving rise to Arnold webs (or transition chains).

The dynamics that we will discuss here concerns the Hamiltonian description of the travelling–wave dynamics of polarized, nearly monochromatic, optical laser pulses counterpropagating in a lossless, cubically nonlinear, parity–invariant, anisotropic, homogeneous medium; for instance, polarized beams in a straight optical fiber constitutes a realization of such a system. Our approach will use the method of dimensional reduction for Hamiltonian systems with a continuous Lie point-group of symmetries, together with the Melnikov–Arnold method to show the existence of complex behaviour under certain small perturbations, induced by modulating the optical parameters characterizing the medium of propagation.

Nonlinear polarization dynamics of optical laser pulses is a subject which has been under investigation for a few decades. Some 25 years ago, Maker *et al.* [1964] analyzed and demonstrated the precession of the polarization ellipse for a single beam propagating in a nonlinear medium. Spatially stable solution configurations for the problem of the interaction of two counterpropagating beams in an isotropic medium were considered in Kaplan [1983] and Lytel [1984]. The phenomenon of optical multi-stability is is also studied in Lytel [1984], and in Otsuka*et al.* [1985]. Reports of numerical evidence for chaotic behaviour and interpretations of experimental data can be found in Trillo *et al.* [1986] and Gaeta *et al.* [1987]. Previous work on special solutions for the one-beam and the two-beam problems appear in Tratnik and Sipe [1987]. Additional references and more detailed treatment of Hamiltonian chaos in nonlinear optical polarization dynamics can be found in David [1989] and in David *et al.* [1988a, 1988b, 1990].

The plan of this paper is as follows. In Section 2 we formulate the polarization dynamics of travelling-wave optical pulses in Hamiltonian form; this is done in terms of two complex 2-component envelope electric field variables. In Section 3, after a convenient change of variables, we use the method of reduction for Hamiltonian systems with symmetry to effectively recast the problem as a more tractable system on a lower-dimensional phase space, namely the sphere. This is actually achieved in two steps, each involving a rotational invariance. The first step transports the system from the initial phase space $C^2 \times C^2$ to $S^2 \times S^2$; the second step takes advantage of the isotropy property of the propagation medium and finally brings the problem down to S^2. Specifically, these reduction steps are taken by restricting to level sets of first integrals of the systems, and then by projecting to the space obtained by quotienting by the residual Hamiltonian invariance group. Section 4 is devoted to the qualitative analysis of the reduced system for a few special cases. Specifically, we classify the fixed points of the reduced dynamics on the sphere and describe the various bifurcations involving these points that take place as material parameters and intensities of the beams are varied. Some degeneracies occur during the second reduction step and lead to peculiar bifurcations; we describe these in detail. Special soliton-like and kink-like solutions are also depicted. In Section 5, we use the Melnikov method to demonstrate how homoclinic (or heteroclinic) orbits tangle and break up to yield complex behaviour when the dynamics is perturbed by modulated material inhomogeneities, modelled as spatially periodic variations of the optical parameters of the medium. Conclusions of this study are summarized in the final Section 6.

2. PHYSICAL FORMULATION OF THE PROBLEM

The travelling-wave evolution of the polarization dynamics of our system of two counterpropagating beams involves complex electric field amplitudes $e_i(\tau)$ and $\bar{e}_i(\tau)$, appearing in the following eikonal form,

$$E_i(z,\tau) = \left[e_i(z,\tau)e^{ikz} + \bar{e}_i(z,\tau)e^{-ikz}\right] e^{-i\omega t} + c.c. \tag{2.1}$$

where $\tau = z - ct$ is the travelling-wave variable and where $i = 1, 2$ is the polarization index. We note that each of $\boldsymbol{e}$ and $\bar{\boldsymbol{e}}$ are elements of C^2. Also, here and below, the "bars" are used to distinguish between quantities associated with one beam or the other. The third-order nonlinear polarizability is written as

$$P_i(z,t) = \chi_{ij}^{(1)} * E_j + \chi_{ijk\ell}^{(3)} * E_j E_k E_\ell \tag{2.2}$$

where $*$ indicates convolution with respect to time and we sum over repeated indices. Using the rotating wave approximation in Maxwell's equations and assuming a lossless medium, small nonlinear anisotropy $\chi^{(3)}$ and null $\chi^{(1)}$ (corresponding to considering an anisotropic and parity–invariant medium), slowly varying amplitudes, and a far–away–from–resonance situation, we obtain the following set of equations governing the dynamics of the electric field amplitudes:

$$\begin{aligned} i\dot{e}_j &= \frac{i\overline{\kappa}}{\overline{r}}\chi^{(3)}_{jk\ell m} e_k \left(e_\ell e^*_m + 2\overline{e}_k \overline{e}^*_\ell\right), \\ i\dot{\overline{e}}_j &= \frac{i\kappa}{r}\chi^{(3)}_{jk\ell m} \overline{e}_k \left(\overline{e}_\ell \overline{e}^*_m + 2e_k e^*_\ell\right). \end{aligned} \tag{2.3}$$

where the constant susceptibility tensor $\chi^{(3)}$ satisfies the following involutions:

$$\chi^{(3)}_{ijk\ell} = \chi^{(3)*}_{ji\ell k}, \qquad \chi^{(3)}_{ijk\ell} = \chi^{(3)}_{\ell jki} = \chi^{(3)}_{ikj\ell}. \tag{2.4}$$

These equations constitute a quasi–canonical Hamiltonian system on $C^2 \times C^2$. The Hamiltonian function is

$$H(\boldsymbol{e},\overline{\boldsymbol{e}}) = \tfrac{1}{2}\chi^{(3)}_{ijk\ell}\left[e^*_i e_j e_k e^*_\ell + 4e^*_i e_j \overline{e}_k \overline{e}^*_\ell + \overline{e}^*_i \overline{e}_j \overline{e}_k \overline{e}^*_\ell\right], \tag{2.5}$$

with the associated Hamiltonian vector field

$$X_H = \frac{i\overline{\kappa}}{\overline{r}}\left(\nabla_{e^*}H \cdot \nabla_e - \nabla_e H \cdot \nabla_{e^*}\right) + \frac{i\kappa}{r}\left(\nabla_{\overline{e}^*}H \cdot \nabla_{\overline{e}} - \nabla_{\overline{e}}H \cdot \nabla_{\overline{e}^*}\right). \tag{2.6}$$

so that any dynamical quantity Q evolves according to the prescription

$$\dot{Q} = X_H Q \qquad (= \{Q, H\}). \tag{2.7}$$

This endows the phase space, $\left(C^2 \times C^2, \{\cdot,\cdot\}\right)$, with a Lie–Poisson structure.

3. REDUCTION TO S^2

Generically, Hamiltonian systems defined on high–dimensional manifolds are somewhat difficult to study because, for one thing, it proves virtually impossible to characterize the geometry of the phase space in any way that is easy to grasp. It is often very convenient to use any continuous symmetry the system may possess in order to redefine it onto a new space with less dimensions. Such a technique is used, for instance, to help one to find solutions of difficult nonlinear partial differential equations by reducing them to equations with fewer variables; these yield so–called (sub) group–invariant solutions. Along the same lines, a Hamiltonian dynamical system can also be reduced, to a system on a lower–dimensional phase space (i.e., to a set of equations with fewer dependent variables) if it is possesses symmetries. This allows for a study of the dynamics on the reduced phase space, after which one may revert to the original space to deduce results pertinent to the variables in terms of which the problem was defined. The machinery lying behind the method of Hamiltonian symmetry reduction is based upon the following theorem (see ,e.g., Chapter 6 of Olver [1986] for details). Let $\mathcal{G}$ be a Hamiltonian symmetry group of transformations acting on $\mathcal{M}$ and generated by a set $\{P_i\}$ of first integrals. We define the momentum map $\mathcal{P} : \mathcal{M} \longrightarrow g^* : x \longmapsto \sum_i P_i(x)\omega_i$, where $\{\omega_i\}$ is a basis of g^*, the dual to the Lie algebra of the group $\mathcal{G}$. We denote by $\mathcal{S}_\alpha$ the common α–level set of $\mathcal{P}$. The residual symmetry group $\mathcal{G}_\alpha$ on $\mathcal{S}_\alpha$ is the isotropy subgroup of $\alpha \in g^*$.

Reduction Theorem. *Let $\mathcal{P}$ be of maximal rank everywhere on $\mathcal{S}_\alpha$ and let $\mathcal{G}_\alpha$ act regularly on the submanifold $\mathcal{S}_\alpha$. Then there is a natural induced immersion $\phi : \mathcal{S}_\alpha/\mathcal{G}_\alpha \hookrightarrow \mathcal{M}/\mathcal{G}$ such that $\mathcal{S}_\alpha/\mathcal{G}_\alpha$ is a Poisson submanifold so that the diagram*

$$\begin{array}{ccc} \mathcal{S}_\alpha & \overset{i}{\hookrightarrow} & \mathcal{M} \\ \Big\downarrow \pi_\alpha & & \Big\downarrow \pi \\ \mathcal{S}_\alpha/\mathcal{G}_\alpha & \underset{\phi}{\hookrightarrow} & \mathcal{M}/\mathcal{G} \end{array}$$

commutes. π and π_α are the natural projections and i the immersion realizing $\mathcal{S}_\alpha$ as a submanifold of $\mathcal{M}$. Moreover, any Hamiltonian system on $\mathcal{M}$ with $\mathcal{G}$ as a symmetry group naturally restricts to Hamiltonian systems on $\mathcal{S}_\alpha/\mathcal{G}_\alpha$ and $\mathcal{M}/\mathcal{G}$.

In practice, in order to prevent reconstructibility problems, the above theorem suggests that one may reduce to the space $\mathcal{S}_\alpha/\mathcal{G}_\alpha$ by first restricting to $\mathcal{S}_\alpha$ and then projecting via π_α. For the system (2.3) under consideration here, two S^1 invariances will allow us to perform reduction. First, it proves convenient to make a change of variables on $C^2 \times C^2$ by reparametrizing this space with Stokes variables, as follows:

$$\begin{aligned} \boldsymbol{u} &= \boldsymbol{e}^\dagger \vec{\sigma} \boldsymbol{e}, & u_0 &\equiv r = \boldsymbol{e}^\dagger \cdot \boldsymbol{e}, \\ \overline{\boldsymbol{u}} &= \overline{\boldsymbol{e}}^\dagger \vec{\sigma} \overline{\boldsymbol{e}}, & \overline{u}_0 &\equiv \overline{r} = \overline{\boldsymbol{e}}^\dagger \cdot \overline{\boldsymbol{e}}. \end{aligned} \tag{3.1}$$

where $\vec{\sigma} = (\sigma_3, \sigma_1, \sigma_2)$ are the usual Pauli matrices. One observes that both u_0 and $\overline{u}_0$ are first integrals of the motion and are associated with the invariance of the system under the rotation transformation $(\boldsymbol{e}, \overline{\boldsymbol{e}}) \mapsto e^{i\theta}(\boldsymbol{e}, \overline{\boldsymbol{e}})$. Our first step, in accord with the above theorem, is to restrict to an immersed common level set $u_0 = r, \overline{u}_0 = \overline{r}$. Let us write $\boldsymbol{e}$ and $\overline{\boldsymbol{e}}$ as

$$\boldsymbol{e} = (e_1, e_2) = (x_1 + ix_2, x_3 + ix_4), \quad \overline{\boldsymbol{e}} = (\overline{e}_1, \overline{e}_2) = (\overline{x}_1 + i\overline{x}_2, \overline{x}_3 + i\overline{x}_4) \tag{3.2}$$

It is then easy to see that $u_0 = x_1^2 + x_2^2 + x_3^2 + x_4^2$ and $\overline{u}_0 = \overline{x}_1^2 + \overline{x}_2^2 + \overline{x}_3^2 + \overline{x}_4^2$; thus the level set is simply $S^3 \times S^3$. On this space, the Hamiltonian function, the Hamiltonian vector field, and the equations of motion become

$$H(\boldsymbol{u}, \overline{\boldsymbol{u}}) = \tfrac{1}{2}(\boldsymbol{u} \cdot W \cdot u + \overline{\boldsymbol{u}} \cdot W \cdot \overline{u}) + 2\overline{u} \cdot W \cdot \boldsymbol{u}, \tag{3.3a}$$

$$X_H = \frac{\overline{\kappa}}{\overline{r}} (\nabla_u H \times \boldsymbol{u}) \cdot \nabla_u + \frac{\kappa}{r} (\nabla_{\overline{u}} H \times \overline{\boldsymbol{u}}) \cdot \nabla_{\overline{u}}, \tag{3.3b}$$

$$\dot{\boldsymbol{u}} = \frac{\overline{\kappa}}{\overline{r}} [\boldsymbol{W} \cdot (\boldsymbol{u} + 2\overline{\boldsymbol{u}})] \times \boldsymbol{u}, \quad \dot{\overline{\boldsymbol{u}}} = \frac{\kappa}{r} [\boldsymbol{W} \cdot (\overline{\boldsymbol{u}} + 2\boldsymbol{u})] \times \overline{\boldsymbol{u}}, \tag{3.3c}$$

where

$$\boldsymbol{W} = \vec{\sigma}_{ij}\, \chi^{(3)}_{ijk\ell}\, \vec{\sigma}_{k\ell} = \mathrm{Diag}\,(\lambda_1, \lambda_2, \lambda_1)\,. \tag{3.4}$$

One can always perform a transformation such that the matrix **W** will be diagonal. The components of **W** are interaction terms of the following types: W_{11}, W_{13}, W_{33} are linear–linear polarization interaction terms, W_{12}, W_{23} are linear–circular polarization interaction terms, and W_{22} is a circular–circular polarization interaction term. Note that the isotropy condition mentioned earlier makes the first and third components equal. Now, using the representation (3.2), we notice that the components of $\boldsymbol{u}$ and $\overline{\boldsymbol{u}}$ write as

$$\begin{aligned} u_3 &= 2(x_1x_3 + x_2x_4), & \overline{u}_3 &= 2(\overline{x}_1\overline{x}_3 + \overline{x}_2\overline{x}_4), \\ u_1 &= 2(x_1x_4 - x_2x_3), & \overline{u}_1 &= 2(\overline{x}_1\overline{x}_4 - \overline{x}_2\overline{x}_3), \\ u_2 &= x_1^2 + x_2^2 - x_3^2 - x_4^2, & \overline{u}_2 &= \overline{x}_1^2 + \overline{x}_2^2 - \overline{x}_3^2 - \overline{x}_4^2. \end{aligned} \tag{3.5}$$

The above formulae (3.5) constitute the transformation $S^3 \mapsto S^2$ due to Hopf (see, e.g., Crampin and Pirani [1987]). We then notice that the components of $\boldsymbol{u}$ and $\overline{\boldsymbol{u}}$ are not all independent; indeed, we have

$$u_1^2 + u_2^2 + u_3^2 = r^2, \qquad \overline{u}_1^2 + \overline{u}_2^2 + \overline{u}_3^2 = \overline{r}^2. \tag{3.6}$$

Thus, we realize that we are dealing with the manifold $S^2 \times S^2$, since each equation in (3.6) actually defines a sphere. Each copy of S^2 in the above product is sometimes referred to as a *Poincaré sphere* (see Figure 1). Points on this sphere represent polarization states: the poles are circularly polarized states, equatorial points correspond to linearly polarized states, and all other points describe elliptical states. We are choosing the north and south poles to lie along the 2–axis to conform with optics notation.

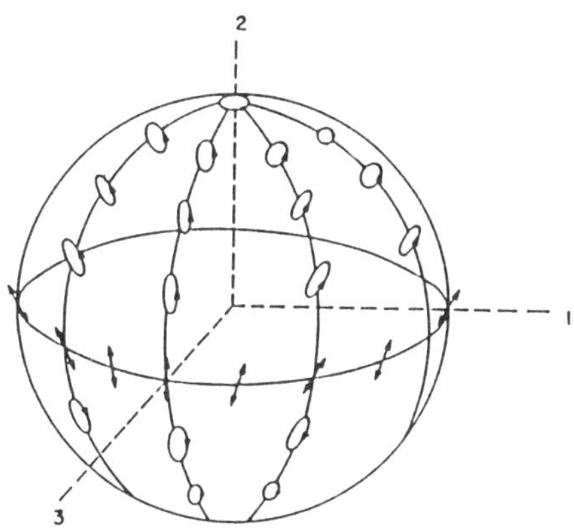

FIGURE 1. The Poincaré sphere.

To actually reduce on $S^2 \times S^2$, we use spherical coordinates,

$$\begin{aligned} u_1 &= r\sin\theta\sin\phi, & \overline{u}_1 &= \overline{r}\sin\overline{\theta}\sin\overline{\phi}, \\ u_2 &= r\cos\theta, & \overline{u}_2 &= \overline{r}\cos\overline{\theta}, \\ u_3 &= r\sin\theta\cos\phi, & \overline{u}_3 &= \overline{r}\sin\overline{\theta}\cos\overline{\phi}, \end{aligned} \tag{3.7}$$

In terms of these coordinates, the Hamiltonian function, X_H, and the equations of motion take the following form:

$$\begin{aligned} H(\theta,\phi,\overline{\theta},\overline{\phi}) =& \frac{1}{2}r^2\left[\lambda_1\sin^2\theta + \lambda_2\cos^2\theta\right] + \frac{1}{2}\overline{r}^2\left[\lambda_1\sin^2\overline{\theta} + \lambda_2\cos^2\overline{\theta}\right] \\ &+ 2r\overline{r}\left[\lambda_1\sin\theta\sin\overline{\theta}\cos(\phi-\overline{\phi}) + \lambda_2\cos\theta\cos\overline{\theta}\right], \end{aligned} \tag{3.8a}$$

$$X_H = \frac{\overline{\kappa}}{r\overline{r}\sin\theta}\left[H_\phi\partial_\theta - H_\theta\partial_\phi\right] + \frac{\kappa}{r\overline{r}\sin\overline{\theta}}\left[H_{\overline{\phi}}\partial_{\overline{\theta}} - H_{\overline{\theta}}\partial_{\overline{\phi}}\right], \tag{3.8b}$$

$$\dot{\theta} = -2\lambda_1\overline{\kappa}\sin\overline{\theta}\sin(\phi-\overline{\phi}), \tag{3.8c}$$

$$\dot{\overline{\theta}} = 2\lambda_1\kappa\sin\theta\sin(\phi-\overline{\phi}), \tag{3.8d}$$

$$\dot{\phi} = \frac{\overline{\kappa}r}{r}(\lambda_2-\lambda_1)\cos\theta - 2\overline{\kappa}\left(\lambda_1\cos\theta\frac{\sin\overline{\theta}}{\sin\theta}\cos(\phi-\overline{\phi}) - \lambda_2\cos\overline{\theta}\right), \tag{3.8e}$$

$$\dot{\overline{\phi}} = \frac{\kappa\overline{r}}{r}(\lambda_2-\lambda_1)\cos\overline{\theta} - 2\kappa\left(\lambda_1\cos\overline{\theta}\frac{\sin\theta}{\sin\overline{\theta}}\cos(\phi-\overline{\phi}) - \lambda_2\cos\theta\right). \tag{3.8f}$$

We now proceed with the second step of the reduction. Notice that the quantity $\sigma = \kappa\cos\theta + \overline{\kappa}\cos\overline{\theta}$ is also a constant of the motion. A level set is thus a three–dimensional manifold specified by a constant value of σ and is coordinatized, for instance, by triplets $(\phi, \overline{\phi}, \omega)$ where $\omega = \kappa\cos\theta - \overline{\kappa}\cos\overline{\theta}$. At this point, we make the following remark. Recall that the reduction theorem necessitates that the momentum be of maximal rank. Here, this means that we will have degenerate situations when the function $\sigma(\theta,\overline{\theta})$ will not be of maximal rank, i.e., when $\sin\theta = \sin\overline{\theta} = 0$. This corresponds to $\sigma = \pm(|\kappa| + |\overline{\kappa}|)$ and $\sigma = \pm(|\kappa| - |\overline{\kappa}|)$. The first two cases are venial; they correspond to the extreme values possible for σ and the dynamics then reduces to a point (physically, this is achieved when the polarizations of the beams are aligned alond the polar direction: this configuration is forever the same). As for the two other situations, they physically arise when the polarizations of the beams are anti–aligned along the polar direction and will yield degeneracies in the phase portraits, as we will observe later in Section 4. Now, we also note that the residual symmetry group $\mathcal{S}_\sigma$ is again S^1 and its Lie algebra is generated by the vector field

$$X_\sigma = \frac{\kappa\overline{\kappa}}{r\overline{r}}\left(\partial_\phi - \partial_{\overline{\phi}}\right). \tag{3.9}$$

This suggests that we pick the following new coordinates:

$$\alpha = \phi - \overline{\phi}, \quad \beta = \phi + \overline{\phi}, \quad \omega = \omega_0 + R\cos\psi \tag{3.10}$$

where ψ is defined through three different formulae according to the relative magnitude of the constant σ:

$$1)\ \sigma \geq |\overline{\kappa}| - |\kappa| : \qquad \cos\psi = \frac{|\overline{\kappa}|\left(1-\cos\overline{\theta}\right) - |\kappa|(1-\cos\theta)}{|\overline{\kappa}|\left(1-\cos\overline{\theta}\right) + |\kappa|(1-\cos\theta)} \tag{3.11a}$$

$$2)\ |\kappa| - |\overline{\kappa}| \leq \sigma \leq |\overline{\kappa}| - |\kappa| : \qquad \cos\psi = \cos\theta, \tag{3.11b}$$

$$3)\ \sigma \leq |\kappa| - |\overline{\kappa}| : \qquad \cos\psi = \frac{|\kappa|(1+\cos\theta) - |\overline{\kappa}|\left(1+\cos\overline{\theta}\right)}{|\kappa|(1+\cos\theta) + |\overline{\kappa}|\left(1+\cos\overline{\theta}\right)}. \tag{3.11c}$$

The constants ω_0 and R are determined by using (3.10). Writing down the expressions for the Hamiltonian function of the system and the Hamiltonian vector field gives

$$H(\alpha,\omega(\phi)) = \frac{1}{2}\lambda_1\{r^2 + \overline{r}^2 + \frac{r\overline{r}}{\kappa\overline{\kappa}}\left[\Gamma\omega^2 + \Delta\sigma^2 + E\sigma\omega + f(\psi)\overline{f}(\psi)\cos\alpha\right]\}, \tag{3.12a}$$

$$X_H = \frac{2\kappa\overline{\kappa}}{r\overline{r}R\sin\psi}\left[H_\alpha\partial_\psi - H_\psi\partial_\alpha\right] + \frac{2\kappa\overline{\kappa}}{r\overline{r}}H_\sigma\partial_\beta, \tag{3.12b}$$

where

$$\Gamma = \tfrac{1}{4}(L-1)\rho_+ - L, \quad \Delta = \tfrac{1}{4}(L-1)\rho_+ + L, \quad E = \tfrac{1}{4}(L-1)\rho_-, \tag{3.12c}$$

$$L = \lambda_2/\lambda_1, \qquad \rho_\pm = (r\overline{\kappa}/\overline{r}\kappa \pm \overline{r}\kappa/r\overline{\kappa}), \tag{3.12d}$$

$$f(\psi) = \sqrt{4\kappa^2 - (\sigma+\omega)^2}, \quad \overline{f}(\psi) = \sqrt{4\overline{\kappa}^2 - (\sigma-\omega)^2}. \tag{3.12e}$$

As for the equations of motion, we find them to be

$$\dot{\psi} = \frac{\lambda_1 \sin\alpha}{R\sin\psi} f(\psi)\overline{f}(\psi), \tag{3.13a}$$

$$\dot{\alpha} = \lambda_1(2\Gamma\omega + E\sigma) + \lambda_1[(\sigma-\omega)f(\psi)/\overline{f}(\psi) - (\sigma+\omega)\overline{f}(\psi)/f(\psi)]\cos\alpha, \tag{3.13b}$$

$$\dot{\beta} = \lambda_1(2\Delta\sigma + E\omega) - \lambda_1[(\sigma-\omega)f(\psi)/\overline{f}(\psi) + (\sigma+\omega)\overline{f}(\psi)/f(\psi)]\cos\alpha, \tag{3.13c}$$

A few remarks are in order. First, we observe that the variable β does not appear in the equations for the variables α and ψ, thus confirming that we reduced to S^2; indeed we can solve for the two latter quantities and then get the former through a quadrature of (3.13c). A second remark concerns the degeneracies mentioned earlier. When $\sigma = \pm(\kappa - \overline{\kappa})$, either one of the poles on the sphere will be singular, or both of them if, in addition, $\sigma = 0$. This indicates that the phase space could actually be viewed as a pierced sphere with one or both removed. A last comment is to the effect that the unreduced system is completely integrable; this is a consequence of the fact that the reduced phased space is two-dimensional, i.e., that the reduced system possesses a single degree of freedom. Moreover, the solution manifold of the unreduced system (2.3) is completely determined from that of the reduced system (3.13a,b) on S^2.

4. PHASE PORTRAIT ANALYSIS

In this section, we will investigate the nature of the phase portrait for the reduced system (3.13a,b). Specifically, we will limit ourselves to a few particularly interesting subcases (more details about the existence and the description of bifurcation sequences for both the one–beam and the two–beam problems can be found in David *et al.* [1990]) and wish to determine the fixed points and determine their type; since we are dealing with a Hamiltonian system, the fixed points can only be stable *centers* or unstable *saddle points*, although some exotic points may arise as *pseudo critical points* when degenerate bifurcations take place. An additional feature of the phase space is the possible existence, aside from periodic trajectories, of infinite–period orbits. These, on one hand, provide us with especially interesting solutions, and on the other hand, are a prime ingredient when looking for chaotic behaviour.

CASE 1: $\overline{r} = r, |\overline{\kappa}| = |\kappa|$. We also mention that setting $\sigma = 0$ and $|\overline{\kappa}| \neq |\kappa|$ also yields the same dynamics. The geometry of the phase portrait depends on two essential parameters, Γ and σ and is depicted in Figure 2. For sufficiently large magnitudes of $|\Gamma|$, and for $\sigma \neq 0$, the sphere is partitioned into three distinct families of periodic orbits, with three limit stable fixed points, separated by a pair of homoclinic loops connected to a hyperbolic saddle point; we thus have a figure-eight pattern. Asymptotically, as $|\Gamma| \to \infty$, the above loops merge to a great circle of fixed points on the equator. As $|\Gamma|$ gets smaller, these homoclinic loops shrink and eventually collapse to the saddle point as $|\Gamma| \to \Gamma_\pm(\sigma)$, at which point a reverse bifurcation takes place, i.e., (Saddle + 2 Centers) $\to$ (Center). The curves on the (Γ, σ)-plane on which the bifurcation occurs are parabolas given by

$$\Gamma_\pm(\sigma) = \frac{\sigma^2 + 4\kappa^2}{\sigma^2 - 4\kappa^2} \tag{4.1}$$

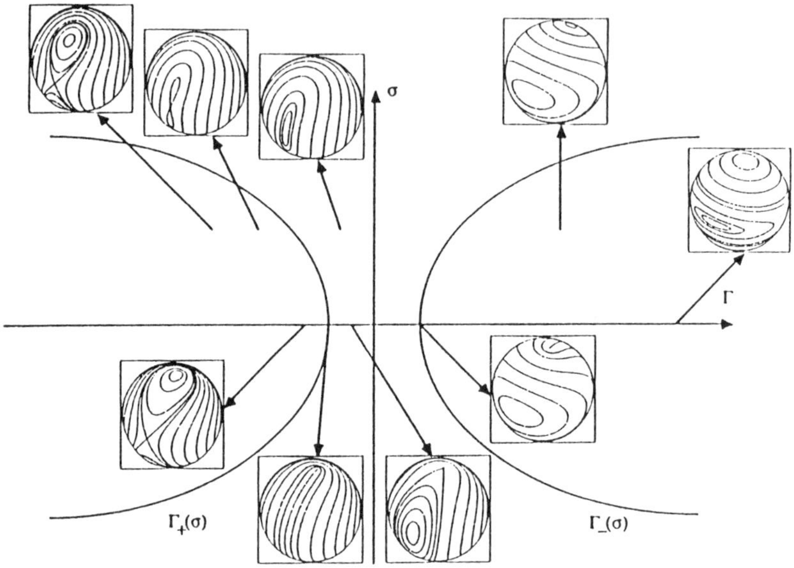

FIGURE 2. The phase portrait for case 1. A degenerate bifurcation takes place when $\sigma = 0$.

The limit $\sigma \to 0$ is special and corresponds to a degeneracy mentioned earlier, where the reduced phase space actually consists of S^2 without its two poles. When $\sigma = 0$, we observe that the poles behave as fixed points, irrespective of the value of Γ. Examining Figure 2, we then see that the homoclinic loops in the phase portrait cannot shrink to the saddle point because the two centers they enclose are constrained to remain at the poles. So these loops actually collapse to form a half great-circle of fixed points from one pole to the other. As the critical parabolic curve is traversed, the nature of the bifurcation is naturally also degenerate: the two *pseudo fixed points* at the poles see their type change to saddles, whereas the saddle point on the equator becomes a stable point. In addition, the line of fixed points opens up into two geodesics between the poles that act as heteroclinic trajectories (infinite period).

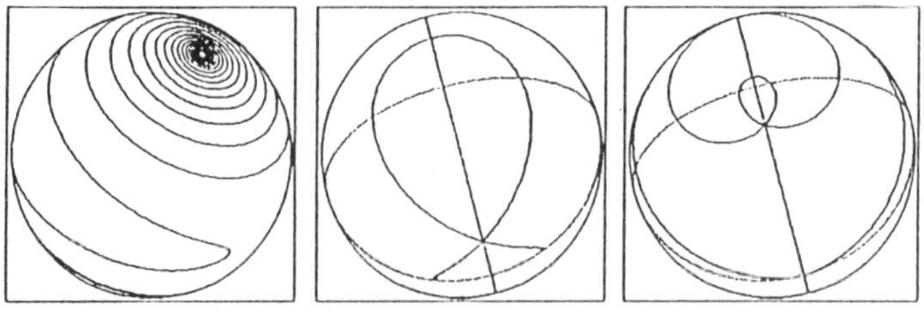

FIGURE 3. Some special kink-like and soliton-like solutions associated with infinite-period orbits appearing in Figure 2.

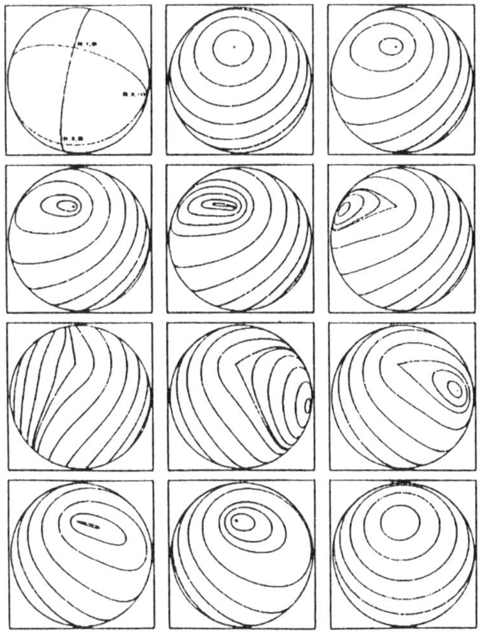

FIGURE 4. Bifurcation behaviour for case 2. Note the presence of a single homoclinic loop connected with one of the poles.

In Figure 3, we illustrate some special solutions associated with trajectories of infinite period appearing in Figure 2. These curves are drawn on the **u**–Poincaré sphere and show how the polarization evolves with time. The left picture corresponds to any of the two geodesic heteroclinic lines connecting the poles: the polarization ellipse describing the beam starts as a circle and then sees its eccentricity increase as it rotates about its center, until it ends up as a line when the trajectory gets to the equator on the Poincaré sphere. The two other pictures correspond to the homoclinic loops appearing in Figure 2 for the non-degenerate situation. Typically, on the Poincaré sphere, these orbits correspond to trajectories that originate on the equator (thus an initial linearly polarization state) and spiral around one of the poles for any number of times before going back to the equator. These solutions are reminiscent of kinks; in the particular case when they get back to their initial location, they are identified as solitary waves.

CASE 2: $\overline{\sigma} = \pm(|\overline{\kappa}| - |\kappa|)$. This second case also exhibits interesting features. The singular situation ($\sigma = 0$) from the previous case is also obtained as a limit when $|\overline{\kappa}| = |\kappa|$. Again, the equations of motion imply a special fixed point, constrained to be located at one pole, irrespective of the parameters. We illustrate the bifurcation behaviour in Figure 4. When the magnitude of $|\Gamma|$ is sufficiently large the phase portrait consists of a family of periodic orbits with two stable fixed points, one exactly at the north pole and the other one nearby the south pole. Nothing happens qualitatively to this description until $|\Gamma|$ reaches a critical value, at which point the north pole becomes unstable and this is accompanied by the creation of a single loop homoclinic to the pole. As Γ crosses zero and changes sign, this loop stretches, goes under the sphere, and eventually shrinks to the north pole, but from the opposite direction to which it emerged from. We also note that during the evolution of this loop, the stable center it initially enclosed was left to stay nearby the south pole, while it dragged the center initially located near the south pole to collapse with it at the north pole.

Let us make the following closing remark for this Section. The Euler index on a surface is definedas the difference between the number of stable and unstable fixed points present in the phase portrait; this number is an invariant. For the two–sphere, S^2, this number is equal to two (it is unity for the real plane). We observe that the index is indeed 2 for all the phase space configurations except when a singular situation occurs, which is just another indication that tells us that the reduced phase space is singular. For the degenerate subcase of case 1, the index is zero (two extra pseudo saddle points at the poles) and it is one for case 2 (one extra hyperbolic point at the north pole).

5. HOMOCLINIC CHAOS

In this section, we will investigate the question of generating complex dynamics for our system when it is subject to some classes of perturbations. These consist in modifying the material parameters of the optical medium in certain specific ways. These perturbations are of physical relevance in applied fields such as fiber optics communications and polarization switching. We will use the Melnikov technique (see Melnikov [1963], Guckenheimer and Holmes [1983], and Wiggins [1988]) to demonstrate the existence of chaos for our optical system, in the form of either horseshoe structures or Arnold diffusion when the dimension is high enough. This method essentially relies on showing that the stable and unstable manifolds of certain hyperbolic manifolds do intersect transversely. This is actually done by calculating so–called Melnikov integral functions, which approximate the separation between these manifolds in some appropriate way, and showing that they possesse countable infinities of simple zeroes, independently of the magnitude of the perturbation. For systems with a single degree of freedom, this Melnikov function has a relatively simple form. Let H^0 denote our previous Hamiltonian function H for the unperturbed system, and let H^1 be the perturbation Hamiltonian function. Then the Melnikov function is the line integral of the symplectic Poisson bracket of H^0 and H^1 along an unperturbed homoclinic or heteroclinic orbit and, for our system, is

$$M(\tau_0) = \int_R \{H^0, H^1\} [\omega(\tau + \tau_0), \alpha(\tau + \tau_0)] \, d\tau \, , \tag{5.1}$$

where ω is defined as in (3.10) and is a more convenient variable to use than ψ. The horseshoe chaoticity is relatively simple to vizualise on the plane. Typically, consider a homoclinic loop connected to a hyperbolic fixed point on the phase plane (see Figure 5).

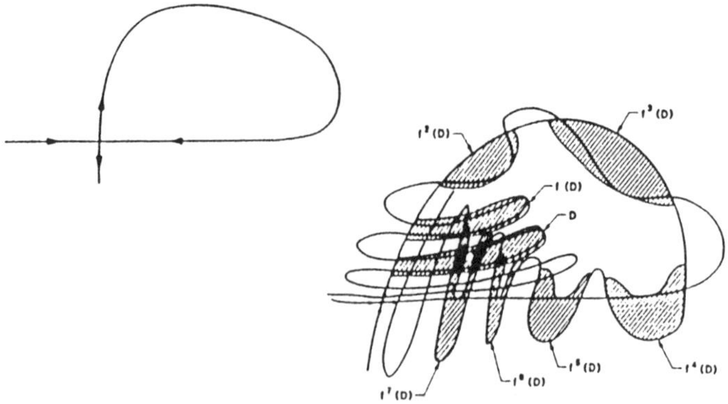

Figure 5. Unperturbed (left) and perturbed (right) dynamics nearby a hyperbolic saddle point.

We consider a perturbation that will preserve the fixed point p, up to a slight displacement. The loop is actually both the stable manifold $W^s(p)$ and the unstable manifold $W^u(p)$ of p. Generically, under a perturbation, the homoclinic loop will break up so that $W^s(p)$ can no longer be identified with $W^u(p)$, thus giving rise to regions of stochasticity. Two fates are then possible. A first possibility is that these manifolds will miss each other; this does not generate any chaos. The second possibility is that they will intersect transversely (tangential intersection is an intermediate situation). Horseshoe chaos is the consequence of the intersection of these manifolds. In particular, transversal intersection in the Poincaré map induces stretching and folding, as a rectangular region of the phase space is mapped away from and then back into the vicinity of the hyperbolic fixed point. These effects are sufficient to cause horseshoe tangles (see Figure 5 and Wiggins [1988]). A second type of chaotic behaviour is characteristic of dynamics ocuring in higher dimensions and is called Arnold diffusion (see Holmes and Marsden [1982b]). It is characterized by the fact that the phase space is less likely to be partitioned into disconnected regions of stochasticity; indeed this necessitates invariant subspaces of co–dimensions separating the regions. If this condition is not met, then solutions do diffuse from region to region.

The application of the Melnikov technique actually consists in calculating the Melnikov integral and ascertaining whether it has simple zeroes. The existence of chaos generating structure is then a consequence of theorems by Melnikov and Poincaré–Birkhoff–Smale. For systems whose dynamics take place in more than two dimensions, the Melnikov function is no longer given by (5.1). Our perturbations are such that the perturbed systems we are considering here fall within certain classes which are investigated in great detail in Wiggins [1988] and we will present the form of the Melnikov intergal as we go along.

We will now proceed to illustrate the method for three types of perturbations corresponding to the following deformations of the matrix $\boldsymbol{W}$:

$$\boldsymbol{W} = \mathrm{Diag}\{\lambda_1, \lambda_2 + \epsilon\cos[\nu(\tau - \tau_0)], \lambda_1\}, \tag{5.2a}$$

$$\boldsymbol{W} = \mathrm{Diag}\{\lambda_1 + \epsilon, \lambda_2, \lambda_1\}, \tag{5.2b}$$

$$\boldsymbol{W} = \mathrm{Diag}\{\lambda_1 + \epsilon\cos[\nu(\tau - \tau_0)], \lambda_2, \lambda_1\}, \tag{5.2c}$$

where ϵ is small. We will examine the consequence of these perturbations in the neighbourhood of the heteroclinic lines connecting the poles depicted in Figure 2.

Perturbations of type (5.2a) preserve S^2 as the reduced phase space and for this case, the Melnikov integral is given by (5.1). For the particular case where $\overline{\kappa} = -\kappa$, we find

$$M(\tau_0) = \frac{3r\nu^2\pi}{8\lambda_1^2\sin^2\alpha_0}\operatorname{csch}\left(\frac{\nu\pi}{4\lambda_1 r\sin\alpha_0}\right)\sin(\nu\tau_0), \tag{5.3}$$

where $\cos\alpha_0 = -(1+L)/2$. Clearly, this expression has simple zeroes located at $\tau_0 = n\pi/\nu$, $n \in Z$. This implies the existence of a Smale horseshoe structure in the Poincaré map, obtained as a recursive folding and stretching of regions nearby the poles. The physical effect implied by the Horseshoe structure is that of an intermittent switching dynamics between two states of circular polarization.

Perturbations of the type (5.2b) break the symmetry that alllowed the second reduction step and thus lift the phase space back to $S^2 \times S^2$. This yields a perturbed system falling within Category III of Wiggins [1988], and for which the Melnikov function is

$$M(\beta_0) = \int_R \frac{\partial H^1}{\partial \beta}(\omega, \alpha, \sigma, \beta + \beta_0)\, d\tau \tag{5.4}$$

Again choosing $\overline{\kappa} = -\kappa$, we find

$$M(\beta_0) = \frac{-r^2(2 - \cos\alpha_0)\sin\beta_0}{\lambda_1 \sin\alpha_0}. \tag{5.5}$$

As in the previous case, this function obviously has simple zeroes and, once again, horseshoe chaos is implied; the difference is that, here, the stable and unstable manifolds are toroidal objects embedded in the space $S^2 \times S^2$. For both of the above types of perturbations, the phase space is partitioned into stochastic dynamical layers separated by invariant tori that form inpenetrable barriers for the polarization.

Finally, perturbations of the type (5.2c) also break symmetries. In contrast with the first two types, it yields Arnold diffusion; the phase space is the five-dimensional manifold $S^2 \times S^2 \times R$ and it can no longer be partitioned in disconnected chaotic regions. Here again, we obtain a system belonging to Category III; however, the Melnikov function is now a vector whose two components are given by

$$\begin{aligned} M_1(\tau_0, \beta_0) &= \int_R \left(\frac{\partial H^0}{\partial \omega}\frac{\partial H^1}{\partial \alpha} - \frac{\partial H^0}{\partial \alpha}\frac{\partial H^1}{\partial \omega} - \frac{\partial H^0}{\partial \sigma}\frac{\partial H^1}{\partial \beta} \right) d\tau + \frac{\partial H^0}{\partial \sigma}\int_R \frac{\partial H^1}{\partial \beta} d\tau, \\ M_2(\tau_0, \beta_0) &= -\int_R \frac{\partial H^1}{\partial \beta} d\tau. \end{aligned} \tag{5.6}$$

Again, choosing $\overline{\kappa} = -\kappa$, we obtain

$$\begin{aligned} M_1(\tau_0, \beta_0) &= \frac{3\nu^2 r\pi(L - \cos\alpha_0 \cos\beta_0)}{16\lambda_1^2 \sin^2\alpha_0} \operatorname{csch}\left(\frac{\nu\pi}{4\lambda_1 r \sin\alpha_0} \right) \sin(\nu\tau_0), \\ M_2(\tau_0, \beta_0) &= \frac{-\nu\pi(1 - \frac{1}{2}\cos\alpha_0)\sin\beta_0}{4\lambda_1^2 \sin^2\alpha_0} \operatorname{csch}\left(\frac{\nu\pi}{4\lambda_1 r \sin\alpha_0} \right) \cos(\nu\tau_0). \end{aligned} \tag{5.7}$$

Here, this pair of functions possesses two distinct sets of zeroes and this is sufficient to ensure the formation of transition chains giving rise to the phenomenon of Arnold diffusion. Physically, this diffusion expresses itself through a back and forth transfer of the polarization among the nonlinear modes of the system in a slow and erratic manner.

6. CONCLUSIONS

We presented an investigation of the dynamics of a pair of laser pulses counterpropagating as travelling waves through a nonlinear polarizable medium. We have seen that the geometry of this system is remarkable and that it reduces, for isotropic media, to motion on the sphere, due to continuous Hamiltonian symmetries of the system. We have discovered a set of degenerate bifurcations that can take place in the phase portrait on the reduced space S^2 as material parameters are varied, as shown in Figures 2 and 4.

Chaotic behavior under certain classes of spatially periodic perturbations of the optical medium has also been demonstrated by using the Melnikov method. The instance of horseshoe chaos which we examined corresponds to sensitive dynamics on the Poincaré sphere in the form of an intermittent switching from one circular polarization state to the other. We also have shown that Arnold diffusion is another possible chaoticity behaviour under some perturbations that are sufficiently symmetry-breaking.

Outlooks for the work presented here include the addition of dissipative and dispersive effects. In this case, the system must be examined as a set of partial differential equations, thus an infinite-dimensional dynamical system, for which the search for homoclinic structures and mechanisms for the onset of chaotic, or complex, dynamics remains to be done.

REFERENCES

V.I. Arnold [1964], Instability of dynamical systems with several degrees of freedom, Sov. Math. Dokl. **156**, 9–12.

M. Crampin and F.A.E. Pirani [1987], *Applicable Differential Geometry*, London Mathematical Society Lecture Notes Series **59**, Cambridge University Press, Cambridge (U.K.).

D. David [1989], Horseshoe chaotic dynamics in a polarized optical beam subject to periodic and dissipative perturbations, Preprint, Los Alamos National Laboratory.

D. David, D.D. Holm, and M.V. Tratnik [1988a], Integrable and chaotic polarization dynamics in nonlinear optical beams, Phys. Lett. **137A**, 355–364.

D. David, D.D. Holm, and M.V. Tratnik [1988b], Horseshoe chaos in a periodically perturbed polarized optical beam, Phys. Lett. **138A**, 29–36.

D. David, D.D. Holm, and M.V. Tratnik [1990], Hamiltonian chaos in nonlinear optical polarization dynamics, Physics Reports (in press).

A.L. Gaeta, R.W. Boyd, J.R. Ackerhalt, and P.W. Milonni [1987], Instabilities and chaos in the polarizations of counterpropagating light fields, Phys. Rev. Lett. **58**, 2432–2435; Instabilities in the propagation of arbitrarily polarized counterpropagating waves in a nonlinear Kerr medium, in *Optical Bistability III*, H.M. Gibbs, P. Mandell, N. Peyghambarian, S.D. Smith (eds.), Springer–Verlag, Berlin.

J. Guckenheimer and P.J. Holmes [1983], *Nonlinear Oscillations, Dynamical Systems, and Bifurcations of Vector Fields*, Applied Mathematical Sciences, Vol. **42**, Springer–Verlag, New–York.

P.J. Holmes and J.E. Marsden [1982a], Horseshoes in perturbations of Hamiltonian systems with two degrees of freedom, Comm. Math. Phys. **82**, 523–544.

P.J. Holmes and J.E. Marsden [1982b], Melnikov's method and Arnold diffusion for perturbations of integrable Hamiltonian systems, J. Math. Phys. **23**, 669–675.

A.E. Kaplan [1983], Light–induced nonreprocity, field invariants, and nonlinear eigenpolarizations, Opt. Lett. **8**, 560–562.

R. Lytel [1984], Optical multistability in collinear degenerate four–wave mixing, J. Opt. Soc. Am. **B1**, 91–94.

P.D. Maker, R.W. Terhune, and C.M. Savage [1964], Intensity–dependent changes in the refractive index of liquids, Phys. Rev. Lett. **12**, 507–509.

V.K. Melnikov [1963], On the stability of the center for time periodic perturbations, Trans. Moscow Math. Soc. **12**, 1–57.

P.J. Olver [1986], *Applications of Lie Groups to Differential Equations*, Graduate Texts in Mathematics **107**, Springer–Verlag, New York.

K. Otsuka, J. Yumoto, and J.J. Song [1985], Optical bistability based on self–induced polarization–state change in anisotropic Kerr–like media, Opt. Lett. **10**, 508–510.

M.V. Tratnik and J.E. Sipe [1987], Nonlinear polarization dynamics. I. The single–pulse equations, Phys. Rev. **A35**, 2965–2975; II. Counterpropagating beam equations: new simple solutions and the possiblity for chaos, Phys. Rev. **A35**, 2976–2988; III. Spatial polarization chaos in counterpropagating beams, Phys. Rev. **A36**, 4817–4822.

S. Trillo, S. Wabnitz, and R.H. Stolen [1986], Experimental observation of polarization instability in a birefringent optical fiber, Appl. Phys. Lett. **49**, 1224–1226.

S. Wiggins [1988], *Global Bifurcations and Chaos – Analytical Methods*, Applied Mathematical Sciences **73**, Springer–Verlag, New York.

IV. QUANTIZATION AND R–MATRICES

C. DUVAL, J. ELHADAD, M.J. GOTAY, and G.M. TUYNMAN

Louis H. KAUFFMAN

B.A. KUPERSHMIDT

NONUNIMODULARITY AND THE QUANTIZATION OF THE PSEUDO-RIGID BODY

C. DUVAL[1], J. ELHADAD[1], M.J. GOTAY[2*] and G.M. TUYNMAN[1**]

[1]CPT-CNRS, Luminy Case 907
F-13288 Marseille cedex 9, France
E-mail: Duval@FRCPTM51.bitnet;
GMT@FRCPTM51.bitnet

[2]Mathematics Department, U.S. Naval Academy
Annapolis, MD21402, USA
E-mail: Gotay@nl.usna.navy.mil

ABSTRACT. We discuss the pseudo-rigid body as an example of a classical system with nonunimodular gauge symmetries. We show that the standard Dirac quantization prescription must be modified for such systems, and illustrate how this can be accomplished by means of "unimodularization."

1. Introduction

When quantizing a constrained classical system with nonunimodular gauge symmetries, the usual Dirac prescription for selecting the gauge invariant states of the system must be modified [2, 11]. Otherwise, one may lose the correlation between the classical and quantum gauge invariant states, and the quantization of the original system may lead to incorrect results.

To be more precise, let (M,ω) be a symplectic manifold, H a Lie group acting symplectically on M on the left, and J an Ad^*-equivariant momentum mapping for this action Φ. The invariance of the system with respect to the gauge group H demands that J vanish [5]. On the quantum level, these constraints are traditionally implemented by means of the *Dirac conditions*

$$\mathcal{Q}J_\xi\,[\Psi] = 0 \tag{1}$$

for all $\xi \in \mathfrak{h}$, the Lie algebra of H, where $\mathcal{Q}J_\xi$ is the quantum operator corresponding to the function $J_\xi = \langle J, \xi\rangle$. The space $\mathcal{H}_0$ of all Ψ is supposed to comprise the physically admissible quantum states of the system.

[1] Recall that a group H is unimodular if it carries a bi-invariant volume. Infinitesimally, this is equivalent to $tr(ad_\xi) = 0$ for all $\xi \in \mathfrak{h}$.

* Partially supported by NSF grant #DMS-8805699.
** Partially supported by the Netherlands Organization for Scientific Research (NWO).

In general, one expects the Hilbert space $\overline{\mathcal{H}}$ obtained by quantizing the classical reduced phase space $\overline{M} = H\backslash J^{-1}(0)$ to coincide with $\mathcal{H}_0$. Put another way, the processes of reduction and quantization should commute. Provided the reducing group H is unimodular[1] this is indeed the case, at least under mild restrictions which are satisfied by most systems of physical interest. (See [2, 4] and references cited therein.)

When H is not unimodular this equivalence is no longer guaranteed, and in fact can be shown to fail [2, 3, 11]. In practical terms, however, this has not been a matter of concern, as nonunimodular symmetries are rare. Nonetheless, there are bona fide mechanical systems which have such symmetries, and for which the Dirac prescription (1) leads to demonstrably erroneous conclusions. In this paper we will study one such system: the *pseudo-rigid body*.

To regain the correspondence between the original and reduced phase space quantizations, a theory has been developed in [2] which for our purposes can be regarded as a means of "unimodularizing" a classical system with nonunimodular symmetries. One thereby obtains a new unimodular system which is "classically equivalent" to the original system and to which the results cited above apply. Thus reduction and quantization will commute when one considers this new system. This theory is in essence a "bosonic" counterpart to the classical BRST formalism, but is phrased purely in terms of the underlying symplectic geometry of the system.

A comprehensive discussion of these matters is given in [2]. It is enough here to sketch the results. One replaces the original Hamiltonian system (M,ω,H,Φ,J) by an "extended" Hamiltonian system $\left(\tilde{M},\tilde{\omega},\tilde{H},\tilde{\Phi},\tilde{J}\right)$, where:

$$
\begin{aligned}
\tilde{M} &= M \times (\mathfrak{h}^* \oplus \mathfrak{h}) \\
\tilde{\omega} &= \omega + \omega_{\mathfrak{h}^*} \\
\tilde{H} &= T^*H \approx H \ltimes \mathfrak{h}^* \\
\tilde{\Phi}_{(h,\nu)}(m,\eta,\rho) &= \left(\Phi_h(m), Ad^*_{h^{-1}}(\eta) + \nu, Ad_h\rho\right) \\
\tilde{J}_{(\xi,\nu)}(m,\eta,\rho) &= J_\xi(m) - \langle ad^*_\xi(\eta), \rho\rangle + \langle \nu, \rho\rangle .
\end{aligned}
\tag{2}
$$

In these expressions $\omega_{\mathfrak{h}^*}$ is the canonical symplectic structure on $\mathfrak{h}^* \oplus \mathfrak{h} \approx T^*\mathfrak{h}^*$ and we have used the canonical left trivialization to endow $T^*H \approx H \times \mathfrak{h}^*$ with a semi-direct product structure. (H acts on $\mathfrak{h}^*$ by the coadjoint representation.)

This construction has three salient features: (*i*) The reduced spaces

$$\tilde{H}\backslash\tilde{J}^{-1}(0,0) \approx H\backslash J^{-1}(0),$$

so that this new extended system is classically equivalent to the original one. (*ii*) The extended reducing group $\tilde{H} = T^*H$ is *always unimodular*, so that the quantization of $\tilde{M}$ will be compatible with that of the reduced space $\overline{M}$. (*iii*) It is canonical and symplectically natural. That is, the definition of $\left(\tilde{M}, \tilde{\omega}, \tilde{H}, \tilde{\Phi}, \tilde{J}\right)$ in (2) is not arbitrary, but rather is the result of applying symplectic induction to H viewed as a subgroup of $T^*H \approx H \ltimes \mathfrak{h}^*$.

Since the extended system is unimodular, we may apply the Dirac prescription (1) to it, obtaining

$$\tilde{Q}\tilde{J}_{(\xi,\nu)}[\tilde{\Psi}] = 0 \tag{3}$$

for all $(\xi, \nu) \in \tilde{\mathfrak{h}} \approx \mathfrak{h} \ltimes \mathfrak{h}^*$. Thus the physically admissible quantum states of the system are those which are $\tilde{H}$-invariant. In terms of the original system, (3) effectively reduces to the *modified Dirac prescription*

$$QJ_\xi[\Psi] = -\frac{i}{2} tr\left(ad_\xi\right) \Psi. \tag{4}$$

It follows that from the standpoint of the nonuniumodular group H, the physically admissible quantum states are actually *quasi*-invariant.

We use the pseudo-rigid body to illustrate these results as well as the formalism developed in [2]. We show that our unimodularization technique (or, equivalently, the modified Dirac prescription (4)) yields the expected — and correct — results, whereas the original Dirac prescription (1) goes awry.

Our conclusions here may have interesting implications regarding the quantization of field theories, as one would not expect their (infinite-dimensional) gauge groups to be unimodular in any reasonable sense. This issue has only begun to be explored [7].

2. The Classical Pseudo-Rigid Body

Consider a pseudo-rigid body, that is, an elastic body moving in $\mathbb{R}^3$ which is allowed to undergo linear deformations. The configuration space for such a body is the group $G = GL(3, \mathbb{R})_+ \ltimes \mathbb{R}^3$, where the identity component $GL(3, \mathbb{R})_+$ of $GL(3, \mathbb{R})$ consists of shears, dilations and rigid rotations, and the $\mathbb{R}^3$ factor comprises rigid translations. The Lie algebra $\mathfrak{g}$ of G consists of pairs (a,b) with $a \in \mathfrak{gl}(3, \mathbb{R})$ and $b \in \mathbb{R}^3$; similarly, we may represent its dual $\mathfrak{g}^*$ by pairs (α,β) with $\alpha \in \mathfrak{gl}(3, \mathbb{R})$ and $\beta \in \mathbb{R}^3$. The pairing between $\mathfrak{g}$ and $\mathfrak{g}^*$ is defined by

$$\langle(\alpha, \beta), (a, b)\rangle = tr(\alpha a) + \beta b, \tag{5}$$

where we regard β and b as row and column vectors, respectively.

The "true" configuration space (or the *space of shapes*) of the body is clearly the homogeneous space $S = H \backslash G$, where $H = (\mathbf{R}_+^* \times SO(3)) \ltimes \mathbf{R}^3$. Geometrically, this means that we identify configurations that differ only by rigid rotations, homogeneous dilations and spatial translations. The crucial point for the present discussion is that *the reducing group H is nonunimodular:* for $(c,d) \in \mathfrak{h}$, we have $tr(ad_{(c,d)}) = tr\ c$. Nonunimodularity actually comes from the dilations via the semi-direct product structure on H.

Let us now describe the details of the classical reduction. We use the canonical left trivialization to identify the phase space T^*G with $M = G \times \mathfrak{g}^*$. (That is, we work in "body coordinates" or the "material representation.") The canonical 1–form on M is $\theta(g,\mu) = \langle \mu, g^{-1}\mathrm{d}g \rangle$, where $g = (A,B)$, $g^{-1}\mathrm{d}g$ denotes the Maurer-Cartan form of G and $\mu = (\alpha,\beta) \in \mathfrak{g}^*$. We thus have

$$\theta = tr(\alpha A^{-1}\mathrm{d}A) + \beta A^{-1}\mathrm{d}B. \tag{6}$$

The Hamiltonian action of H on M is $(h,(g,\mu)) \rightsquigarrow (hg,\mu)$ with momentum mapping $J : M \rightarrow \mathfrak{h}^*$ given by

$$\begin{aligned} J(A,B,\alpha,\beta) &= i^*\left(Ad^*_{(A,B)^{-1}}(\alpha,\beta)\right) \\ &= i^*\left(Ad^*_{A^{-1}}(\alpha + A^{-1}B\beta), \beta A^{-1}\right), \end{aligned} \tag{7}$$

where $i : \mathfrak{h} \rightarrow \mathfrak{g}$ is the inclusion (*cf.* [1], §4.4). From (7) and (5) we see that the vanishing of $\langle J(A,B,\alpha,\beta),(c,d)\rangle$ for all $(c,d) \in \mathfrak{h}$ implies $\beta = 0$ and

$$tr(Ad^*_{A^{-1}}(\alpha)\,c) = 0. \tag{8}$$

Since $c \in Lie(\mathbf{R}_+^* \times SO(3))$, c must be of the form $rI + K$, where $r \in \mathbf{R}$, I is the 3×3 identity and K is skew-symmetric. Taking $K = 0$, (8) yields $tr\ \alpha = 0$. Similarly, taking $r = 0$ we find that $Ad^*_{A^{-1}}(\alpha) = A\alpha A^{-1}$ must be symmetric, so that

$$\alpha^t = (A^tA)\,\alpha(A^tA)^{-1}. \tag{9}$$

The constraint submanifold $J^{-1}(0)$ therefore consists of triples (A,B,α), where α is traceless and satisfies (9).

Now every matrix $A \in GL(3,\mathbb{R})_+$ admits a unique polar decomposition $A = RQ$, where $Q = \sqrt{A^t A}$ is positive-definite symmetric and $R \in SO(3)$. Set $\bar{p} = Q\alpha Q^{-1}$ and $\bar{q} = Q/(det\ Q)^{\frac{1}{3}}$. From this and (9) we deduce that the reduced manifold $\overline{M} = H\backslash J^{-1}(0)$ consists of pairs $(\bar{q},\bar{p})$ of symmetric 3×3 matrices with $\bar{q} > 0$ & $det\ \bar{q} = 1$ and $tr\ \bar{p} = 0$. The symplectic form on $\overline{M}$ is $\bar{\omega} = -\mathbf{d}\bar{\theta}$, where from (6)

$$\bar{\theta} = tr\left(\bar{q}^{-1}\bar{p}\mathbf{d}\bar{q}\right). \tag{10}$$

Evidently

$$\overline{M} \approx T^*S \approx \mathbb{R}^5 \times \mathbb{R}^5.$$

Remarks: (*i*) These results could also have been obtained by means of the cotangent bundle reduction theorem [8]. In this regard we point out that the symplectic structure on $\overline{M} \approx T^*S$ *is* canonical, although the coordinates $(\bar{q},\bar{p})$ are not (*cf.* (10)). This is because we are working in the material representation of T^*S.

(*ii*) The reduced symplectic manifold T^*S also arises in the liquid drop model of the nucleus [6, 9, 10]. In this context, the cotangent bundle of $S = SO(3)\backslash SL(3,\mathbb{R})$ appears as a coadjoint orbit of the 15–dimensional group $CM(3)$ of (lower) block triangular matrices in $Sp(3,\mathbb{R})$. The fact that $\overline{M}$ can be viewed as a reduction of $T^*SL(3,\mathbb{R})$ has proved quite useful in deriving the general form of "collective" Hamiltonians. Here, however, the reducing group $SO(3)$ is unimodular.

3. Unimodularization and Quantization

We now unimodularize the pseudo-rigid body and quantize it. For the necessary background regarding quantization theory in this context, see [2]. Since we will be working exclusively in the cotangent category, it suffices to consider only the half-density quantization.

Let $\tilde{G} = G \times \mathfrak{h}^*$ be the "extended" configuration space. Then the smooth quantum state space associated to the extended phase space $\tilde{M} \approx T^*\tilde{G}$ is $\Delta^{\frac{1}{2}}(\tilde{G})$, the linear space of 1/2–densities on $\tilde{G}$. We may explicitly describe these half-densities as follows.

Fix an everywhere nonvanishing *right* G–invariant half-density δ_G on G and let $\varepsilon_{\mathfrak{h}^*}$ denote the Lebesgue volume on $\mathfrak{h}^*$. Then

$$\delta_{\tilde{G}} = \delta_G \otimes |\varepsilon_{\mathfrak{h}^*}|^{\frac{1}{2}} \tag{11}$$

is a half-density on the extended configuration space which is invariant under the *left* action of $\tilde{H}$ on $\tilde{G}$ given by[2]

$$((h,\nu),(g,\eta)) \rightsquigarrow (hg, Ad^*_{h^{-1}}(\eta)+\nu).$$

Indeed, invariance follows from the observations that (*i*) for $h \in H$, $det(Ad_h)$ is the same regardless of whether one computes the adjoint action with respect to G or its subgroup H,[3] and (*ii*) $[det(Ad_h)]\,[det(Ad^*_{h^{-1}})] = 1$. Thus wave functions $\tilde{\Psi} \in \Delta^{\frac{1}{2}}(\tilde{G})$ take the form $\tilde{\Psi} = \tilde{\psi}(A,B,\eta)\,\delta_{\tilde{G}}$.

Since $\delta_{\tilde{G}}$ is $\tilde{H}$-invariant it follows that $\mathfrak{L}_{(\xi,\nu)_{\tilde{G}}}\delta_{\tilde{G}} = 0$ for all $(\xi,\nu) \in \tilde{\mathfrak{h}}$. (Here $(\xi,\nu)_{\tilde{G}}$ denotes the fundamental vector field on $\tilde{G}$ corresponding to $(\xi,\nu) \in \tilde{\mathfrak{h}}$.) Thus $\tilde{J}_{(\xi,\nu)}$ quantizes as

$$\tilde{Q}\tilde{J}_{(\xi,\nu)}[\tilde{\Psi}] = -i\left[(\xi,\nu)_{\tilde{G}}\,\tilde{\psi}\right]\delta_{\tilde{G}}. \tag{12}$$

Setting $\xi = 0$, the Dirac prescription (3) yields $\tilde{\psi} = \tilde{\psi}(A,B)$ only. Similarly, setting $\nu = 0$ we obtain $\tilde{\psi} = \tilde{\psi}\left(Q/(det\ Q)^{\frac{1}{3}}\right)$. Thus the space $\tilde{\mathcal{H}}_0$ of $\tilde{H}$–invariant quantum states of the pseudo-rigid body consists of wave functions of the form

$$\tilde{\Psi} = \tilde{\psi}\left(Q/(det\ Q)^{\frac{1}{3}}\right)\delta_{\tilde{G}}. \tag{13}$$

We remark that elements of $\tilde{\mathcal{H}}_0$ are not square integrable over $\tilde{G}$.

It is instructive to compare (12) and (13) with what one would obtain by directly quantizing the original system. Every element $\Psi \in \Delta^{\frac{1}{2}}(G)$ can be written $\Psi = \psi(A,B)\,\delta_G$. Note, however, that δ_G is *not* invariant under the action of H on G; in fact,

$$\mathfrak{L}_{\xi_G}\delta_G = \frac{1}{2}\,tr(ad_\xi)\,\delta_G \tag{14}$$

for all $\xi \in \mathfrak{h}$. Thus we have

$$QJ_\xi[\Psi] = -i\left[\xi_G\psi + \frac{1}{2}\,tr(ad_\xi)\,\psi\right]\delta_G. \tag{15}$$

---

[2] Note that the action $\tilde{\Phi}$ in (2) is just the cotangent lift of this action to $\tilde{M} \approx T^*\tilde{G}$.

[3] The same can then be said for $tr(ad_\xi)$ for all $\xi \in \mathfrak{h}$. These coincidences result from the particular forms of G and H.

Consequently the modified quantization condition (4) reduces to $\xi_G \psi = 0$, since the right hand side of (4) exactly cancels the half-density contribution in (15). (This is, of course, why we chose δ_G as we did. Here we are again making crucial use of the coincidence that $tr\left(ad_\xi\right)$ for $\xi \in \mathfrak{h}$ is the same regardless of whether ad_ξ is computed relative to H or G.) The space $\mathcal{H}_*$ of quasi H–invariant states therefore consists of wave functions of the form $\Psi = \psi\left(Q/\left(det\ Q\right)^{\frac{1}{3}}\right)\delta_G$, consistent with (13) and (11).[4]

Next we turn to the quantization of the reduced phase space. The corresponding wave functions $\bar{\Psi} \in \Delta^{\frac{1}{2}}(S)$ may be expressed as $\bar{\Psi} = \bar{\psi}\left(\bar{q}\right)\delta_S$, where δ_S is the natural half-density on S induced from $\delta_{\tilde{G}}$ on $\tilde{G}$ as follows. Since $\tilde{H}$ is unimodular, $\tilde{\mathfrak{h}}$ carries an Ad–invariant volume.[5] Fix a frame $\tilde{\vartheta}$ for $\tilde{\mathfrak{h}}$ of unit volume. Now $\tilde{H}$ acts freely on $\tilde{G}$, so that $\tilde{G} \to S$ is a principal $\tilde{H}$–bundle. Thus if $\bar{\zeta}$ is a frame for TS, we may define

$$\delta_S\left(\bar{\zeta}\right) = \delta_{\tilde{G}}(\zeta, \tilde{\vartheta}_{\tilde{G}}), \tag{16}$$

where ζ is any lift of $\bar{\zeta}$ to TG. Since $\delta_{\tilde{G}}$ is left $\tilde{H}$–invariant and $\tilde{H}$ is unimodular, we see that this definition makes sense. Moreover, δ_S so defined is independent of the choice of frame $\tilde{\vartheta}$ for $\tilde{\mathfrak{h}}$.

The association $\widetilde{\mathcal{H}}_0 \approx \overline{\mathcal{H}}$ given by

$$\tilde{\psi}\left(Q/\left(det\ Q\right)^{\frac{1}{3}}\right)\delta_{\tilde{G}} \leftrightarrow \bar{\psi}\left(\bar{q}\right)\delta_S \tag{17}$$

therefore provides a *canonical* identification of the $\tilde{H}$–invariant quantum states of the extended system with the quantum states of the reduced system, as guaranteed by the Smooth Equivalence Theorem of [2, 4].

Using this result we can endow the space $\widetilde{\mathcal{H}}_0$ with an inner product. (Recall that elements of $\widetilde{\mathcal{H}}_0$ are not square integrable over $\tilde{G}$.) We simply use the inner product on $\overline{\mathcal{H}} \approx L^2(S)$ along with (17) to define

$$\langle\tilde{\Psi}, \tilde{\Phi}\rangle = \int_S \bar{\psi}\left(\bar{q}\right)^* \bar{\phi}\left(\bar{q}\right)\delta_S^{\ 2}.$$

Thus our unimodularization technique enables us to make the space of $\tilde{H}$–invariant states into a Hilbert space.

---

[4] Although the coefficient function ψ is H–invariant, the total wave function $\Psi = \psi\delta_G$ is *not*, since δ_G is not.

[5] In fact, since $\tilde{H}$ is a cotangent bundle there is a *canonical* choice for this volume. Simply demand that it give one when evaluated on frames of the form (ϑ, ϑ^*) for $\tilde{\mathfrak{h}} \approx \mathfrak{h} \ltimes \mathfrak{h}^*$, where ϑ is any frame for $\mathfrak{h}$ and ϑ^* is the dual frame for $\mathfrak{h}^*$.

4. On the Dirac Quantization Conditions

The computations above illustrate how one unimodularizes (and then quantizes) a system with nonunimodular gauge symmetries. We wish to emphasize here the fact that one *must* unimodularize the original system if there is to be an intrinsically defined correspondence between the gauge invariant and reduced quantum states of the system. Why this is so is explained in detail in [11]; see also [2]. But one can gain an appreciation for what goes wrong in the presence of nonunimodular gauge symmetries in general by again examining the specific case of the pseudo-rigid body. One basic problem is that there is no *intrinsic* way to associate H–invariant (or even quasi H–invariant) half-densities on G with half-densities on S; indeed, there is no strict analogue of (16) in either of these contexts. And it is this relation which is the key ingredient in the canonical equivalence (17) of the extended and reduced phase space quantizations.

In the absence of unimodularity — as in the quantization of the original phase space — one can try to mimic the approach of the last section. For instance, fix a frame ϑ for $\mathfrak{h}$; then one may verify from (16) and (11) (scaling ϑ as appropriate) that

$$\delta_S\left(\bar{\zeta}\right) = \delta_G\left(\zeta, \vartheta_G\right). \tag{18}$$

This result gives rise to an isomorphism $\mathcal{H}_* \approx \overline{\mathcal{H}}$, but it is not intrinsically defined as — in contrast to (16) — it depends on the choice of frame ϑ for $\mathfrak{h}$. Even worse, one must resort to an ad hoc construction in order to identify the space $\mathcal{H}_0$ of H–invariant states with $\overline{\mathcal{H}}$. (An example of such an ad hoc isomorphism is provided by (24) below.)

Having extolled the virtues of quantizing the extended unimodular system, let us now study the quantization of the original system. Our analysis thus far does not distinguish in any significant way between the original and modified Dirac quantization conditions; as we have indicated above, both $\mathcal{H}_*$ and $\mathcal{H}_0$ can be (noncanonically) identified with $\overline{\mathcal{H}}$. Thus it may seem that either quantization prescription will lead to results which are (again, noncanonically) equivalent to those obtained by quantizing the reduced space. In fact this is true if one uses the modified Dirac prescription (4) as was demonstrated in [11]. But this is certainly not the case if one uses instead the *uncorrected* Dirac prescription (1): we will show that the resulting quantization is actually *inconsistent* with the quantization of the reduced phase space. The simplest way to do this is via the quantization of observables on M.

If the quantizations of the original and reduced systems are to be equivalent, there must exist an isomorphism $\sigma : \mathcal{H}_0 \to \overline{\mathcal{H}}$ which intertwines the quantizations of (a certain class of) observables. That is, suppose that f is an observable on M which is H–invariant and so reduces to an observable $\bar{f}$ on $\overline{M}$; then we require

$$\sigma(\mathcal{Q}f[\Psi]) = \bar{\mathcal{Q}}\bar{f}[\sigma(\Psi)] \tag{19}$$

for all $\Psi \in \mathcal{H}_0$. The class of observables to which (19) applies must include those which are at most linear in the momenta. We now prove, in the case of the pseudo-rigid body, that there does not exist such an isomorphism σ.

For suppose there did. Set $\delta = \sigma^{-1}(\delta_S)$, where δ_S is given by (18). Since $\delta \in \mathcal{H}_0$ it is left H–invariant.

On the other hand, let us consider the action of G on itself by *right* translations. Denote by K the momentum map for the lifted action on M; observe that K is linear in the momenta. Since right translations commute with the left H–action, the G–action descends to S with corresponding momentum map $\bar{K}$ on $\overline{M}$. Using (18) and the fact that δ_G is right–invariant, it follows that δ_S is right-invariant. Thus $\bar{\mathcal{Q}}\bar{K}_\mu[\delta_S] = 0$ for all $\mu \in \mathfrak{g}^*$. Then (19) implies that $\mathcal{Q}K_\mu[\delta] = 0$ so that δ is right-invariant as well. Then δ must be a constant multiple of δ_G. But according to (14), δ_G is *not* left H–invariant unless $tr(ad_\xi)$ vanishes for all $\xi \in \mathfrak{h}$, which it does not. Thus we have a contradiction.

As a concrete illustration, consider the right action of $GL(3,\mathbb{R})$ on G given by $(W,(A,B)) \rightsquigarrow (AW,B)$. The lifted action on M is

$$(W,(A,B,\alpha,\beta)) \rightsquigarrow (AW,B,W^{-1}\alpha W,\beta W).$$

The corresponding momentum map $K : M \to \mathfrak{gl}(3,\mathbb{R})^*$ can be interpreted as the "material momentum." Analogous to (7) we obtain from (5) that

$$K_\Omega(A,B,\alpha,\beta) = tr(\alpha\Omega) \tag{20}$$

for $\Omega \in \mathfrak{gl}(3,\mathbb{R})$. Since K_Ω is linear in the momenta, its quantization is routine; using the fact that δ_G is right–invariant, we compute

$$\mathcal{Q}K_\Omega[\Psi] = -i\left[tr\left((A\Omega)^t\frac{\partial\psi}{\partial A}\right)\right]\delta_G. \tag{21}$$

Extend the $GL(3,\mathbf{R})$–action to $\tilde{G} = G \times \mathfrak{h}^*$ by letting it act trivially on the second factor. Then the momentum map $\tilde{K} : \tilde{M} \to \mathfrak{gl}(3,\mathbf{R})^*$ for the lifted action on $\tilde{M}$ is given by the same formula (20), and it quantizes according to essentially the same formula (21), *viz.*,

$$\tilde{Q}\tilde{K}_\Omega[\tilde{\Psi}] = -i\left[tr\left((A\Omega)^t \frac{\partial\tilde{\psi}}{\partial A}\right)\right]\delta_{\tilde{G}}. \tag{22}$$

On the other hand, as it is H–invariant, K_Ω descends to $\overline{M} \approx T^*S$ as $\bar{K}_\Omega(\bar{q},\bar{p}) = tr(\bar{q}^{-1}\bar{p}\bar{q}\Omega)$. We then find that

$$\bar{Q}\bar{K}_\Omega\left[\bar{\Psi}\right] = -i\left[tr\left(\kappa\frac{\partial\bar{\psi}}{\partial\bar{q}}\right)\right]\delta_S, \tag{23}$$

where the symmetric matrix κ is the unique solution of

$$\kappa\bar{q} + \bar{q}\kappa = \Omega^t\bar{q}^2 + \bar{q}^2\Omega^t - \frac{2}{3}(tr\,\Omega)\,\bar{q}^2.$$

It is then straightforward, albeit tedious, to directly verify that the isomorphism $\widetilde{\mathcal{H}}_0 \approx \overline{\mathcal{H}}$ unitarily intertwines the operators $\tilde{Q}\tilde{K}_\Omega$ and $\bar{Q}\bar{K}_\Omega$. (Alternatively, this follows from the Smooth Equivalence Theorem.) Similarly one shows using [11] that the isomorphism $\mathcal{H}_* \approx \overline{\mathcal{H}}$ induced by the association (18) intertwines the operators (21) and (23), and that the isomorphism $\widetilde{\mathcal{H}}_0 \approx \mathcal{H}_*$ induced by the association (11) intertwines the operators (21) and (22).

Now suppose that we use instead the uncorrected quantization condition (1). Furthermore assume that the isomorphism $\sigma : \mathcal{H}_0 \to \overline{\mathcal{H}}$ is chosen in such a way that

$$\begin{aligned}\delta(A,B) &= \left[det\left(Ad_{(A,B)}\right)\right]^{-\frac{1}{2}}\delta_G(A,B)\\ &= (det\,A)^{-\frac{1}{2}}\,\delta_G(A,B)\,.\end{aligned} \tag{24}$$

Note that δ is *the* left-invariant helf-density on G associated to the right-invariant half-density δ_G. Then from (21) we compute

$$QK_\Omega[\delta] = \frac{i}{2}(tr\,\Omega)\,\delta,$$

and this does not correspond under σ to $\bar{Q}\bar{K}_\Omega(\delta_S)$, which vanishes. Thus this particular isomorphism σ is not an equivalence of the original and reduced quantum systems. Of course, our result above shows that this is the case for *any* $\sigma : \mathcal{H}_0 \rightarrow \overline{\mathcal{H}}$.

We close by presenting an entirely different argument as to why the uncorrected Dirac prescription (1) is untenable. Suppose that we fix the center of mass of the body at the outset, so that the spatial translations in G and H are eliminated. Then H is unimodular! If we now quantize this system and its reduction, we obtain results which coincide exactly with those presented here using the modified quantization condition (4). As one does not expect that fixing the center of mass of the body will substantially alter its quantum behavior, one must again conclude that the "$-\frac{i}{2}\, tr\left(ad_\xi\right)$" correction is essential.

References

[1] R. Abraham and J. Marsden, *Foundations of Mechanics*, Benjamin-Cummings, Reading, MA, second ed., 1978.

[2] C. Duval, J. Elhadad, M. J. Gotay, J. Śniatycki, and G. M. Tuynman, *Quantization and bosonic BRST theory*, to appear, (1990).

[3] C. Duval, J. Elhadad, and G. M. Tuynman, *The BRS method and geometric quantization: some examples*, Commun. Math. Phys., to appear, (1990).

[4] M. J. Gotay, *Constraints, reduction and quantization*, J. Math. Phys., **27** (1986), pp. 2051–2066.

[5] M. J. Gotay, J. A. Isenberg, J. E. Marsden, R. Montgomery, J. Śniatycki, and P. B. Yasskin, *Momentum Mappings and the Hamiltonian Structure of Classical Field Theories with Constraints*, to appear, 1990.

[6] V. Guillemin and S. Sternberg, *Symplectic Techniques in Physics*, Cambridge U. P., Cambridge, 1984.

[7] B. Kostant and S. Sternberg, *Symplectic reduction, BRS cohomology and infinite dimensional Clifford algebras*, Ann. Phys. (N.Y.), **176** (1987), pp. 49–113.

[8] M. Kummer, *On the construction of the reduced phase space of a Hamiltonian system with symmetry*, Indiana U. Math J., **30** (1981), pp. 281–291.

[9] C. Quesne, *The nuclear collective WSp(6,*$\mathbf{R}$*) model*, Ann. of Phys., **185** (1988), pp. 46–85.

[10] G. Rosensteel, *Hamiltonian formulation of Riemann ellipsoid dynamics*, Lett. Math. Phys., **17** (1989), pp. 79–86.

[11] G. M. Tuynman, *Reduction, quantization and non-unimodular groups*, J. Math. Phys., to appear, (1990).

FROM KNOTS TO QUANTUM GROUPS (AND BACK)

LOUIS H. KAUFFMAN

Department of Mathematics, Statistics and Computer Science,
The University of Illinois at Chicago,
Chicago, Ill 60680, USA

1. INTRODUCTION

This paper traces how the Jones polynomial leads naturally to the notion of quantum group, and how quantum groups give rise to invariants of links via solutions to the Yang-Baxter equation. Section 5 is an original treatment of the construction of the universal R–matrix. All the other material has, or will appear elsewhere in similar form.

2. KNOT THEORY

Let's begin by recalling the Reidemeister moves:

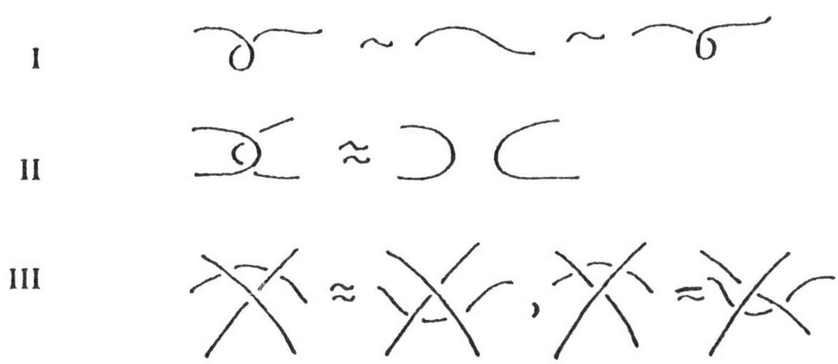

These moves can be performed on a link diagram. A link diagram is a locally four-valent plane graph with extra structure at the vertices in the form of crossings

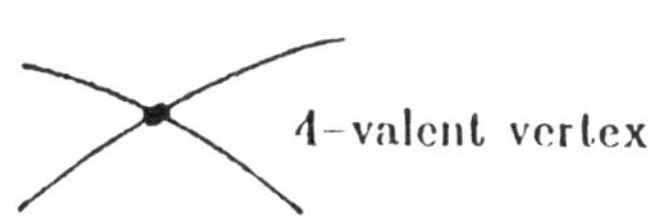

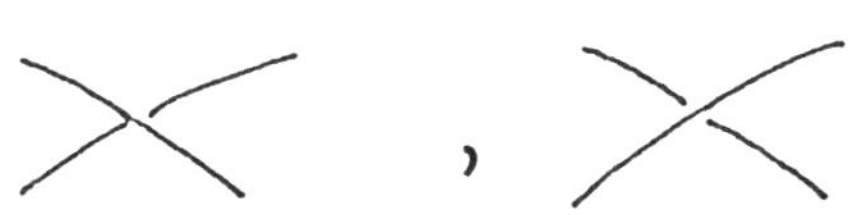

These crossings are taken to indicate the projection of arcs embedded in a three-space, and projected to the plane. The broken arc pair at a crossing indicates the arc that passes underneath the other arc in space. Any link (a link is a collection of circles embedded in a three-sphere or Euclidean three space.) has a point of projection to the surface of a two dimensional sphere or to a plane, so that the projection (with under and over-crossing indications) becomes a diagram for that link.

Two links are said to be **ambient isotopic** if there is a continuous time-parameter family of embeddings starting with one link and ending with the other one. The theory of knots and links is the theory of link embeddings under the equivalence relation of ambient isotopy. (A **knot** is a link with one component. That is, a knot is an embedding of a single circle into three-space.)

It is assumed that all the embeddings are represented (up to ambient isotopy) by an embedding that is a differentiable curve(s) in the three-space. Links that do not admit such a representation are called **wild**, and must be treated separately.

The Reidemeister moves generate the theory of knots and links in three-dimensional space in the sense of the following theorem:

THEOREM (REIDEMEISTER [18]). *Let K and K' be two links embedded in three-dimensional space (either the three-dimensional sphere or the Euclidean space R^3). Then, K and K' are ambient isotopic if and only if diagrams for K and K' are related by a finite sequence of the moves I, II, III.*

REMARK: In Reidemeister's day the notion of ambient isotopy was also combinatorial. Let R denote Reidemeister. For R, ambient isotopy was generated by a *single* move type called an **elementary combinatorial isotopy** (or elementary isotopy for short). The knots and links for R are **piecewise linear** – meaning that they consist of interconnections of straight line segments embedded in Euclidean space. Vertices are regarded as the endpoints of these segments, and any straight segment can be regarded as the connection of two segments, by adding a vertex at an interior point. The elementary combinatorial isotopy has two directions: **expansion**, and **contraction**. In an expansion, one takes two vertices on the link, and a new vertex in the complement such that the (two dimensional) triangle spanned by these vertices intersects the link only at one of its three edges. Expansion consists in replacing this edge in the link by the two remaining edges in the triangle. Contraction is the opposite of expansion – three points on the link span a triangle intersecting the link only along two edges; these edges are replaced by the third edge of the triangle.

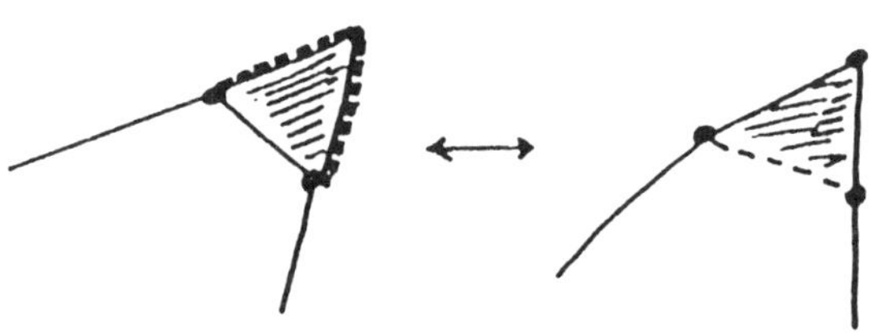

The Reidemeister moves come about via examination of the forms of planar projections of the elementary isotopies. For example, the diagram below shows how a type I Reidemeister move is the shadow of an elementary isotopy.

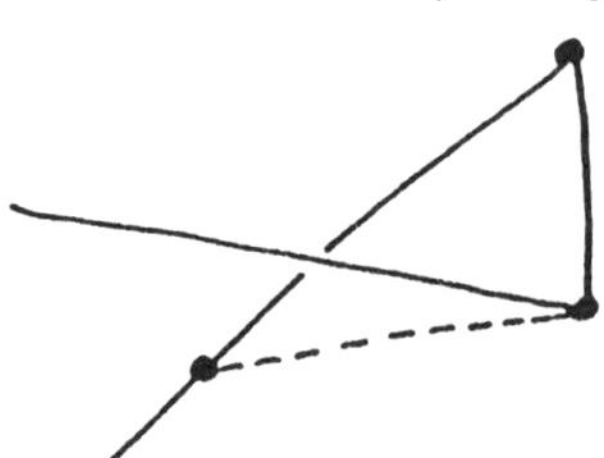

Reidemeister's approach to his theorem is a good way to get a geometric feel for the situation. For a modern treatment of the Theorem, using the continuous (or differentiable) notion of isotopy, see [5].

While Reidemeister's Theorem is an excellent starting point for a combinatorial theory of knots and links, it does not make life easy. The easiest way to illustrate this is to exhibit a demon such as the one shown below. (This demon – shown to the author by Ken Millet – improves over previous culprits, and is the smallest possible for projections on a sphere.)

This demon D is unknotted, but does not admit any simplifying Reidemeister moves, nor does it admit any type three moves. (A Reidemeister move is said to be **simplifying** if it reduces the number of crossings in the diagram.) In order to unknot D it is necessary to first make the diagram more complex before it can become simpler. Examples of this sort show that the equivalence relation generated by the Reidemeister moves is subtle, and that the matter of constructing invariants is non-trivial.

There are many accounts of the classical construction of knot and link invariants ([1], [8], [4], [11], [21]). In the next section, I shall go directly to a model for the Jones polynomial and discuss its physical interpretations. For these purposes it does make sense to make one remark about the process of abstraction leading to mathematical knots. If we were to make a knot or link from rope or other material, then the amount of twisting on the rope would make a difference in the behaviour or the resulting knotted form. Such twisting has been abstracted when we go to the diagram or the mathematical curve embedded in space. We can recover some of this structure by considering **framed links**. A framed link is a link such that each component has a continuous normal vector field. This is equivalent to thinking about embeddings of **bands** rather than circles. Thus the figure below indicates a framed trefoil, with standard framing inherited from its planar embedding.

If we keep track of the framing then one no longer has invariance under the type I move:

For this reason it is useful to have the concept of **regular isotopy**. Two links are said to be regularly isotopic if one can be obtained from the other by a sequence of type II and type III moves only. Regular isotopy is the equivalence relation generated by the type II and type III moves. Note:

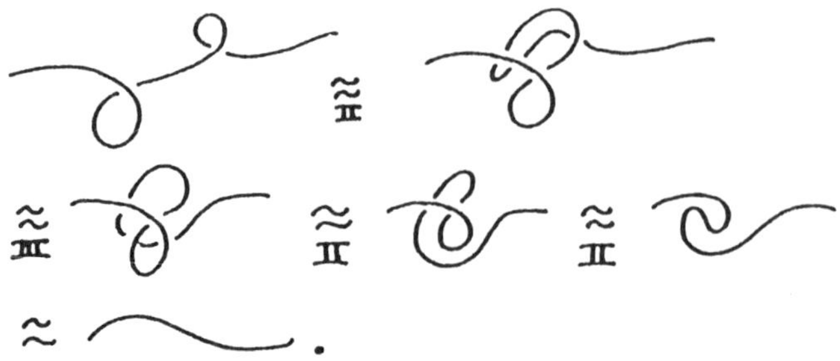

Opposite curls cancel. This regular isotopy is the knot theoretic version of the Whitney trick [24]. Actually, regular isotopy is a bit subtler than simple framing. The bands shown below

are isotopic, but the corresponding string diagrams are not regularly isotopic (they have different Whitney degree [24].).

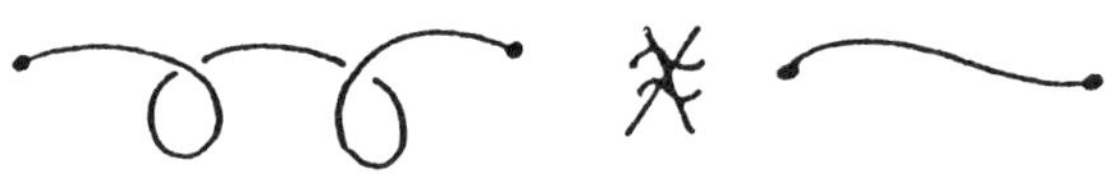

A useful invariant of regular isotopy is the **writhe**, $w(K)$. The writhe is the sum of the crossing signs

in a given diagram. Thus

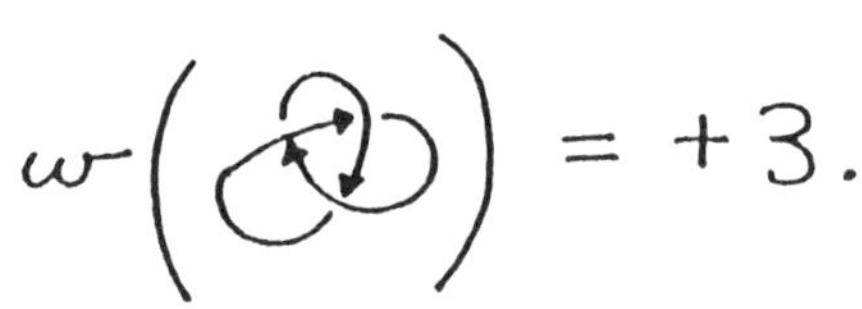

It is easy to see that the writhe is a regular isotopy invariant for diagrams. It is very useful for normalizing other invariants of regular isotopy. It turns out that most of the invariants we shall discuss need such a normalization.

III. LINK INVARIANTS AS VACUUM-VACUUM AMPLITUDES

First, a quick description of the bracket model [12] of the Jones polynomial [9]: We give a method of associating a well-defined polynomial in three variables, $\langle K\rangle(A, B, d)$, to an unoriented link K. This polynomial is defined recursively by the formulas:

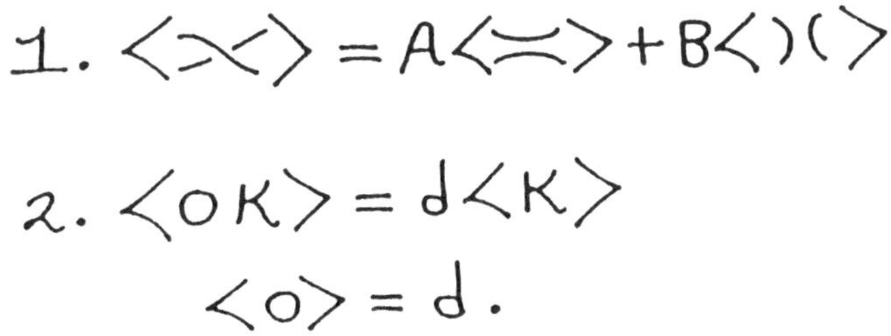

The first formula asserts that the polynomial for a given diagram is obtained as an additive combination of the polynomials for the diagrams obtained by splicing away the given crossing in two possible ways. Thus the small diagrams indicate larger diagrams that differ only as indicated. The second formula says that the value of a loop (simple closed curve in the plane) is d, and that if a loop occurs (isolated) inside a larger diagram, then the value of the polynomial acquires a factor of d from this loop. In particular, the value of a disjoint union of N simple closed curves is d^N.

Together, the two formulas completely determine $\langle K\rangle$, and $\langle K\rangle$ is well-defined just so long as A, B and d commute with one another. Thus this polynomial takes values in the ring $Z[A, B, d]$ of polynomials in three variables with integer coefficients.

As it stands, $\langle K\rangle$ is not an invariant of any of the Reidemeister moves. However, the following formula is an easy consequence of 1. and 2. above.

PROPOSITION.

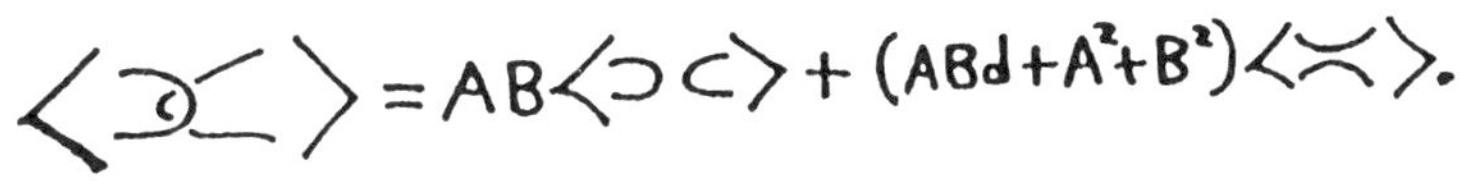

As a result we see that $\langle K\rangle$ is invariant under the type II move if we select $B = A^{-1}$ and $d = -A^2 - A^{-2}$. Furthermore, it now follows directly that $\langle K\rangle$ is invariant under the move III:

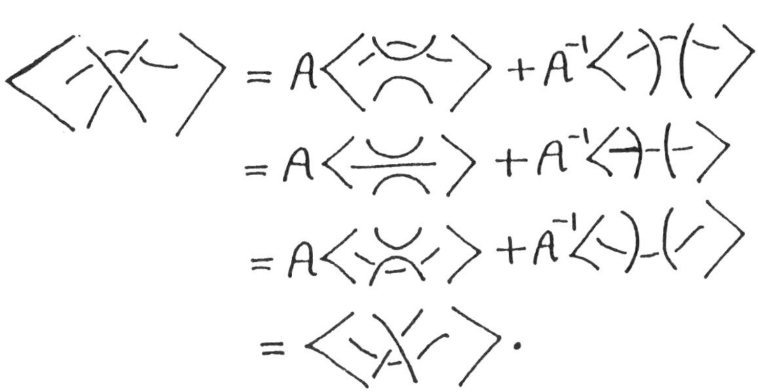

Thus, with $B = A^{-1}$ and $d = -A^2 - A^{-2}$, we have that $\langle K\rangle$ is an invariant of regular isotopy. To obtain an invariant of ambient isotopy for oriented links, we form

$$f_K(A) = (-A^3)^{-w(K)}\langle K\rangle/\langle O\rangle$$

where K is oriented, $w(K)$ is the writhe of K as defined in the previous section, and $\langle K\rangle$ is the bracket evaluated on the unoriented link underlying K. The reason for this

factor of $-A^3$ is that

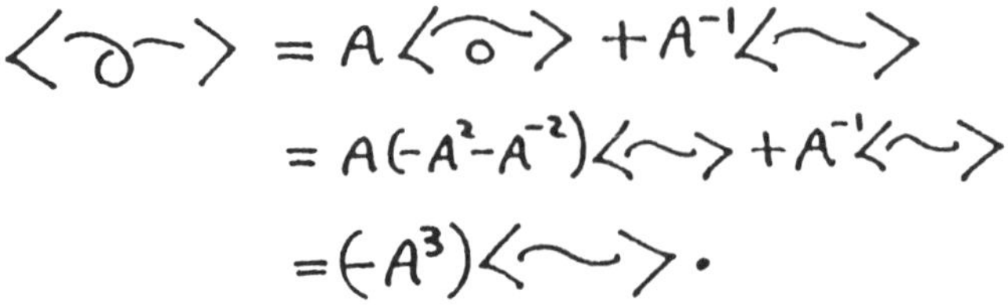

One then has the

THEOREM [12]. *For any oriented link $K, V_K(t) = f_K(t^{-1/4})$ where V_K denotes the original one-variable Jones polynomial.*

Thus, the bracket, suitably normalized, gives a direct model for the Jones polynomial.

THE VACUUM-VACUUM AMPLITUDE.

In the rest of this section I shall stick to the bracket, and show how it can be seen as a "vacuum-vacuum amplitude" in a combinatorial version of topological quantum field theory [**25**]. More generally, we can consider an amplitude associated to a given diagram by regarding the plane as 1 + 1 spacetime. By convention, let time run vertically up the page, and space proceed from left to right. (This is the convention of the reader of English.) Position the link diagram so that it is transversal to the space levels except at critical points corresponding to maxima, minima and crossings.

Each minimum can be regarded as a creation of two particles from the vacuum, each maximum an annihilation, and each crossing is an interaction (thought of as involving braiding in the extra spatial dimension orthogonal to the page). To each of these events we associate a matrix whose indices go over (say) the spins of the particles, and whose values are the amplitudes for each of these processes.

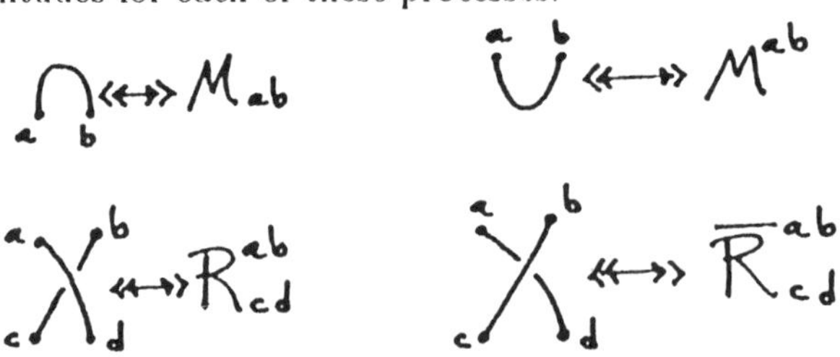

The amplitudes used here are a generalization of amplitudes in quantum mechanics, suitably generalized for the purposes of topology (In the process we take leave of the usual interpretations of observation in quantum mechanics. In the topology the amplitude itself is a real property of the system. There is no "collapse of the wave function"). Therefore the amplitudes will take values in a commutative ring (e.g. in $Z[A, A^{-1}]$), and the spins will run over an arbitrary finite index set (e.g. $\{-1,+1\}$). Amplitudes are calculated according to the principles of quantum mechanics [7]:

(1) If an event occurs in a way that can be decomposed into a set of individual steps (e.g. creations, annihilations, interactions), then the amplitude of the given event is the product of the amplitudes of the individual steps.

(2) If an event may occur in several disjoint alternative ways, then the amplitude of this event is the sum of the amplitudes of the ways.

Given a diagram K, and a set of matrices as above, we can calculate the amplitude for particles to be created from the vacuum, interact in the pattern of the link diagram, and return to the vacuum. This amplitude decomposes as a sum of the amplitudes for configurations of the diagram. (I take configuration as a neutral term here. For amplitudes one may prefer the term *history*. For a purely spatial interpretation one may prefer the term *state*.) Each configuration σ is an assignment of spins to the nodes of the diagram. (The nodes are the input and output nodes of the small diagrams corresponding to the matrices). Given a configuration, each matrix has a well-defined value, and the amplitude of this configuration is the product of these values. Thus the vacuum-vacuum amplitude, $T(K)$, for a given diagram K is the sum (over the configurations) of the product of the matrix values for each configuration.

Symbolically, this works out in accord with the usual Einstein convention for repeated indices: Write down a product of all the matrices for the given diagram, in indices, with one index for each node. The amplitude is then the value of this expression interpreted as a sum over all cases of repetitions of an index in lower and upper positions.

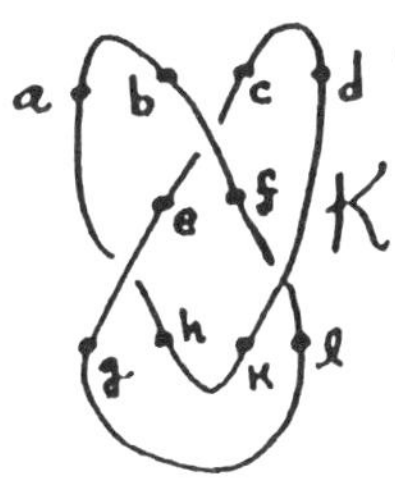

$$T(K) = M_{ab} M_{cd} R^{bc}_{ef} \overline{R}^{ae}_{gh} \overline{R}^{fd}_{kl} M^{gl} M^{hk}.$$

Having defined the vacuum-vacuum amplitude $T(K)$, we must see when it will be an invariant of regular isotopy, and when it will model the bracket. In order for T (K) to be an invariant of regular isotopy, we need the following restrictions on the matrices [15]:

1. 

$\longleftrightarrow M_{ai} M^{ib} = \delta^b_a.$

2. 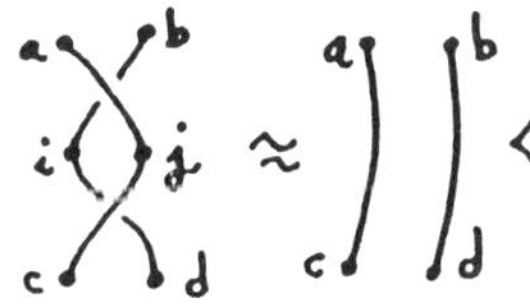

$\longleftrightarrow R^{ab}_{ij} \overline{R}^{ij}_{cd} = \delta^a_c \delta^b_d$

$(R\overline{R} = I).$

3. 

$$\overline{R}^{ab}_{cd} = M_{ci} R^{ia}_{dj} M^{jb}.$$

4. 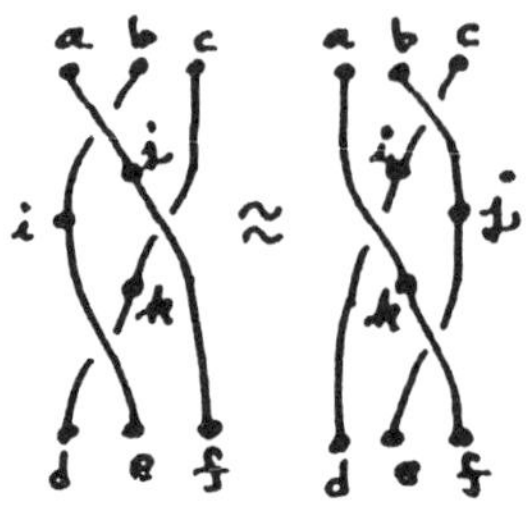

$$R^{ab}_{ij} R^{jc}_{kf} R^{ik}_{de} = R^{bc}_{ij} R^{ai}_{dk} R^{kj}_{ef}.$$

(Yang–Baxter Equation)

(In 3. there is a corresponding left-hand twist, and in 4. there is also the same equation for all crossings reversed.) Equation 4. is called the Yang-Baxter Equation [3] (here given without rapidity parameter).

REMARK: It is interesting to speculate about the physical meaning of these restrictions. The twist condition 3. is the most mysterious since it relates R and R^{-1} via creations and annihilations. A simpler physical situation may lend some insight here. In the simplified scenario, we assume that 1. holds, and that parallel identity lines are interchangeable with pairs of creations and annihilations:

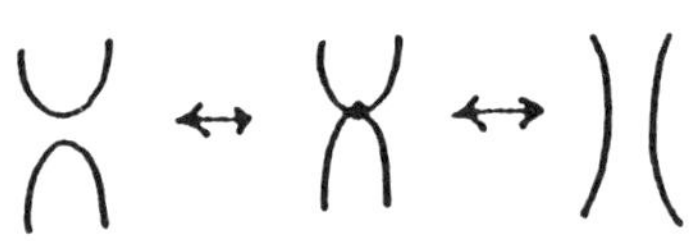

Then one has the sequence of identifications

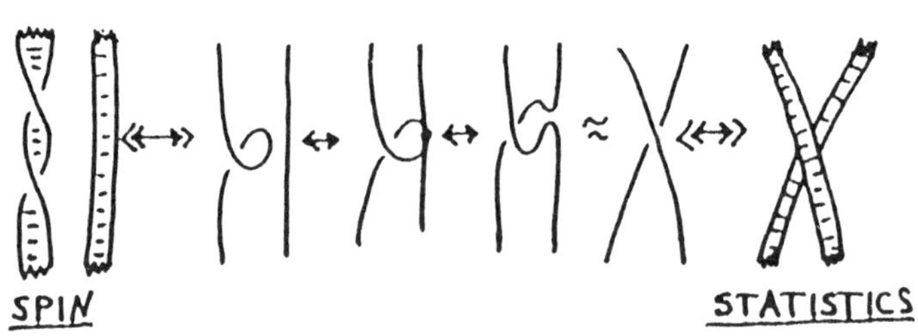

This has been interpreted as a depiction of the *equivalence of spin and statistics* (see e.g. [22] and references therein) where spin is regarded as catalogued by the twist of framing (become curl of diagram) and statistics corresponds to the braiding of the two lines. This shows that part of our diagrammatics correspond to ordinary physical interpretations, and that where the topology begins the equivalence of spin and statistics leaves off. In this sense, the topology is an index of the non-standard statistics.

MODELLING THE BRACKET.

In order to model the bracket with a vacuum–vacuum amplitude we need to find creation and annihilation matrices that are inverse to one another, and that give a loop value of $-A^2 - A^{-2}$. Here is an answer to that puzzle:

$$M_{ab} = M^{ab}$$

$$M = \begin{bmatrix} 0 & \sqrt{-1}A \\ -\sqrt{-1}A^{-1} & 0 \end{bmatrix}.$$

Note that the matrix M has square the identity, and that the loop value is therefore the sum of the squares of the entries of M. (See [13],[14],[15] for motivations for this construction.)

With a given choice for the creations and annihilations, there is one choice for the R matrix to give the bracket:

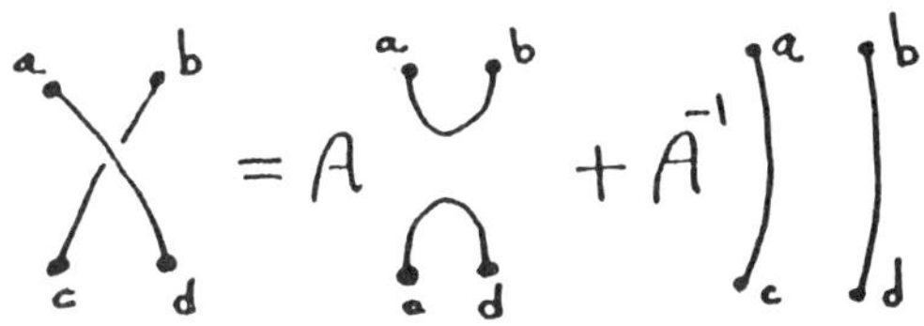

With this choice, $T(K)$ will satisfy the defining equations of the bracket and therefore $\langle K \rangle = T(K)$ (since we have correctly adjusted the loop value).

REMARK: In fact it is interesting to note that if the creation and annihilation are inverse matrices, and R is defined as above, then 2. follows easily, while 3. goes as below,

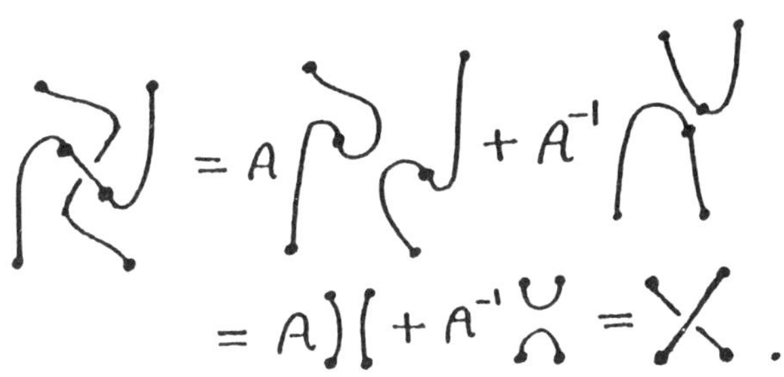

and 4. is proved by first checking

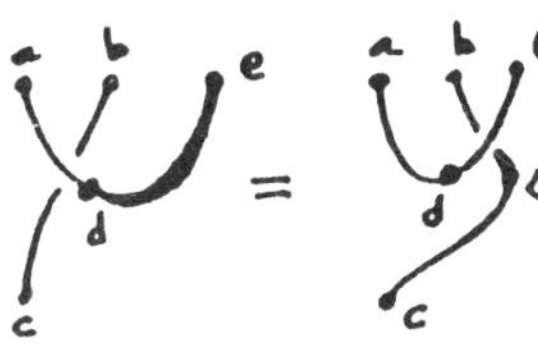

$$\Longleftrightarrow R^{ab}_{cd} M^{de} = M^{ad} \bar{R}^{be}_{dc}.$$

and then performing the following variation on our bracket derivation of the invariance under the III move

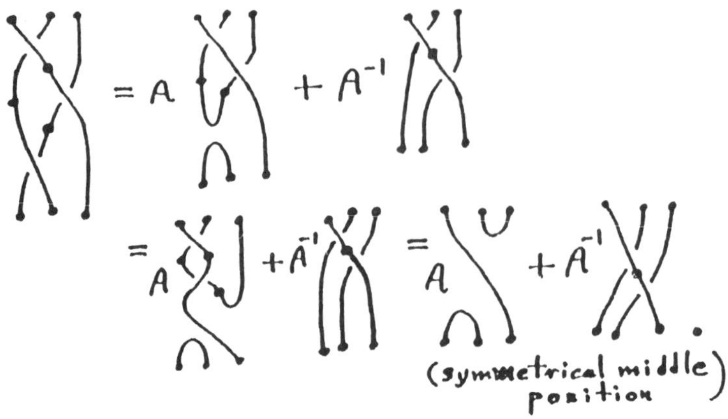

The upshot of this discussion is that by simply adjusting the creation and annihilation matrices correctly, we automatically produce a model of the bracket and a solution to the Yang-Baxter equation. This is the simplest instance of a solution to the Yang-Baxter equation appearing naturally from the knot theory. This is the well-known [19] R-matrix corresponding to the SL(2) quantum group. In fact, the structure that we have created so far will now enable us to see one motivation for the construction of the quantum group.

IV. THE SL(2) QUANTUM GROUP

Note that we can write

$$M = \begin{bmatrix} 0 & \sqrt{-1}A \\ -\sqrt{-1}A^{-1} & 0 \end{bmatrix} = \sqrt{-1}\,\tilde{\varepsilon},$$

$$\tilde{\varepsilon} = \begin{bmatrix} 0 & A \\ -A^{-1} & 0 \end{bmatrix}.$$

and that, as A approaches 1, the matrix $\tilde{\varepsilon}$ approaches

$$\varepsilon = \begin{bmatrix} 0 & 1 \\ -1 & 0 \end{bmatrix}.$$

The matrix ε is significant in linear algebra because it expresses the determinant of a 2×2 matrix. That is, let P be a 2×2 matrix with commuting entries. Then

$$P\varepsilon P^{T} = \mathrm{Det}\,(P)\varepsilon.$$

(T denotes transpose.)

Now SL(2) (over a commutative ring) is the set of matrices of determinant one, and can therefore be characterized as the set of matrices leaving the epsilon invariant:

$$SL(2) = \{P \mid P\varepsilon P^T = \varepsilon\}$$

Here T denotes matrix transpose.

At $A = 1$ the bracket does not discriminate between under and overcrossings, and the identity

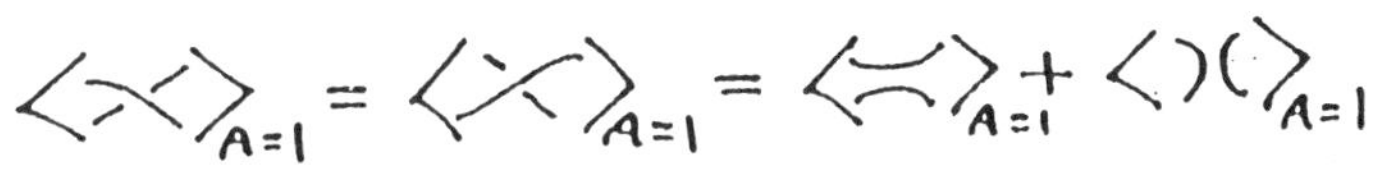

corresponds directly ot the Fierz identity

$$\epsilon^{ab}\epsilon_{cd} = \delta^{ac}\delta_{bd} - \delta^{ad}\delta_{bc}$$

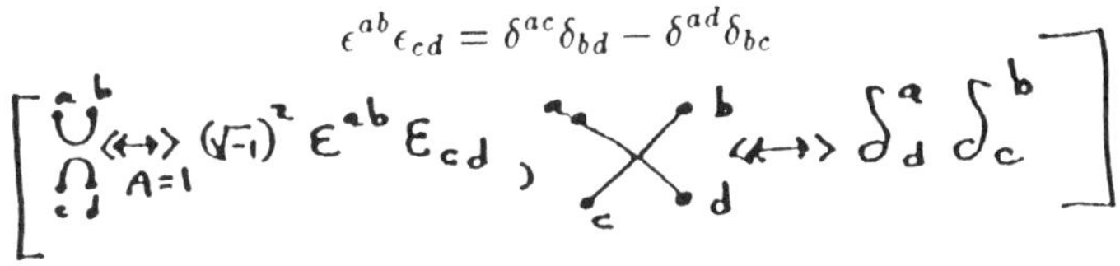

Thus at $A = 1$ (and also at $A = -1$) the diagrams become interpreted as tensor diagrams for SL(2) invariant expressions.

It is then natural to ask whether there is a generalization of this symmetry for the topology and link diagrams. Specifically, we ask whether $\tilde{\varepsilon}$ has a symmetry group analogous to SL(2). Some experimentation shows that the way to ask this question is to consider

$$P = \begin{bmatrix} a & b \\ c & d \end{bmatrix}$$

with associative, possibly non-commutative entries, and ask for the invariances:

$$P\tilde{\varepsilon}P^T = \tilde{\varepsilon}$$

$$\underline{\text{and}} \quad P^T\tilde{\varepsilon}P = \tilde{\varepsilon}.$$

It is then an exercise in elementary algebra to see that these conditions are equivalent to the equations: $(q = \sqrt{A})$:

$$ca = qac \qquad db = qbd$$
$$ba = qab \qquad dc = qcd$$
$$bc = cb$$
$$ad - da = (q^{-1} - q)bc$$
$$ad - q^{-1}bc = 1$$

These are the defining relations for the algebra $U^* = SL(2)_q$ ([6], [17]) , sometimes called the SL(2) quantum group. It is not a group, but rather a Hopf algebra. The co-algebra structure is given by the map

$$\Delta = U^* \longrightarrow U^* \otimes U^*$$

where

$$\Delta(P^i_j) = \sum_K P^i_K \otimes P^K_j .$$

Thus

$$\Delta(a) = a \otimes a + b \otimes c$$

$$\Delta(b) = a \otimes b + b \otimes d$$

$$\Delta(c) = c \otimes a + d \otimes b$$

$$\Delta(d) = c \otimes b + d \otimes d$$

In this case the Hopf algebra has an antipode and this is directly related to the fact that the matrix P has an inverse $\gamma(P)$:

$$\gamma(P) = \begin{pmatrix} d & -qb \\ -q^{-1}c & a \end{pmatrix} .$$

Recall that the antipode is a map $\gamma : U^* \longrightarrow U^*$ such that the following diagram commutes

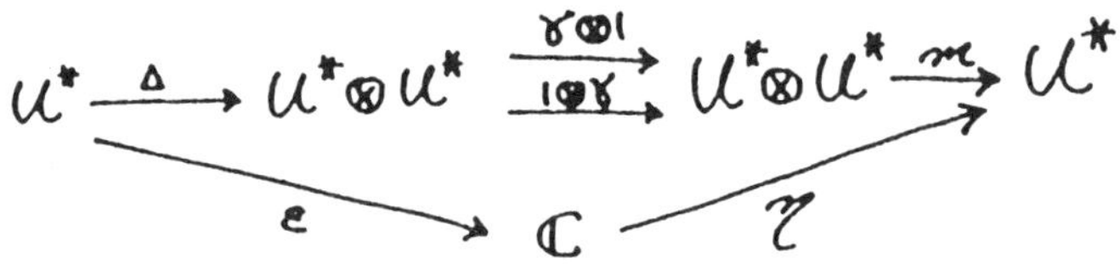

where ε and η are the co-unit and unit respectively and m denotes the multiplication in the algebra. Here

$$\varepsilon(P^i_j) = \delta^i_j \quad \text{and} \quad \eta(\delta^i_j) = \delta^i_j .$$

Thus the condition that γ be an antipode is just that

$$\sum_K \gamma(P^i_K) P^K_j = \delta^i_j$$

$$\sum_K P^i_j \gamma(P^K_j) = \delta^i_j .$$

And this is the same as saying that P and $\gamma(P)$ are inverse matrices.

We could now discuss a number of things about the relationship of this quantum group to solutions to the Yang-Baxter equation, and to its dual form as a deformation of the Lie Algebra for $SL(2)$ (see [6], [19]). But here there is not space for this. The purpose of this section has been to show how the quantum group arises naturally from a combination of the topology and a desire to extend the algebraic symmetry inherent in a significant special case of the vacuum-vacuum expectation model.

V. AND BACK

In order to indicate how the trail looks going back from quantum groups to link invariants I shall make a leap to the formalism behind the so-called quantum double construction of Drinfeld [6]. We shall then see how a Hopf algebra structure can give rise, through its matrix representations, to invariants of links.

We begin with an algebra with generators $e_0, e_1, ..., e_n$ and $e^0, e^1, e^n, ..., e^n$ and the following relations describing multiplication in the algebra

(A) $e_s e_t = m^i_{st} e_i$

(B) $e^s e^t = \mu^{st}_i e^i$

Diagrammatically, I shall write  $\longleftrightarrow e^s$,  $\longleftrightarrow e_s$

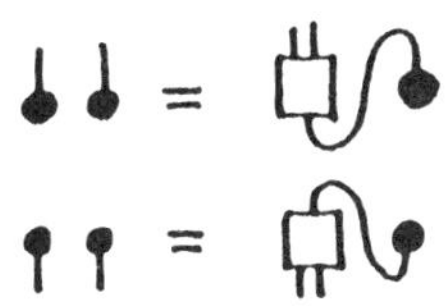

where it is understood that the boxes denote the product expansion coefficients, and hence the boxes commute with the e-nodes (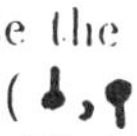) and with each other. We further assume the following relationship between multiplying upper and lower e's:

(C)

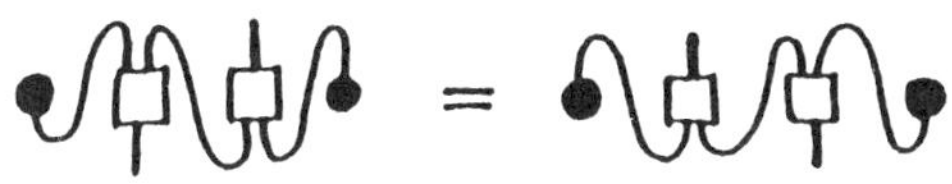

$$e_i \mu^{ij}_a m^b_{jk} e^k = e^k m^b_{kj} \mu^{ji}_a e_i .$$

Then we have

THEOREM. *With the above assumptions, the element*

$$\rho = \sum_s e_s \otimes e^s =$$ 

satisfies the algebraic form of the Yang-Baxter equation:

$$\rho_{12}\,\rho_{13}\,\rho_{23} = \rho_{23}\,\rho_{13}\,\rho_{12}.$$

$$[\rho_{12} = \sum_s e_s \otimes e^s \otimes 1, \quad \rho_{13} = \sum_s e_s \otimes 1 \otimes e^s, ...]$$

PROOF: (See [15]).

In a matrix representation the e-nodes sprout indices, and an algebraic solution to the Yang-Baxter Equation becomes an knot theorist's matrix solution via an added permutation. Thus if

denotes the algebraic solution, then

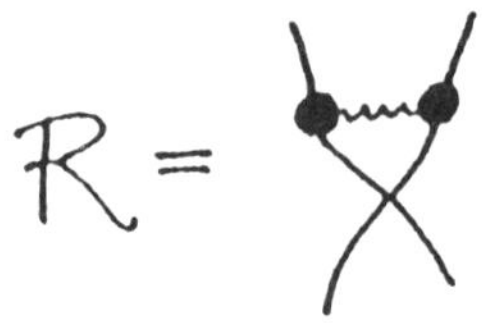

denotes the corresponding knot theoretic R-matrix in some representation (the indices of this representation correspond to the new lines).

As we know from the previous section, the knot theory demands a relationship between the creation and annihilation matrices and the R-matrix. This "twist relation" is given diagrammatically as follows:

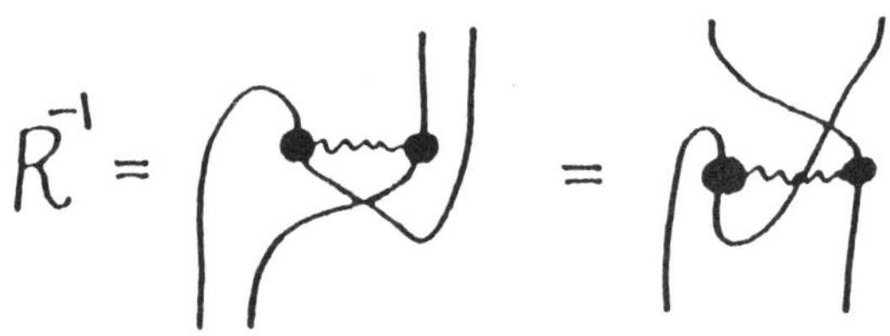

Thus

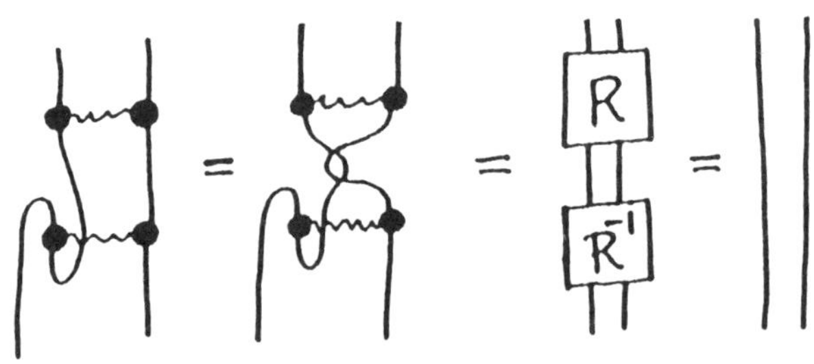

We conclude that there should be an automorphism

$$x \longrightarrow \gamma(x)$$

of the abstract algebra that corresponds to the map in the representation

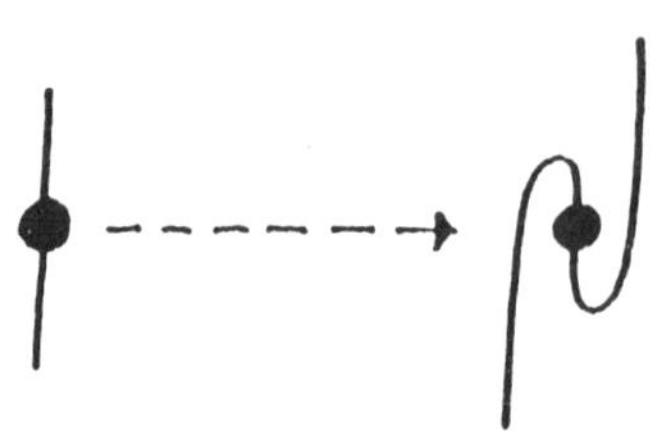

and we need that $\gamma(e_s)\otimes e^s$ is the inverse of $e_s \otimes e^s$ (with summation on repeated lower and upper indices). It is then not hard to see that if we were to make the algebra into a Hopf algebra such that the co-multiplication for the lower index e's is the multiplication for the upper index e's and vice versa (this is the double construction), then this inverse requirement is equivalent to γ being an antipode. (See the appendix.) The twist conditions of the knot theory are intimately tied with the Hopf algebra structure for the quantum group. This completes the journey back, albeit in an abstract mode.

APPENDIX

Here we verify that γ, as defined in section 5, is an antipode in the Hopf algebra. Recall that we demanded that $\gamma(e_s)\otimes e^s$ be inverse to $e_s \otimes e^s$. Multiplying this out, we find

$$\begin{aligned}
[\gamma(e_i)\otimes e^i][e_j \otimes e^j] &= \gamma(e_i)e^j \otimes e_i e^j \\
&= \gamma(e_i)e^j \otimes \mu_k^{ij} e_k \\
&= \mu_k^{ij}\gamma(e_i)e^j \otimes e_k \\
&= m(\mu_k^{ij}\gamma(e_i)\otimes e^j)\otimes e_k \\
&= m((\gamma\otimes 1)\ (\mu_k^{ij} e_i \otimes e^j\ ))\otimes e_k \\
&= m((\gamma\otimes 1)(\Delta(e_k)))\otimes e_k
\end{aligned}$$

We are summing on repeated indices, and in the last step we used the fact that in the Drinfeld double construction, the diagonal in the algebra of the e's with lower subscripts is given by the multiplication coefficients in the algebra of the upper e's. Note also that m denotes multiplication, and the formula for γ to be an antipode is

$$\begin{aligned}
m((\gamma\otimes 1)(\Delta(e_k))) &= \ (\Delta(e_k)) = e_0 \text{ if } k=0 \\
m((\gamma\otimes 1)(\Delta(e_k))) &= \ 0 \qquad\qquad \text{if } k \neq 0.
\end{aligned}$$

Here we take e_0 to be 1, the identity element in the Hopf algebra. Thus we have shown that if γ is an antipode for the Drinfeld double construction, then $\gamma\ (e_s)\ \otimes\ e^s$ and $e_s\ \otimes\ e^s$ are inverses. This, in turn, shows that the diagrammatic twist condition for the vacuum-vacuum expectation corresponds, through representation of the Drinfeld construction, to the existence of an antipode in the Hopf algebra. In this sense, the abstract quantum group defined by the formalism of the Drinfeld double construction gives a universal link invariant.

Acknowledgements. The author is pleased to be able to thank the organizers of the October 1989 Montreal Conference on Hamiltonian Systems, Transformation Groups and Spectral Transform Methods for their hospitality, encouragement and stimulation.

References

1. Alexander, J.W., *Topological invariants of knots and links*, Amer. Math. Soc. **20** (1928), 275–306.
2. Akutsu, Y., Deguchi, T., Wadati, M., "The Yang-Baxter Relation: A New Tool for Knot Theory," Braid Group, Knot Theory and Statistical Mechanics, Ed. Yang and Ge, World Scientific, 1989.
3. Baxter, R.J., "Exactly Solved Models in Statistical Mechanics," Academic Press, 1982.
4. Birman, J.S., *Braids, Links and Mapping Class Groups*, Annals of Math. Study **82** (1976).
5. Burde, G. and Zieschang, H., "Knots," De Gruyter, 1985.
6. Drinfeld, V.G., "Quantum Groups," Proc. Intl. Congress Math., Berkeley, Calif., USA, 1986, pp. 798–820.
7. Feynman, R.P., "Lectures on Physics," Addison Wesley, 1964.

7.1. Finkelstein, D. and Rubenstein, J., J. Math. Phys. **9** (1968), 1762.

8. Fox, R.H., "A quick trip through knot theory: Topology of 3-Manifolds," M.K. Fort Ed. Prentice Hall, 1962, pp. 120–167.

8.1. Hennings, M., "Invariants of links and three manifolds obtained from Hopf algebras," preprint 1989.

9. Jones, V.F.R., *A polynomial invariant for links via von Neumann algebras*, Bull. A.M.S. **12** (1985), 103–112.
10. Jones, V.F.R., *On knot invariants related to some statistical mechanics models*, Pacific J. Math. **137** (1989), 311–334.
11. Kauffman, L.H., "On Knots," Annals of Mathematics Study 115, Princeton Univ. Press, 1987.
12. Kauffman, L.H., *State models and the Jones polynomial*, Topology **26** (1987), 395–407.
13. Kauffman L.H., *Statistical mechanics and the Jones polynomial*, AMS Contemp. Math. **78** (1988), 263–297.
14. Kauffinan L.H., *Spin networks and knot polynomials*, Int. J. Mod. Phys. A. No 1, **5** (Jan.1990), 93–115.
15. Kauffman L.H., *Knots, abstract tensors and the Yang-Baxter equation.*
16. Lawrence, R., *A universal link invariant using quantum groups*, (preprint 1988).

16.1. Lee, H.C., *Q-deformation of* $sl(2,C) \times Z_N$ *and link invariants.*

17. Manin, Y.I., *Quantum groups and non-commutative geometry*, CRM- Université de Montéal (1988).
18. Reidemeister K., "Knotentheorie," Chelsea Pub. Co, New York, Copyright 1932, Julius Springer, Berlin, 1948.
19. Reshetikhin, N.Y., "Quantized universal enveloping algebras, The Yang-Baxter equation and invariants of links, I and II," LOMI reprints E-4-87 and E-17-87, Steklov Institute, Leningrad, USSR.
20. Reshetikhin, N.Y. and Turaev, V.G., "Invariants of 3-manifolds via link polynomials and quantum groups," preprint 1989.
21. Rolfsen, D., "Knots and Links," Publish or Perish Press, 1976.
22. Sorkin, R., *A general relation between kink-exchange and kink-rotation*, Comm. Math. Phys. **115** (1988), 421–434.
23. Turaev, V.G., *The Yang-Baxter equations and invariants of links*, Invent. Math. **92** (1988), 527–553.
24. Whithney, H., *On regular closed curves in the plane*, Comp. Math. 4 (1937), 276–284.
25. Witten, E., *Quantum field theory and the Jones polynomial*, Comm. Math. Phys. **121** (1989), 351–399.

CONFORMALLY MULTIPLICATIVE POISSON STRUCTURES IN LINEAR ALGEBRA AND ALGEBRAIC GEOMETRY

B.A. KUPERSHMIDT

The University of Tennessee Space Institute
Tullahoma, TN 37388, USA

ABSTRACT. This is a review of multiplicative and conformally multiplicative Poisson brackets on matrix spaces and on commutative and other rings. Topics discussed include: Quantum linear spaces, Quantum $GL_q(n)$, and their quasiclassical limits; super– and coloured matrix groups; triangular decomposition as a Poisson isomorphism; Affine groups and supergroups; Poisson brackets on: Grassmann algebras, group rings of $\mathbf{Z}_2$ and $\mathbf{Z}_3$, $\mathbf{C}^{\times}$ and general quadratic extensions, rings of polynomials and formal power series and their factor–rings. A list of open problems concludes the review.

This is a report on:

(1) multiplicative Poisson structures on matrix groups and supergroups viewed as the quasiclassical limits of the corresponding quantum objects;

(2) conformally multiplicative Poisson brackets in polynomial and formal power series rings and their factor–rings, and in other rings, such as the Grassmann algebra $\Lambda(N)$ and group rings of finite groups. No doubt it will be dated by the time it is finished.

1. WHAT IS A QUANTUM VECTOR SPACE?

Let k be a field or a commutative ring. In the free associative ring $k < x > := k < x_1, \dots, x_n >$ on noncommuting generators x_i's, consider the ideal $I(\lambda)$ generated by the relations

$$x_i x_j - q^{\lambda(i,j)} x_j x_i = 0, \quad 1 \le i, j \le n, \tag{1.1}$$

where $\lambda(i,j) = -\lambda(j,i)$ is a set of fixed constants from k, and q is a formal parameter commuting with everything. The factor–ring $k < x > / I(\lambda) =: A_q^n(\lambda)$ is called the function ring of a quantum n–dimensional vector space [1]. The dual to (1.1) relations

$$\xi_i \xi_j + q^{\lambda(j,i)} \xi_j \xi_i = 0, \quad 1 \le i, j \le n, \tag{1.2}$$

generate an ideal $\tilde{I}(\lambda)$ in $k < \xi >$, the duality being with respect to the pairing

$$(\xi_i, x_j) = \delta_{ij}; \tag{1.3}$$

the factor–ring $k < \xi > / \tilde{I}(\lambda)$ describes a quantum Grassmann algebra [1]. Let $A = (a_{ij})$ be a matrix with elements in $k < a_{ij} >$. We get commutations relations on a_{ij} by demanding that, acting on x and ξ, A preserves the relations (1.1) and (1.2); a_{ij}'s commute with x_k's and ξ_k's [1]. The resulting relations are

$$a_{j\beta} + a_{i\beta} a_{j\alpha} q^{\lambda(\beta,\alpha)} = q^{\lambda(i,j)} (a_{j\alpha} a_{i\beta} + a_{j\beta} a_{i\alpha} q^{\lambda(\beta,\alpha)}) \tag{1.4}$$

$$a_{i\alpha} a_{j\beta} - a_{i\beta} a_{j\alpha} q^{\lambda(\alpha,\beta)} = -q^{\lambda(j,i)} (a_{j\alpha} a_{i\beta} - a_{j\beta} a_{i\alpha} q^{\lambda(\alpha,\beta)}). \tag{1.5}$$

We now pass to the "quasiclassical limit", i.e., consider $q = e^{\varepsilon} \approx 1 + \varepsilon, \varepsilon^2 = 0$, and replace the commutator [,] by (ε times the Poisson bracket { , }). Then formulae (1.4), (1.5) become

$$\{a_{i\alpha}, a_{j\beta}\} = [\lambda(i,j) + \lambda(\alpha,\beta)] a_{i\beta} a_{j\alpha}. \tag{1.7}$$

THEOREM 1.8. *The Poisson brackets (=: P.B.) (1.7) satisfy the Jacobi identities iff, after a possible permutation of indices* $\{1, 2, \ldots, n\}$,

$$\lambda(i,j) = \text{const } \theta(j-i), \tag{1.9}$$

where θ *is the sign function:* $\theta(s) = +1, 0, -1$ *for* $s > 0, s = 0, s < 0$ *respectively. With this* $\lambda(i,j)$ *from now on, the relations (1.1) and (1.2) turn into the popular choice*

$$x_i x_j = q x_j x_i, \quad i < j,$$

$$\xi_i \xi_j = -q^{-1} \xi_j \xi_i, \quad i \leq j,$$

[1,2], while the P.B. (1.7) takes the form

$$\{a_{i\alpha}, a_{j\beta}\} = [\theta(j-i) + \theta(\beta - \alpha)] a_{i\beta} a_{j\alpha}, \tag{1.10}$$

which I had extracted [3] from Drinfeld's formulae for the quantum group $SL_q(n)$ *[4].*

REMARK 1.11: Taking the quasiclassical limit of the defining relations (1.1) and (1.2),

$$\{x_i, x_j\} = \lambda(i,j) x_i x_j, \tag{1.12}$$

$$\{\xi_i, \xi_j\} = \lambda(i,j) \xi_j \xi_i, \tag{1.13}$$

we can forget about quantum spaces and groups and simply demand that the multiplication maps

$$A \times x \to x, \quad A \times \xi \to \xi, \tag{1.14}$$

be Poisson maps, with the Poisson structure on the product space being the direct sum. This demand results in (1.7). (More generally, formula (1.10) gives a Poisson structure to each vector space of rectangular matrices, so that matrix multiplication becomes a Poisson map; formula (1.14) is a particular case of this general property.)

REMARK 1.15: If we make an illegitimate (from the point of view of quantum groups) operation by using q^ν instead of q in (1.2), ν being a constant:

$$\xi_i \xi_j + q^{\nu \lambda(j,i)} \xi_j \xi_i = 0, \tag{1.16}$$

then formulae (1.13) and (1.7) become

$$\{\xi_i, \xi_j\} = \nu \lambda(i,j) \xi_j \xi_i, \tag{1.17}$$

$$\{a_{i\alpha}, a_{j\beta}\} = \frac{1+\nu}{2}[\theta(j-i) + \theta(\beta-\alpha)] a_{i\beta} a_{j\alpha} + \frac{1-\nu}{2}[\theta(j-i) - \theta(\beta-\alpha)] a_{i\alpha} a_{j\beta}. \tag{1.18}$$

Everything is fine for the bracket (1.18): it satisfies the Jacobi identities and is multiplicative (i.e., it makes the matrix multiplication into a Poisson map); except, the $det(A)$ is no longer a Casimir for $\nu \neq 1$:

$$\{a_{i\alpha}, det(A)\} = \frac{1-\nu}{2} a_{i\alpha} \, det(A) \sum_{j=1}^{n} [\theta(j-i) - \theta(j-\alpha)]. \tag{1.19}$$

REMARK 1.20: The bracket (1.18) allows self-consistent constrains:

$$U := \{a_{i\alpha} = 0 | i > \alpha\}, \quad \text{uppertriangular subgroup,} \tag{1.21}$$

$$L := \{a_{i\alpha} = 0 | i < \alpha\}, \quad \text{lowertriangular subgroup,} \tag{1.22}$$

$$Deg := \{det(A) = 0\}, \quad \text{degenerate semigroup,} \tag{1.23}$$

$$GL(n) := \{det(A) \neq 0\}, \tag{1.24}$$

$$SL(n) := \{det(A) = 1, \nu = 1\}. \tag{1.25}$$

REMARK 1.26: In the R-matrix language, the quasiclassical limit consists in replacing the familiar formulae

$$RA_1A_2 = A_2A_1R, \tag{1.27}$$

$$R_{12}R_{13}R_{23} = R_{23}R_{13}R_{12}, \tag{1.28}$$

$$A_1 = A \otimes \mathbf{1}, A_2 = \mathbf{1} \otimes A, \quad \text{etc., by} \quad (R = \mathbf{1} \otimes \mathbf{1} + \varepsilon r + O(\varepsilon^2))$$

$$\{A_1, A_2\} = [A_2A_1, r], \tag{1.29}$$

$$[r_{12}, r_{13}] + [r_{12}, r_{23}] + [r_{13}, r_{23}] = 0. \tag{1.30}$$

REMARK 1.31: For $\nu = 1$ and $n = 2$, the ring of Casimirs is generated by

$$det(A) \quad \text{and} \quad a_{12}/a_{21}. \tag{1.32}$$

CONJECTURE 1.33. *For $\nu = 1$ and $n > 2$, the $det(A)$ is the generator of the ring of invariants.*

REMARK 1.34: In view of Theorem 1.8, it is not suprising that, given a constant $n \times n$ matrix b, the maps

$$A \mapsto bA, \qquad A \mapsto Ab, \tag{1.35}$$

$$A \mapsto bAb^{-1}, \tag{1.36}$$

are not Poisson, while the map ($\mu_1, \ldots, \mu_n$ are arbitrary invertible constants)

$$a_{ij} \mapsto \mu_i a_{ij} \mu_j^{-1} \tag{1.37}$$

is, since it results from the only automorphism of the defining relations (1.12), (1.13):

$$x_i \mapsto \mu_i x_i, \quad \xi_i \mapsto \mu_i \xi_i. \tag{1.38}$$

The map (1.36), by the way, induces a new $(n^2 - n)$ - parameter family of multiplicative Poisson structures on Mat_n.

2. SUPERGROUPS AND COLOURED GROUPS

Let Γ be a discrete abelian group and $\mathcal{X} : \Gamma \times \Gamma \to k^{\times}$ be a map, satisfying the relations

$$\mathcal{X}(\gamma_1, \gamma_2) = \mathcal{X}(\gamma_2, \gamma_1)^{-1}, \tag{2.1}$$

$$\mathcal{X}(\gamma_1, \gamma_2 + \gamma_3) = \mathcal{X}(\gamma_1, \gamma_2)\mathcal{X}(\gamma_1, \gamma_3), \tag{2.2}$$

$$\mathcal{X}(0, \gamma) = 1. \tag{2.3}$$

Let $p : \{1, \dots, n\} \longrightarrow \Gamma$ be the grading map. On the space $Mat_n(\mathcal{R})$, where $\mathcal{R}$ is an associative Γ -graded space, the induced grading is

$$p(A) = p(a_{ij}) + p(j) - p(i), \tag{2.4}$$

so that the matrices of grading zero ("even") are those for which

$$p(a_{ij}) = p(i) - p(j). \tag{2.5}$$

We consider the ring of only such matrices.

THEOREM 2.6. *Set*

$$(\bar{i}, \overline{\alpha}, \bar{j}, \overline{\beta}) := \mathcal{X}(\overline{\alpha}, \overline{\beta})\mathcal{X}(\overline{\beta}, \bar{j})\mathcal{X}(\bar{j}, \overline{\alpha}), \tag{2.7}$$

where $\overline{\alpha}$ etc. stands for $p(\alpha)$ etc. Then the P.B.

$$\{a_{i\alpha}, a_{j\beta}\} = \frac{1+\nu}{2}[\theta(j-i) + \theta(\beta - \alpha)]a_{i\beta}a_{j\alpha} \sqcap (\bar{i}, \overline{\alpha}, \bar{j}, \overline{\beta}) + \frac{1-\nu}{2}[\theta(j-i) - \theta(\beta-\alpha)]a_{i\alpha}a_{j\beta} \tag{2.8}$$

are multiplicative and satisfy the (graded) Jacobi identities.

THEOREM 2.9. *For $\nu = 1$, the $\Gamma det(A)$ $[= sdet(A) = Ber(A)$ for $\Gamma = \mathbf{Z}_2]$ is a Casimir. The coloured Poisson groups U, L, Deg, GL, and $SL|_{\nu=1}$ are defined thereby, as in Remark 1.20.*

REMARK 2.10: The recipe of "adding missing relations" of [1] §6, does not seem to work in the case of general grading: given the relations (1.1) with a fixed Γ-grading of the x's, there is no unique choice of the Γ-gradings of the ξ's in (1.2). Formula (2.8) is the unique Γ-graded generalization of the formula (1.18).

3. TRIANGULAR DECOMPOSITION

Denote by D the ring of diagonal matrices, and by $\tilde{L}$ $(\tilde{U})$ those matrices in L (U) whose diagonal entries are all $\mathbf{1}$. The multiplication map

$$\tilde{L} \times D \times \tilde{U} \to Mat_n \tag{3.1}$$

is regular, and its inverse is rational. Define the Poisson bracket on the space $\tilde{L} \times D \times \tilde{U}$ by the rules:

$$D \text{ is commutative;} \tag{3.2a}$$

$$\tilde{L} \text{ and } \tilde{U} \text{ commute between themselves;} \tag{3.2b}$$

the commutation relations on $\tilde{L} \times D$ and $D \times \tilde{V}$ are such that the multiplication maps $\tilde{L} \times D \to L$ and $D \times \tilde{V} \to U$ are Poisson. (3.2c)

Thus, the nonzero commutation relations on the space $\tilde{L} \times D \times \tilde{U}$ (for simplicity, for the case $\nu = 1$ and $\Gamma = \{0\}$) are:

$$\{x_{i\alpha}, x_{j\beta}\} = [\theta(j-i) + \theta(\beta-\alpha)]x_{i\beta}x_{j\alpha} + [\theta(i-\beta) + \theta(\alpha-\beta)]x_{i\beta}x_{j\beta}x_{\beta\alpha} \\ - [\theta(j-\alpha) + \theta(\beta-\alpha)]x_{i\alpha}x_{j\alpha}x_{\alpha\beta}, \quad x_{ii} = 1, \tag{3.3}$$

$$\{d_i, x_{j\beta}\} = [\theta(j-i) + \theta(\beta-i)]d_i x_{i\beta}x_{ji}, \tag{3.4}$$

$$\{y_{i\alpha}, y_{j\beta}\} = [\theta(j-i) + \theta(\beta-\alpha)]y_{i\beta}y_{j\alpha} + [\theta(i-j) + \theta(\alpha-j)]y_{j\beta}y_{j\alpha}y_{ij} \\ - [\theta(j-i) + \theta(\beta-i)]y_{i\alpha}y_{i\beta}y_{ji}, \quad y_{ii} = 1, \tag{3.5}$$

$$\{d_i, y_{j\beta}\} = [\theta(j-i) + \theta(\beta-i)]d_i y_{i\beta}y_{ji}, \tag{3.6}$$

where

$$\tilde{L} = (x_{ij}), \quad D = \text{diag } (d_1, \ldots, d_n), \quad \tilde{U} = (y_{ij}). \tag{3.7}$$

THEOREM 3.8. *The birational isomorphism (3.1) is a Poisson isomorphism.*

Since

$$\Gamma det(\tilde{L}D\tilde{U}) = \Gamma det(D) = \sqcap_{i=1}^{n} d_i^{\chi(\bar{i},\bar{i})} \tag{3.9}$$

commutes with $\tilde{L}, D$, and $\tilde{U}$, Theorem 2.9 follows. Also, from (3.2c) it follows that the natural projections

$$Mat_n \longrightarrow Mat_n \; / \; \tilde{U} \approx L, Mat_n \longrightarrow \tilde{L} \setminus Mat_n \approx U, \tag{3.10}$$

are Poisson maps.

REMARK 3.11: The Poisson structures (3.3) and (3.5) on unipotent subgroups are <u>not</u> multiplicative.

REMARK 3.12: Besides Poisson maps (3.10), there are many other Poisson projections, such as from Mat_n unto its subgroup $\{a_{i+1,1} = 0, 1 \leq i < n\}$ (onto which the *P.B.* (1.10) properly restrict) resulting from the left action on Mat_n by the subgroup of lower triangular unipotent matrices whose only nonzero elements are in the first column.

4. AFFINE GROUPS

These act as

$$\mathbf{x} \mapsto A\mathbf{x} + \mathbf{p}, \tag{4.1}$$

$$\xi \mapsto A\xi + \mathbf{w}, \tag{4.2}$$

and do not seem to come in the quasiclassical limit from anything quantum. To find a multiplicative $P.B.$ for the group of affine transformations Aff_n, we realize these transformations as projectivisations of the group of <u>linear</u> transformations of k^{n+1} (or $\mathcal{R}^{n+1}$ in the Γ-graded case): 1) the matrices $(\overline{a}_{ij}), 1 \leq i,j \leq n+1$, with $\{\overline{a}_{n+1,j} = 0 | 1 \leq j \leq n\}$ form a subgroup on which the $P.B.$ (1.18) restrict properly; 2) then the submatrices of ratios $\{\overline{a}_{ij}/\overline{a}_{n+1,n+1} | 1 \leq i,j \leq n+1\}$ are isomorphic to the desired group Aff_n; 3) the $P.B.$ (1.18) for these ratios split out a separate multiplicative $P.B.$ on Aff_n. With the identification

$$a_{ij} = \overline{a}_{ij}/\overline{a}_{n+1,n+1}, \quad p_i = \overline{a}_{i,n+1}\overline{a}_{n+1,n+1}, \quad 1 \leq i,j \leq n, \tag{4.3}$$

the $P.B.$ on Aff_n have the form (for the gradeless case $\Gamma = \{0\}$, the general case is similar):

$$\{a_{i\alpha}, a_{j\beta}\} = \frac{1+\nu}{2}[\theta(j-i) + \theta(\beta-\alpha)]a_{i\beta}a_{j\alpha} + \frac{1-\nu}{2}[\theta(j-i) - \theta(\beta-\alpha)]a_{i\alpha}a_{j\beta}, \tag{4.4a}$$

$$\{p_i, p_j\} = \theta(j-i)p_ip_j, \tag{4.4b}$$

$$\{a_{i\alpha}, p_j\} = \frac{1+\nu}{2}[\theta(j-i)+1]a_{j\alpha}p_i + \frac{1-\nu}{2}[\theta(j-i)-1]a_{i\alpha}p_j + a_{i\alpha}p_j. \tag{4.4c}$$

REMARK 4.5: Aff_n is isomorphic to the subring of those matrices in Mat_{n+1} which satisfy the conditions

$$\{a_{n+1,j} = 0 \mid 1 \leq j \leq n\}, \tag{4.6}$$

$$a_{n+1,n+1} = 1. \tag{4.7}$$

The $P.B.$ (1.18) (or even (1.13) for Mat_{n+1} restrict properly on the subset (4.6) but <u>cannot</u> be restricted onto (4.7).

5. CONFORMAL POISSON STRUCTURES

Multiplicative $P.B.$ do <u>not</u> require the <u>group</u> structure, but only the <u>semigroup</u> one. Thus, if $\mathcal{M}_1, \mathcal{M}_2, \mathcal{M}_3$ are Poisson manifolds then we can talk about a map

$$\mathcal{M}_1 \times \mathcal{M}_2 \to \mathcal{M}_3 \tag{5.1}$$

being Poisson, and even <u>conformally</u> Poisson, with the conformal factor λ, if the map

$$\mathcal{M}_1 \times \mathcal{M}_2 \to \lambda\mathcal{M}_3 \tag{5.2}$$

is Poisson; here $\lambda\mathcal{M}$ stands for the manifold $\mathcal{M}$ whose new $P.B.$ equals λ times the old $P.B.$ [5]. Naturally, the most interesting case is when $\mathcal{M}$ is a group or a semigroup. Let us consider a few examples

(a) $\mathcal{M} = \mathbf{R}^2 = \{u_1\mathbf{e}_1 + u_2\mathbf{e}_2\}$, with multiplication of the $2-d$ solvable Lie algebra: $[\mathbf{e}_1, \mathbf{e}_2] = \mathbf{e}_2$. Then

$$\{u_1, u_2\} = \varphi(u_1, u_2), \quad \varphi \text{ is arbitrary with } \varphi(0,u) = 0, \quad \lambda \text{ is arbitrary;} \tag{5.3}$$

(b) $\mathcal{M} = \mathbf{R}^3 \approx so(3)$, with multiplication of the cross product. Here $\lambda = -1$ and there are two types of quadratic $P.B.$:

$$\{u_1, u_2\} = \delta_1 u_1 u_2, \quad \{u_2, u_3\} = \delta_2 u_2 u_3, \quad \{u_3, u_1\} = \delta_3 u_3 u_1,$$
$$\delta_1 + \delta_2 + \delta_3 = 0, \quad \delta_i \in \mathbf{R}, \tag{5.4}$$

$$\{u_1, u_2\} = u_2^2 - u_1^2, \quad \{u_2, u_3\} = u_3 u_1, \quad \{u_3, u_1\} = -u_3 u_2, \tag{5.5}$$

and two cyclic permutations of (5.5);

(c) $\mathcal{M} = Fun(\mathbf{Z})$:

$$\{u_n, u_m\} = (n - m) u_n u_m, \quad \lambda = 1; \tag{5.6}$$

(d) $\mathcal{M} = Fun(\mathbf{Q}_{>0}) = Fun(\{p^\sigma := p_1^{\sigma 1} p_2^{\sigma 2} \ldots | \sigma_i \in \mathbf{Z}, \sigma_i = 0 \text{ for } i \gg 0, p_1, p_2, \ldots$ are ordered positive primes $\})$:

$$\{u_{p\sigma}, u_{p\nu}\} = \left[\sum_m c_m(\sigma_m - \nu_m)\right] u_{p\sigma} u_{p\nu}, \quad c_m \text{ are arbitrary constants}, \quad \lambda = 1; \tag{5.7}$$

(e) This is a differential example: $\mathcal{M} = K[u^{(m)}]$ is a differential algebra, $B = u($any skewsymmetric operator with coefficients in $K)u$, is the Hamiltonian (1×1) matrix, and

$$u \times v \longmapsto uv, \quad \lambda = 2; \tag{5.8}$$

(f) $\mathcal{M} = \{x + y\epsilon | \epsilon^2 = c^2\}$:

$$\{x, y\} = x^2 - c^2 y^2, \quad \lambda = 2, \tag{5.9}$$

$$\{x, y\} = (x^2 - c^2 y^2)\left[\eta_1 \frac{tanh^{-1}(cy/x)}{c} + \eta_2 ln(x^2 - c^2 y^2)\right], \quad \lambda = 1, \tag{5.10}$$

where η_1, η_2 are arbitrary constants. Formula (5.10) describes all multiplicative $P.B.$ on $\mathcal{M}$ which are defined for $(x, 0)$ when x is close to 1. For $c = 0$, (5.10) yields a $P.B.$

$$\{x, y\} = xy, \quad \lambda = 1. \tag{5.11}$$

(g) Grassmann algebra $\Lambda(N) = \{\sum_\sigma u_\sigma \theta^\sigma | \sigma \in \{0, 1\}^N\}$:

$$\{u_\sigma, u_\nu\} = (|\sigma| - |\nu|) u_\sigma u_\nu, \quad \lambda = 1, \tag{5.12}$$

where

$$|\sigma| := \sum_{s=1}^{N} \sigma_s; \tag{5.13}$$

(h) {Group ring of $\mathbf{Z}_2$} = $\{x0 + y1\}$:

$$\{x, y\} = y(x + y)[\eta_1 ln y + \eta_2 ln(x + y)], \quad \lambda = 1, \tag{5.14}$$

and these are all $P.B.$ which are defined for $(0, y)$ when y is close to 1;

(i) {Group ring of $\mathbf{Z}_3$ } = $\{\overline{x}0 + \overline{y}1 + \overline{z}2\} = \{x(0 + 1 + 2) + y1 + z2\}$: here $\lambda = 1$ and

$$\{x, y\} = x(z - y)\left[\eta_1 ln y + \eta_2 tan^{-1}(z/y)\right], \tag{5.15a}$$

$$\{z, x\} = x(y - z)\left[\eta_1 ln z + \eta_2 tan^{-1}(y/z)\right], \tag{5.15b}$$

$$\{y, z\} = (y^2 - z^2)\left[\eta_3 ln(y^2 - z^2) + \eta_4 tanh^{-1}(z/y)\right]. \tag{5.15c}$$

6. THE VALUE $\lambda = 2$ IS SPECIAL

A conformally multiplicative *P.B.* on a group G satisfies at the identity element e the equality

$$2\{f,h\}|_e = \lambda\{f,h\}|_e. \tag{6.1}$$

Thus, for $\lambda \neq 2$,

$$\{f,h\}|_e = 0, \quad \forall f,h \in Fun(G). \tag{6.2}$$

Thus [6,7], one defines a Lie bracket on T_e^*G by the rule

$$[df|_e, dh|_e] := d(\{f,h\})|_e. \tag{6.3}$$

For example, the *P.B.* (1.18) leads to the Lie bracket

$$[X,Y]_{k\gamma} = \sum_{s=1}^{n}[\theta(s-k)-\theta(s-\gamma)]\left[\frac{1+\nu}{2}(X_{ks}Y_{s\gamma} - Y_{ks}X_{s\gamma}) + \frac{1-\nu}{2}(X_{k\gamma}Y_{ss} - Y_{k\gamma}X_{ss})\right]. \tag{6.4}$$

REMARK 6.5: Formula (6.3) has a differential analog:

THEOREM 6.6. *Let $C = K[u_i^{(\sigma)}], 1 \leq i \leq n, \sigma \in \boldsymbol{Z}_+^n$, be a differential algebra over a differential algebra K (see [8]) for basic definitions). Let $B \in Mat_n(C[\partial^\sigma])$ be a Hamiltonian matrix. Let $\Phi : C \to K$ be a differential homomorphism over K such that $\Phi(B) = 0$. Then the formula*

$$[X,Y] := \Phi\left(\frac{\delta}{\delta \boldsymbol{u}}[X^t B(Y)]\right), \quad X,Y \in K^n, \tag{6.7}$$

defines the structure of a differential Lie algebra on K^n.

7. COMMUTATIVE RINGS

The ring $\left\{\sum_{i\geq 0} a_i x^i\right\}$ has the following multiplicative

P.B.:

$$\{a_i, a_j\} = (j-i)a_i a_j, \quad \lambda = 1, \tag{7.1}$$

$$\{a_i, a_j\} = a_i a_{j-1} - a_{i-1}a_j, \quad \lambda = 2, \tag{7.2}$$

with the understanding that $a_{(.)} = 0$ every time the symbol $a_{(.)}$ does not make sense. When there is more than one x, these formulae turn into

$$\{a_\sigma, a_\nu\} = (|\nu| - |\sigma|)a_\sigma a_\nu, \quad \lambda = 1, \tag{7.3}$$

$$\{a_\sigma, a_\nu\} = a_\sigma \sum_s a_{\nu-1_s} - \left(\sum_s a_{\sigma-1_s}\right) a_\nu, \quad \lambda = 2. \tag{7.4}$$

In the language of generating functions

$$a(x) = \sum a_i x^i \quad \text{or} \quad \sum a_\sigma x^\sigma, \tag{7.5}$$

formulae (7.1) – (7.4) become [5]

$$\{a(x), a(y)\} = (y\partial_y - x\partial_x)\,(a(x)a(y))\,, \quad \lambda = 1, \tag{7.6}$$

$$\{a(x), a(y)\} =< y - x > a(x)a(y), \quad \lambda = 2, \tag{7.7}$$

where

$$y\partial_y := \sum_s y_s \frac{\partial}{\partial y_s}, \quad x\partial_x := \sum_s x_s \frac{\partial}{\partial x_s}, \tag{7.8}$$

$$< y - x >:= \sum_s (y_s - x_s). \tag{7.9}$$

Formula (7.6) is the first in the infinite series of multiplicative $P.B.$:

$$\{a(x), a(y)\} = a(x)a(y)[\sqcap_x(ln a(x)) - \sqcap_y(ln a(y))], \quad \lambda = 1, \tag{7.10}$$

where

$$\sqcap_x = \sum c_{\sigma\nu} x^\sigma \left(\frac{\partial}{\partial x}\right)^\nu, \quad \sqcap_y := \sum c_{\sigma\nu} y^\sigma \left(\frac{\partial}{\partial y}\right)^\nu \tag{7.11}$$

are arbitrary linear differential operators. Formulae (7.9) are uniquely defined by the additional property that

$$\left\{\frac{a(x)}{a(0)}, \frac{a(y)}{a(0)}\right\} = 0, \tag{7.12}$$

$$\left\{a(0), \frac{a(x)}{a(0)}\right\} = a(0)\mathcal{O}\left(\frac{a(x)}{a(0)}\right), \tag{7.13}$$

where $\mathcal{O}$ is a nonlinear operator.

The situation becomes more interesting when we consider factor–rings $k[[x_i, \dots x_m]]/I$. The $\lambda = 1 - P.B.$ (7.6) (or (7.10)) does not, in general, restrict onto the factors, with the exception of the factors of the form $k[x]/(x^{n+1})$ [or $k[x]/((x+c)^{n+1})$ for some c], in which case the $P.B.$ are of the form (7.10) with

$$\sqcap_x = x\left(\frac{\partial}{\partial x}\right)^r, \quad 1 \le r \le n. \tag{7.14}$$

Thus, e.g., for $\{a_0 + a_1 x + a_2 x^2 | x^3 = 0\}$, one has

$$\{a_0, a_1\} = a_0 a_1, \quad \{a_0, a_2\} = 2a_0 a_2, \quad \{a_1, a_2\} = a_1 a_2, \quad (r = 1), \tag{7.15}$$

like in (7.1), and also

$$\{a_0, a_1\} = a_0\left(a_2 - \frac{a_1^2}{2a_0}\right), \quad \{a_0, a_2\} = a_1\left(a_2 - \frac{a_1^2}{2a_0}\right), \tag{7.16a}$$

$$\{a_1, a_2\} = \left(\frac{a_1^2}{a_0} - a_2\right)\left(a_2 - \frac{a_1^2}{2a_0}\right), \quad (r = 2). \tag{7.16b}$$

Quite likely, one has to deform formula (7.10) with I in some specific way to obtain a multiplicative *P.B.* on $k[\mathbf{x}]/I$; as formula (5.10) shows, the result is going to be not algebraic but transcendental. In contrast, the $\lambda = 2 -$ *P.B.* (7.7) restricts properly onto any $k[\mathbf{x}]/I$. For example, for the case

$$\left\{ \sum_{i=0}^{n} a_i x^i | x^{n+1} = \sum_{i=0}^{n} c_i x^i \right\}, \tag{7.17}$$

we have the same formula (7.7) in the form

$$\{a(x), a(y)\} = (y-x)a(x)a(y)/\left(x^{n+1} = \sum c_i x^i, \quad y^{n+1} = \sum c_i y^i\right), \quad \lambda = 2, \tag{7.18}$$

or, in components,

$$\{a_i, a_j\} = a_i a_{j-1} - a_{i-1} a_j + a_n (c_j a_i - c_i a_j), \quad \lambda = 2. \tag{7.19}$$

More generally, we have

THEOREM 7.20. *Let $\mathcal{A}$ be an affine variety, and R a ring of regular functions on $\mathcal{A}$. Then there exists a basis of monomials in R (making it into an infinite-dimensional linear space and a semigroup) such that the following P.B. between the coordinate functions on R:*

$$\{a(x), a(y)\} = < y - x > a(x)a(y)/(I_x, I_y), \tag{7.21}$$

make R into a conformal Poisson semigroup with the conformal factor 2.

PROOF: By Noether normalization theorem, R can be represented as a finite sequence of finite extensions of the form (7.17), and each of these extensions allows a proper restriction of the *P.B.* (7.21). ∎

As an example, consider a hyperelliptic curve of arbitrary genus:

$$R = k[x_1, x_2]/\left(x_2^2 = \sum_{s \geq 0} c_s x_1^s\right). \tag{7.22}$$

In the basis of R : $\{x_1^i, x_1^i x_2 | i \in \mathbf{Z}_+\}$, the coordinate functions a_i and b_i have the following *P.B.*:

$$\{a_i, a_j\} = a_i \left(a_{j-1} - \sum_s c_s b_{j-s}\right) - a_j \left(a_{i-1} - \sum_s c_s b_{i-s}\right), \tag{7.23a}$$

$$\{a_i, b_j\} = a_i b_{j-1} - a_{i-1} b_j + a_i a_j - b_i \sum_s c_s b_{j-s}, \tag{7.23b}$$

$$\{b_i, b_j\} = b_i(b_{j-1} - a_j) - b_j(b_{i-1} - a_i). \tag{7.23c}$$

REMARK 7.24: In formula (7.21) one can replace $< y - x >$ by arbitrary antisymmetric function $\varphi(x, y)$.

REMARK 7.25: A 1–Poisson analog exists of Theorem 7.20:

$$\{a(x), a(y)\} = [\varphi(x, y)a(x)a(y)/(I_x, I_y)] \sum_s c_s ln a(\xi_s), \tag{7.26}$$

where c_s are arbitrary constants, and ξ_s are general points on $\mathcal{A}$.

8. OPEN PROBLEMS

(A) The most important situations where multiplicative and 2–multiplicative $P.B.$ are not known are:

(a) classical groups $O(n), Sp(n), U(n)$;

(b) unipotent groups and their intersections with the Toeplitz matrices;

(c) the supergroup $Q(n)$, and more generally, groups of block matrices such as

$$\begin{pmatrix} A_1 & A_2 \\ cA_2 & A_1 \end{pmatrix}, \quad c \text{ is a fixed constant;} \tag{8.1}$$

(d) more generally still, $Mat_n(R[x])/I$;

(e) $k[\mathbf{x}]/I$ for $\lambda = 1$;

(f) projective and quasiprojective varieties;

(g) abelian varieties;

(h) Quaternions.

(B) It is an experimental fact that the inversion map: $g \mapsto g^{-1}$ is always:

$$\text{anti-Poisson}, \quad \lambda = 1, \tag{8.2}$$

$$\text{Poisson}, \quad \lambda = 2. \tag{8.3}$$

There must be a general principle behind this observation.

Acknowledgement. Thanks are due to the organizers and the CRM for their hospitality, and to Igor Dolgachev, Alexei Reiman, and George Wilson for useful discussions. This work was partially supported by the National Science Foundation.

REFERENCES

[1] Yu.I. Manin, "Quantum Groups and Non–Commutative Geometry", Publ. CRM–1561, Université de Montréal (1988).

[2] N. Yu. Reshetikhin, L.A. Takhtadjan, and L.D. Faddev, "Quantization of Lie Groups and Lie Algebras", Algebra and Analysis 1 №1 (1989) 178–206.

[3] B.A. Kupershmidt, "Quasiclassical Limit of Quantum Matrix Groups" (Preprint, 1989).

[4] V.G. Drinfel'd, "Quantum Groups", Proc. ICM, Berkeley (1986) 798–820.

[5] B.A. Kupershmidt, "Quadratic Poisson Structures on Polynomial Rings" (Preprint, 1989).

[6] V.G. Drinfel'd, "Hamiltonian Structures on Lie Groups, Lie Bialgebras, and the Geometric Meaning of the Yang–Baxter Equations", Dokl. Akad. Nauk SSSR 268 (1983) 285–287.

[7] M.A. Semenov–Tian–Shansky, "Dressing Transformations and Poisson Group Actions", Publ. RIMS Kyoto Univ. 21 (1985) 1237–1260.

[8] B.A. Kupershmidt, "Discrete Lax Equations and Differential–Difference Calculus", Asterisque (Paris, 1985).

V. OTHER PHYSICAL APPLICATIONS

Gil BOR and Richard MONTGOMERY

Miroslav GRMELA

D.R. LEWIS, J.E. MARSDEN, T.S. RATIU, and J.C. SIMO

Jedrzej SNIATYCKI

$SO(3)$ INVARIANT YANG-MILLS FIELDS WHICH ARE NOT SELF-DUAL

GIL BOR AND RICHARD MONTGOMERY

Department of Mathematics, University of California
Berkeley, CA 87545, USA

INTRODUCTION

This note considers $SO(3)$ invariant solutions to the Yang-Mills equations on the standard four-sphere S^4 in five-dimensional Euclidean space $\mathbf{R}^5$, with $SU(2)$ as a gauge group. The $SO(3)$ action on S^4 is induced by the irreducible representation of $SO(3)$ on $\mathbf{R}^5$. Since a generic orbit of this action is three-dimensional, the Yang-Mills equations reduce to a system of ordinary differential equations with certain boundary conditions.

Our work is motivated by the following question: Are all solutions to the $SU(2)$-Yang-Mills equations over S^4 (anti-)self dual? Sibner, Sibner, and Uhlenbeck [**1**] recently answered "no" to this question, by proving the existence of a non self-dual solution to these equations. Their proof is far from constructive. We believe that by using our symmetry we can find more explicit non self-dual connections of any Chern class.

The Yang-Mills (YM) equations are the Euler-Lagrange equations for the Yang-Mills action functional

$$YM(\omega) = \int_{S^4} \|F\|^2.$$

Here F is the curvature of ω, a connection defined on some principal $SU(2)$ bundle P over S^4. Isomorphism classes of such bundles are classified by their second Chern number c_2, an integer. A connection ω is (anti-)self dual if and only if it minimizes YM among all smooth connections defined on P, in which case $YM(\omega) = 8\pi^2|c_2|$.

In order to define invariant connections it is necessary to lift the $SO(3)$ action on S^4 to an action on P by $SU(2)$, the universal cover of $SO(3)$. The decomposition of $SU(2)$ bundles with $SU(2)$ symmetry into (equivariant) isomorphism classes is finer than that of all $SU(2)$ principle bundles. In fact, we will show that the isomorphism classes of these bundles with symmetry are indexed by pairs of integers (n_-, n_+), each equivalent to 1 mod 4, and with

$$c_2 = \frac{n_-^2 - n_+^2}{8}.$$

Thus, there can be a number of distinct bundles with symmetry for a given Chern number. In case of $c_2 = 0$, there are an infinite number, but only one of these contains a self dual connection. So if we can show that the Yang-Mills functional attains its infimum on each of these equivariant components, then we obtain the existence of an infinite family of non (anti-)self dual connections. So far, we have not succeeded.

In this note we set the groundwork for this project by showing how to classify the invariant connections, how to write their Yang-Mills equations as ODE's on an interval, and how to get the correct boundary conditions for this system of ODE's. The trickiest part is the boundary conditions, which encode the smoothness of the connection. In the final section we present several possible approaches for proving the existence of the minimum.

Some of this groundwork can be found in Urakawa [6], and in Harnad [**2**]. These authors work much more generally than we do, but do not work out any details of our specific symmetry.

Our investigation of this specific symmetry was inspired by the paper of Avron et al [3]. For each half-integer J (representing the total spin of a fermionic system) they construct a family of quaternionic line bundles with connections over S^4 with this symmetry. They show that all but one of these connections is not self-dual.

1. EQUIVARIANT BUNDLES

The equivariant YM set-up consists of

(1) a principle G bundle $P \to X$ with a group S (for symmetry) acting on P by bundle automorphisms,

(2) an S invariant connection ω on P such that

(3) ω satisfies the YM equations $D * D\omega = 0$.

In this note $X = S^4$, $G = SU(2)$, $S = SU(2)$ and the S action is required to project to the action on S^4 given by the irreducible representation of S on $\mathbf{R}^5$.

We consider first the lifting problem: classify all possible lifts of the given S action on S^4 to bundle automorphisms of a principal G bundle $P \to S^4$. (The S action on S^4 has an ineffective kernel $\{\pm 1\}$, so it is actually an $S/\{\pm 1\} \cong SO(3)$ action; we look for S actions on P instead of $SO(3)$ actions because there are more of them: every $SO(3)$ action defines an S action via the projection $S \to SO(3)$, but not conversely).

We begin by giving a more detailed description of the S action on S^4. The space of 3×3 traceless symmetric matrices Q forms a real 5-dimensional vector space, $\mathbf{R}^5$, with norm $||Q||^2 = tr(Q^2)$. $SO(3)$ acts orthogonally on this vector space by conjugation of matrices, and hence S acts via its projection to $SO(3)$. The action on S^4 is obtained by restricting this action to the four-sphere $S^4 = \{[tr(Q^2) = 1\} \subset \mathbf{R}^5$.

This action on S^4 has two orbit types. The principal (generic) type consists of matrices $Q \in S^4$ with distinct eigenvalues. The singular orbits, of which there are two, consist of matrices with double eigenvalues. As a principal isotropy group we can take the the subgroup Γ of S which leaves the three coordinate axes (the eigenspaces of a generic matrix) invariant. This is a finite subgroup of S, and so the generic orbit is three dimensional. Each singular orbit is isomorphic to a real two-dimensional projective space, since, given two eigenvalues, one of which is degenerate, the symmetric matrix is uniquely determined by the choice of the nondegenerate eigenspace. On one singular orbit the nondegenerate eigenvalue is positive, and on the other it is negative.

For the rest of the paper we will identify S with the group of unit quaternions. Then

$$\Gamma = \{\pm 1, \pm i, \pm j, \pm k\}.$$

Let

$$\Delta = \{diag(\lambda_1, \lambda_2, \lambda_3) \mid \lambda_1^2 + \lambda_2^2 + \lambda_3^2 = 1, \lambda_1 + \lambda_2 + \lambda_3 = 0\}$$

be the fixed-point set of Γ in S^4. Then Δ is a great circle in S^4 parameterized by

$$c_t = \frac{\sqrt{3}}{2} diag(\cos(t), \cos(t + 2\pi/3), \cos(t - 2\pi/3)), 0 \leq t \leq 2\pi.$$

The arc $c_t, 0 \leq t \leq \pi/3$ is a global slice for the S action: it intersects each orbit precisely once. The two endpoints c_0 and $c_{\pi/3}$ correspond to the two singular orbits, the interior of the arc to the principal orbits. Let S_t be the isotropy subgroup of c_t. We have

$$S_0 = \{e^{i\theta}, je^{i\theta}\}, S_{\pi/3} = \{e^{j\theta}, ie^{j\theta}\}, S_t = \Gamma = S_0 \cap S_{\pi/3}, 0 < t < \pi/3.$$

Note that the two singular orbits intersect Δ at the 6 points $t = 0, \pi/3, 2\pi/3, \ldots$ etc., 3 per orbit.

A lift of the S action to a principle G bundle is described by the following data. Let $u_t, 0 \leq t \leq \pi/3$, be a lift of c_t to P. This defines the holonomy representations $\lambda_t : S_t \to G$ by

$$s \cdot u_t = u_t \lambda_t(s), s \in S_t, 0 \leq t \leq \pi/3.$$

A different choice of u_t differs by a curve $g_t \in G$. This results in the conjugation of λ_t by g_t. This construction is reversible:

Proposition 1. *There is a 1-1 correspondence between*

(1) *Isomorphism classes of S equivariant principle G bundles P over S^4 (with the given S action on S^4), and*

(2) *conjugacy classes of homomorphisms $\lambda_t : S_t \to G$.*

See [5] for details.

A short calculation yields the following:

Proposition 2. *The conjugacy classes of homomorphisms $\lambda_t : S_t \to G$ are given by the following list:*

a) $\lambda_t \equiv 1$;

b) $\lambda_t(\gamma) = \gamma, \gamma \in \Gamma$,
$\lambda_0(e^{i\theta}) = e^{in_+\theta}$,
$\lambda_{\pi/3}(e^{j\theta}) = e^{jn_-\theta}$,
where $n_\pm \equiv 1 \pmod 4$, *i.e.* $n_\pm = 1, -3, 5, -7 \ldots$ *etc.*

Remark: In case (b), the lift u_t can be chosen as follows: consider the action $p \to \gamma \cdot p\gamma^{-1}$ of Γ on $P|_\Delta$. It is easily seen from the definitions that the fixed point set of this action intersects each fiber of $P|_\Delta$ in two antipodal points, and so forms a 2:1 covering space of the circle Δ . Define u_t to be a local section of this 2:1 covering defined over $c_t, 0 \leq t \leq \pi/3$. (A global section may not exist.) Clearly there are two such choices.

This completely solves the lifting problem.

2. INVARIANT CONNECTIONS

In this section we determine an explicit form for the S-invariant connections. This is based on propositions 1 and 2 describing the possible S actions. We will soon see that the only connection having the symmetry of type (a) of proposition 2 is the flat connection on the trivial bundle. Thus, we will be concentrating on connections having symmetry type (b).

If ω is an S invariant connection on P then it is determined by its values along a lift u_t of the section c_t of the S action on S^4. When the symmetry of P is of type (b) we fix this lift according to the remark at the end of the previous section. In case the symmetry is of type (a) we take u_t to be a horizontal lift of c_t.

For a symmetry generator $X \in$ Lie algebra of S we define

$$\phi_t(X) = \omega(X_{u_t}) \in \text{Lie algebra of } G.$$

This is the "vertical" component, or Higgs field, corresponding to the X action at u_t. Note that in case (b) the definition of ϕ_t does not depend on which of the two lifts of c_t we take, for they differ by right multiplication by -1, and this acts trivially on $Lie(G)$.

More generally, define $\phi : P \to Hom(Lie(S), Lie(G))$ by

$$\phi_p(X) = \omega(X_p).$$

If ω is S-invariant, this has the symmetries $\phi_{pg} = g \cdot \phi_p$, and $\phi_{sp} = \phi_p \cdot s$, where $g\cdot$ and $\cdot s$ mean the composition with the adjoint actions of these elements on the appropriate Lie algebra. These imply the following symmetries in our situation:

(1) $\phi_t(\gamma \cdot X) = \lambda_t(\gamma) \cdot \phi_t(X)$

(2) $\omega(\dot{u}_t) = \lambda_t(\gamma) \cdot \omega(\dot{u}_t)$

where λ_t is given by proposition 2. In case (a) of that proposition these equivariance conditions easily imply that $\phi_t \equiv 0$. Since we chose $\omega(\dot{u}_t) \equiv 0$, ω is the flat connection on the trivial bundle. We therefore concentrate on case (b). Now $Lie(G) = Lie(S) = \mathbf{R}^3$, so that $\phi_t \in Hom(\mathbf{R}^3, \mathbf{R}^3)$, and Γ acts on each of these $\mathbf{R}^3$'s by reflections about the i, j, k axes. Thus the equivariance condition (1) implies that ϕ_t fixes these axes. Then

$$\phi_t = a_1(t) \cdot i \otimes \sigma^1 + a_2(t) \cdot j \otimes \sigma^2 + a_3(t) \cdot k \otimes \sigma^3$$

where $\{\sigma^1, \sigma^2, \sigma^3\}$ is the dual basis to $\{i, j, k\}$ and a_i are real valued functions on $[0, \pi/3]$. Similarly, the Γ equivariance property (2) implies that $\omega(\dot{u}_t) = 0$ (i.e. u_t is horizontal after all) since the only Γ invariant vector of $\mathbf{R}^3$ is 0. So a_1, a_2, a_3 completely describe the S invariant connection ω.

The boundary conditions satisfied by the a_i at $t = 0$ and $t = \pi/3$ are determined by using some additional symmetries of $\phi : \Delta \to \mathbf{R}^3 \cong 3 \times 3$ diagonal matrices; namely

(1) W equivariance, where $W = N/\Gamma$ and $N = N(\Gamma)$ is the normalizer of Γ in S,

(2) S_t equivariance for $t = 0$ and $t = \pi/3$.

In (1) $W \cong$ the group of permutations of $\{1, 2, 3\}$. It acts on $\mathbf{R}^3$ by permutation of the coordinates and on the circle Δ by reflections, i.e. (23) acts by $t \to -t$, (13) acts by $t \to \pi/3 - t$, etc. To verify this W equivariance recall the general fact that for any S action, W acts on the fixed point set of Γ. In our case, S acts on P and $Hom(Lie(S), Lie(G))$ by conjugation. The Γ fixed point set in P is a double cover of Δ (see the remark at the end of section 1), and the fixed point set in $Hom(Lie(S), Lie(G))$ is "diagonal matrices" $\cong \mathbf{R}^3$. Since ϕ is ± 1 invariant it descends to a W equivariant map on Δ. Finally, calculate that $N(\Gamma)$ projects to the subgroup of $SO(3)$ generated by 90^0 rotations about the coordinate axes.

The above necessary conditions, (1) and (2), are also sufficient for ϕ_t to represent a smooth S invariant connection (see [5] for details):

PROPOSITION 3. *There is a 1-1 correspondence between S invariant connections on S^4 (of the symmetry type in proposition 2b) and maps $\phi = (a_1, a_2, a_3) : [0, \pi/3] \to \mathbf{R}^3$*

which can be extended smoothly to a map of the circle $\Delta \supset [0, \pi/3]$ subject to the following conditions: At 0:

(1) $a_1(0) = n_+ \equiv 1 \pmod 4$,

(2) $a_1(t) = a_1(-t)$,

(3) $a_2(t) = a_3(-t)$, *and*

(4) *if* $n_+ \neq 1$ *then* $a_2(0) = a_3(0) = 0$.

At $\pi/3$*:*

(1) $a_2(\pi/3) = n_- \equiv 1 \pmod 4$,

(2) $a_2(\pi/3 + t) = a_2(\pi/3 - t)$,

(3) $a_1(\pi/3 + t) = a_3(\pi/3 - t)$, *and*

(4) *if* $n_- \neq 1$ *then* $a_1(\pi/3) = a_3(\pi/3) = 0$.

3 THE REDUCED YM EQUATIONS

The YM equations in terms of a_1, a_2, a_3 are :

$$\frac{1}{4}\frac{d}{dt}(K_1\dot{a_1}) = a_1\left(\frac{1}{K_1} + \frac{a_3^2}{K_2} + \frac{a_2^2}{K_3}\right) - a_2a_3\left(\frac{1}{K_1} + \frac{1}{K_2} + \frac{1}{K_3}\right), \ldots etc.$$

where " ... etc." stands for another 2 similar equations that we get by cyclic permutations of 1,2,3. The K_i are functions of t defined by

$$K_1(t) = \frac{3}{\sin t} - 4\sin t,$$
$$K_2(t) = K_1(t + 2\pi/3),$$
$$K_3(t) = K_1(t - 2\pi/3).$$

The (anti-)self dualequations are

$$\frac{1}{2}K_1\dot{a_1} = \pm(a_1 - a_2a_3), \ldots etc.$$

where "+" stands for the self-dual, and "−" for the anti self-dual equations.

To derive these equations we write the curvature along the cross-section c_t in terms of the a_i 's

$$F = i[2(a_2a_3 - a_1)\sigma^2 \wedge \sigma^3 - \dot{a_1}\sigma^1 \wedge dt] + \ldots etc.$$

which gives the (anti-)self-dual equations (the coefficients K_i arise from the fact that the σ^i are not orthonormal). The YM equations can be derived from variation of the YM action

$$\int_{S^4} \|F\|^2 \propto \int_0^{\pi/3} dt\left[\frac{4(a_2a_3 - a_1)^2}{K_1} + K_1\dot{a_1}^2 + \ldots etc.\right]$$

w.r.t. a_i, or directly by evaluating $D * F = 0$ in terms of a_i.

4. CHERN NUMBERS

We now would like to calculate the second Chern number c_2 corresponding to each of the equivariant bundles given in proposition 2b.

PROPOSITION 4. *The second Chern number of a bundle P with the $n_\pm$ symmetry given in proposition 2b is*

$$c_2 = \frac{n_-^2 - n_+^2}{8}.$$

PROOF: This is a straightforward calculation similar to the one in [3]. We use one of the equivariant connections in proposition 3 and the formula for its curvature in the previous section. We than have

$$\begin{aligned} tr(F^2) &\propto (\dot{a_1}(a_2a_3 - a_1) + \dots)\sigma^1\sigma^2\sigma^3 dt \\ &\propto \frac{d}{dt}(a_1a_2a_3 - \frac{a_1^2 + a_2^2 + a_3^2}{2})\sigma^1\sigma^2\sigma^3 dt, \\ \Rightarrow c_2 \propto \int_{S^4} tr(F^2) &\propto a_1a_2a_3 - \frac{a_1^2 + a_2^2 + a_3^2}{2}\Bigg|_0^{\pi/3} \propto n_-^2 - n_+^2 \end{aligned}$$

We can determine the missing constant by some more care in the above calculation or by doing a special case, say $n_+ = 1, n_- = -3$, corresponding to the lift to the quaternionic Hopf bundle $S^7 \to S^4$ ($c_2 = -1$) by the irreducible representation on $\mathbf{C}^4$. ∎

REMARKS: **1.** It is possible to calculate c_2 somewhat more simply without using the explicit form of an equivariant connection. This is based on a general equivariant integration formula due to Atiyah and Bott [4], but the above calculation is more elementary.

2. The lifts in the general $n_\pm$ case can also be realized concretely. These arise by the "spectral splitting" of the trivial rank-n quaternionic vector bundle $S^4 \times \mathbf{H}^n$ w.r.t. a certain linear operator on its sections (the "quadrupole Hamiltonian" in [3]).

5. EXISTENCE OF NON-DUAL CONNECTIONS

If a particular equivariant class of connections does not posses (anti-)self dual connections than any critical point of the YM functional restricted to this class (e.g. a minimum) is a YM connection which is not (anti-)self dual.

The only (anti-)self dual connections on a trivial bundle are flat. The only equivariant flat connection has $a_1 = a_2 = a_3 \equiv 1$ and $n_+ = n_- = 1$ (by the curvature formula in section 3). Therefore, any equivariant YM connection on a bundle with $n_+ = n_- \neq 1$ is not (anti-)self dual.

There are a number of possible approaches for proving the existence of such a non self-dual solution. We have not yet been able to push any of these through to fruition. We will review one fairly convincing approach in some detail, and discuss some others briefly.

The most convincing approach so far is the direct method of the calculus of variations. What one has to do is show that the set of invariant connections satisfy the Palais-Smale condition. Tom Parker is working on a general theorem to this effect.

In more detail , this approach runs as follows. Take ω^i to be a sequence of invariant connections all lying in one invariant component, and satisfying $YM(\omega^i) \to \inf YM$, where the infimum is taken over this component. Then show

(1) a subsequence of this sequence converges to some ω^∞,

(2) ω^∞ is invariant, lies in the same invariant component as the ω^i, and enjoys some regularity,

(3) $YM(\omega^\infty) \leq YM(\omega^i)$,

(4) the validity of the principle of symmetric criticality [**5**], which states that if an invariant field is extremal among invariant fields, then it is extremal among all fields.

(2) and (3) imply that ω^∞ satisfies the hypothesis of the principle of symmetric criticality, and so, if we can complete these steps we are done. Steps (1), (3), and (4) can be completed.

Step (1) is essentially Uhlenbeck's famed compactness theorem, which says the following. Suppose we are handed a sequence of connections on which YM is bounded. Then there exists a finite set D of points of S^4, a subsequence of the sequence, and a sequence of gauge transformations defined on $S^4\backslash D$, such that after applying the gauge transformations to the connections of the subsequence, the resulting subsequence converges to some ω^∞. This convergence is in the weak topology of the Sobolev space of $L^1_{2,loc}$ connections on $S^4\backslash D$. Moreover, the points of D, so called "bubbling off points", are characterized by concentration of curvature density there. If a connection is invariant, then the curvature density is an invariant function. Consequently, for invariant sequences, blow up of curvature must happen along entire group orbits. But D is a discrete point set, and for our symmetry, no orbits are discrete. Consequently D is automatically empty. This shows (1).

Proving the regularity of ω^∞ in step (2) is usually bound up with showing that it is a critical point. In regard to its invariance, the following observation is in order. Suppose that ω is an invariant connection and that ω' is obtained from ω by gauge transformation. Then ω' is itself invariant if and only if it equals ω. This suggests that in applying the Uhlenbeck compactness result, we need not apply any gauge transformations.

In addition to this four-dimensional approach, there are one- dimensional approaches to existence. One approach is to apply the direct method of the calculus of variations to the one-dimensional action. The role played by the special endpoint conditions of proposition 3 is unclear from this one-dimensional point of view. Another approach is to work directly with the one-dimensional ODE. The endpoints, 0 and $\pi/3$, are the only singular points for this ODE.

A fourth approach is to use the "ambi-twistor" method pioneered by Witten [**7**]. See also Isenberg, Yasskin and Green [**8**]. This is a generalization of the twistor method for self-dual solutions. Of the four approaches this is the most constructive.

6. NUMERICAL EVIDENCE

Notice that the reduced YM action and equations are invariant w.r.t. the reflection $t \to t + \pi$. (This interchanges dual and anti-self-dual solutions and flips the boundary conditions). It is therefore natural to consider reflection invariant solutions on the trivial bundle $n_+ = n_- \neq 1$. Combined with the W symmetries of proposition 3 this means that a solution to the equations on $[0, \pi/6]$ with

$$a_1(\pi/6) - a_2(\pi/6) = a_1'(\pi/6) + a_2'(\pi/6) = a_3(\pi/6) = 0$$

extends to a smooth reflection-invariant solution on $[0, \pi/3]$ with $n_+ = n_-$.

A numerical integration of the equations for the case $n_+ = n_- = -3$ was done as follows: first a (truncated) power series expansion for the solutions was calculated at $t = 0$. This gives a family of solutions starting at $t = 0$ depending on the 3 parameters $a_1''(0), a_2'(0)$, and $a_2''(0)$. For example, the values $a_1''(0) = 2, a_2'(0) = \sqrt{3}, a_2''(0) = -1$ correspond to the anti-self dual solution $a_1(t) = -2\cos t - 1$. After "jumping-off" a small amount off the $t = 0$ singularity by the power series we begin a numerical integration (Runge-Kutta) up to $t = \pi/3$ where we check the above "reflectability" conditions. This method yields a solution for the values

$$a_1''(0) = 7, a_2'(0) = -4, a_2''(0) = -3$$

with approximately 1 precent error.

References

1. L. M. Sibner, R. J. Sibner, and K. Uhlenbeck, *Solutions to YM equations which are not self-dual*, Preprint (1989).
2. J. Harnad, S. Shnider, and J. Tafel, *Group actions on principle bundles and dimensional reduction*, Lett. in Math. Phys. **4** (1980), 107–113.
3. J. E. Avron, L. Sadun, J. Segret, and B. Simon, *Chern numbers and Berry's Phase in Fermi system*, Caltech preprint (1989).
4. M. F. Atiyah and R. Bott, *The moment map and equivariant cohomology*, Topology **23 (No. 1)** (1984), 1–28.
5. R. S. Palais, *The Principle of Symmetric Criticality*, Comm. Math. Phys. **69** (1979), 19–30.
6. H. Urakawa, *Equivariant theory of YM connections over Riemannian manifolds of cohomogeneity one*, Indiana Univ. Math. J. **37 (No. 4188)** (1988), 753–788..
7. E. Witten, *An interpretation of classical Yang-Mills theory*, Phys. Lett. **77B** (1978), p. 394.
8. J. Isenberg, Ph. B. Yasskin, and P. G. Green, *Non-self-dual guage fields*, Phys. Lett. **78B** (1978), p. 462.

THERMODYNAMICAL LIFT OF THE NONLINEAR ONSAGER-CASIMIR VECTOR FIELD

MIROSLAV GRMELA

Département de génie chimique, Ecole Polytechnique de Montréal
CP 6128 -A, Montréal, Québec H3C 3A7, Canada

Contact geometry is shown to provide a setting in which thermodynamics and the dynamics that is compatible with it, are formulated in a unified manner.

1. PHYSICAL CONTEXT

A physical system is called macroscopic if it is composed of a large number of particles. We recall some experimental evidence collected about the macroscopic systems.

(i) Let the macroscopic system be isolated from the external influence. Let moreover the system be observed by performing on it macroscopic measurements. A measurement is called macroscopic if results of the measurement provide an information about gross (overall) characteristics of the macroscopic system and not about details concerning positions and velocities of the particles composing the macroscopic system. For example, a measurement of hydrodynamic fields is a macroscopic measurement. Experience shows that isolated macroscopic systems approach (as the time goes to infinity) states at which results of the macroscopic measurements do not change in time. Such states are called equilibrium states. At the equilibrium states, macroscopic systems are found to be well described by classical equilibrium thermodynamics (see Section 4).

(ii) Even if the macroscopic system are not isolated, there are many well known examples in which an approach (as the time goes to infinity) to a reduced, lower dimensional, description is still observed. This happens for example for the system considered in the Bénard problem (a layer of a fluid subjected to the gravitational force and heated from below). It is a well known experience that this externally forced macroscopic system is well described by hydrodynamic (Boussinesq) equations. This means that results of more microscopic measurements show an approach to a state at which the system is well described by the hydrodynamic description. An example of a measurement that is more microscopic that hydrodynamic measurements is a measurement of the one particle distribution function (recall that five hydrodynamic fields are five moments of the one particle distribution function).

Now we recall some experience collected while trying to describe and understand the above experience. The case by case analysis of many particular macroscopic systems revealed that the time evolution equations whose solutions exhibit the experience (i) share some structure that we collect in the concept of the nonlinear Onsager-Casimir dynamics (Section 2). We then suggest that this dynamics models also the experience (ii), i.e. the approach to lower dymansional dynamical systems in the context of externally forced macroscopic systems.

In this Note, we observe that the nonlinear Onsager-Casimir vector field (introduced in Section 2) can be lifted to a larger state space, equiped with a contact structure, in such a way that:

(i) The flow generated by the lifted vector field is a contact flow (i.e. the flow preserving the contact structure).

(ii) The flow generated by the lifted vector field has an invariant submanifold. The invariant submanifold is a Legendre submanifold of the larger state space. The lifted vector field restricted to the invariant Legendre submanifold coincides with the nonlinear Onsager-Casimir vector field.

(iii) The geometry of the invariant Legendre submanifold is interpreted physically as thermodynamics implied by the nonlinear Onsager-Casimir vector field.

The observation made in this Note clarifies the role that thermodynamics plays in dissipative dynamics. The geometrical formulation of dynamics and thermodynamics introduced in this Note should also be useful for bringing the geometrical methods developed in the context of nondissipative classical mechanics to the assistance in analysing dissipative dynamics and thermodynamics. On the other hand, the new context in which the geometrical methods are applied may stimulate the development of the geometrical methods themselves.

2. NONLINEAR ONSAGER-CASIMIR VECTOR FIELD

Let M, called a state space, be a vector space; $u \in M$ denotes the state variables used to describe states of macroscopic systems. For example, $u \equiv$ five hydrodynamic fields of mass, momentum and energy densities (this is called a hydrodynamic description) or $u \equiv$ one particle distribution function (kinetic theory description). Let in addition M be a bundle, $M = (M, N, \mathcal{M})$, where N is the base space and $\mathcal{M}$ the bundle projection. The space N is also assumed to be a vector space.

Furthermore, we introduce $S : M \to \mathbb{R}$, the function S called entropy (see Section 4) is assumed to be sufficiently regular. We also introduce another function Φ, called a thermodynamic potential (see Section 4), by $\Phi : M \times N^* \to \mathbb{R}$, $\Phi(u,q) = -S(u) + \langle q, \mathcal{M}(u) \rangle$, where $\langle , \rangle$ is pairing in N, N^* is dual to N.

Next, we introduce a vector field in the state space M by

$$u_t = L^c(u)\partial\Phi/\partial u - L^0(\partial\Phi/\partial u)\partial\Phi/\partial u \tag{1}$$

We assume that:

OC1 L^c, called a Casimir tensor, is a Poisson tensor (i.e. L^c is skewsymmetric and $\{A, B\} = \langle \partial A/\partial u, L^c(u)\partial B/\partial u \rangle$ is a Poisson bracket ($\langle , \rangle$ denotes pairing in M (see Ref. [1])).

OC2 L^0, called Onsager tensor, is specified by $L^0(p) = \partial^2\Psi/\partial p\partial p$, where $p = \partial\Phi/\partial u$ and $\Psi : M^* \to \mathbb{R}$, called dissipative potential, is a sufficiently regular function that reaches its minimum at the origin, $\Psi(0) = 0$ and Ψ is a convex function at the origin.

OC3

$$\frac{d\mathcal{M}(u)}{dt} = 0 \tag{2}$$

OC4 Solutions to eq. (1) exist for all times. As $t \to \infty$, the equilibrium states u_{Neq}, defined as solutions of $\partial\Phi/\partial u = 0$, are approached.

We note that due to the assumptions (OC1) and (OC2), $d\Phi/dt \leq 0$. The thermodynamic potential plays thus the role of the Lyapunov function provided it also enjoys the necessary convexity property. We also note that if the Onsager tensor $L^0 \equiv 0$ then the vector field (1) is a hamiltonian vector field, the thermodynamic potential Φ plays the role of the hamiltonian. We shall refer to the first term on the right hand side of eq. (1) as a nondissipative (Casimir) part and to the second term on the right hand side of eq. (1) as a dissipative (Onsager) part of the nonlinear Onsager-Casimir vector field. The vector field (1) satisfying the properties (OC1)-(OC4) is called a nonlinear Onsager-Casimir vector field. Without the property (OC4), we shall call it a formal nonlinear Onsager-Casimir vector field. The reason for using the name Onsager-Casimir is explained in Example 1.

Example 1.

If L^c is a skewsymmetric twice contravariant tensor independent of u and L^o a symmetric twice contravariant tensor independent of $\partial\Phi/\partial u$ (i.e. Ψ is a quadratic function) then the vector field (1) is the vector field introduced by Onsager [2] and Casimir [3]. From the physical point of view, this vector field describes the linear time evolution in a small neighborhood of equilibrium states (defined as solutions of $\partial\Phi/\partial u = 0$). Onsager and Casimir have also distinguished the nondissipative and dissipative parts by introducing an involution I, called a parity mapping, in the state space $M, I : M \to M, I \circ I =$ identity. The parity mapping I induces a mapping among the vector fields. The dissipative (Onsager) part of the vector field remains invariant with respect to this mapping and the nondissipative (Casimir) part of the vector field changes its sign if the mapping is applied on it. From the physical point of view, the parity mapping is the mapping that in the context of the reversible time evolution compensates the reversion of time. For Example, if $u \equiv (\mathbf{r}, \mathbf{v})$, where $\mathbf{r}$ is the position vector and $\mathbf{v}$ the velocity, then $(\mathbf{r}, \mathbf{v}) \overset{\mathbf{I}}{\longmapsto} (\mathbf{r}, -\mathbf{v})$. In this Note, we shall not use the additional structure in the state space M provided by the parity mapping I.

Example 2.

Equation (1) becomes the Cahn-Hilliard equation [4] by making the following choices: $u \equiv \rho(\mathbf{r})$, where ρ is the field of the mass density, $\mathbf{r}$ is the position vector; the entropy S depends also on $\partial\rho/\partial\mathbf{r}; \mathcal{M}(\rho) = \int_\Omega d^3\mathbf{r}\rho(\mathbf{r})$, where $\Omega \subset \mathrm{I\!R}^3$ denotes the region confining the macroscopic system under consideration; $L^c \equiv 0, L^0 \equiv -\frac{\partial}{\partial r_\alpha}\left(D\frac{\partial}{\partial r_\alpha}\right)$, where the summation convention has been used, $D > 0$ is a constant diffusion coefficient. The nonlinear generalization of L^0 appearing in the assumption (OC2) has been first introduced in Refs. [6–8].

Example 3.

All vector fields discussed for Example in Ref. [5] are also Examples of the nonlinear Onsager-Casimir vector field (1) with $L^0 \equiv 0$. Dissipative fluid mechanics equations and dissipative kinetic equations (e.g. the Navier-Stokes-Fourier fluid mechanics equations and the Boltzmann kinetic equation) have been cast into the form of eq. (1) in Refs. [9–11]. In the majority of these Examples the assumption (OC4) is not verified so that the vector fields remain formal. The thermodynamic potentials arising in these Examples will be mentioned in Section 4.

Example 4.

The realizations of the nonlinear Onsager-Casimir vector field introduced in the previous Examples describe the time evolution of isolated macroscopic systems that reach as $t \to \infty$ thermodynamic equilibrium states. This means in particular that the base space N $(M \equiv (M, N, \mathcal{M}))$ is the two dimensional state space of equilibrium thermodynamics (see Section 4). We conjecture that eq. (1) can also be used to describe externally forced macroscopic systems (like e.g. the system considered in the Bénard problem – see Section 1). We shall now briefly explain how we suggest to apply the nonlinear Onsager-Casimir vector field to the Bénard problem. Let $u \equiv f(\mathbf{r}, \mathbf{r}), f : \Omega \times \mathbb{R}^3 \to R_+$ be one particle distribution function, $\mathbf{r}$ is the position vector $\mathbf{r} \in \Omega \subset \mathbb{R}^3$, and $\mathbf{v}$ velocity. The base space N is the state space used in hydrodynamics (i.e. $\mathcal{M}(f) = \left(\int d^3\mathbf{v} f(\mathbf{r}, \mathbf{v}), \int d^3\mathbf{v} \frac{1}{2} m v^2 f(\mathbf{r}, \mathbf{v}), \int d^3\mathbf{v}\mathbf{v} f(\mathbf{r}, \mathbf{v})\right)$. We conjecture that there exists a thermodynamic potential Φ (see eq. (1)) and the Onsager and Casimir tensors L^0 and L^c so that a pair of the time evolution equations, one being eq. (1) and the other the Boussinesq hydrodynamic equation

$$v_t = B(v) \quad ; v \in N \tag{3}$$

represents the kinetic theory description of the Bénard problem. The variable q appearing in Φ is thermodynamically conjugate to the state variable v. This means that there is a one-to-one relation between v and q (provided the thermodynamic potential Φ is appropriately chosen – see Section 4). Note that in the case of the Bénard problem the time evolution equation (1) is non-autonomus since q appearing in Φ evolves in time. The realizations of the nonlinear Onsager-Casimir vector field presented in Examples 1–3 can be regarded as a special case of the pair of equation (1) and (3) in which eq. (3) degenerated to the equation $v_t = 0$. The pair of equations (1) and (3) is interpreted physically as follows. The time evolution equation (1) governs the time evolution of fast modes and the time evolution equation (3) governs the time evolution of slow modes. The formulation (1), (2), (3) of the splitting of the time evolution into fast and slow has the advantage of revealing the associated thermodynamics. The thermodynamics is determined by the thermodynamic potential Φ. We shall consider this point separately in Section 4.

3. THERMODYNAMICAL LIFT

If $L^0 \equiv 0$, i.e. if the dissipative (Onsager) term in eq. (1) is absent, then eq. (1) defines a hamiltonian system and the symplectic geometry (see Ref. [1]) provides an excellent setting for analysing solutions of eq. (1). If however $L^0 \not\equiv 0$ then the flow generated by eq. (1) does not preserve the symplectic structure and consequently the symplectic geometry ceases to provide an appropriate environment for analysing solutions of eq. (1). The question arises of what is the appropriate environment replacing the symplectic geometry if $L^0 \not\equiv 0$. The main result of this note, presented in this section, is the introduction of such environment. Its thermodynamical interpretation is discussed in Section 4.

Starting with the state space M, we construct $T^*M \times \mathbb{R}$. This space has a natural contact structure defined by one form $\omega = dz - pdu$, where $u \in M, p \in T^*_u M, z \in \mathbb{R}$ are local coordinates. We shall call the space $T^*M \times \mathbb{R}$ a thermodynamic space (the terminology is explained in Section 4). We introduce the following vector field in the thermodynamic space $T^*M \times \mathbb{R}$,

$$\begin{aligned} u_t &= -\frac{\partial K}{\partial p} \\ p_t &= \frac{\partial K}{\partial u} - p\frac{\partial K}{\partial z} \\ z_t &= K - \left< p, \frac{\partial K}{\partial p} \right>, \end{aligned} \tag{4}$$

where

$$K = \langle p, \partial\Psi/\partial p \rangle \quad - \langle p, [\partial\Psi/\partial p]_{p=\partial\Phi/\partial u} \rangle \quad - \langle p, L^c(u)\partial\Phi/\partial u \rangle \tag{5}$$

We make the following observations:

(i) The vector field (4) is a canonical formulation of a contact vector field (see Refs. [1] and [12]). This means that the flow generated by the vector field (4) presrves the contact structure, defined by $\omega = dz - pdu$, of the thermodynamic space $T^*M \times \mathbb{R}$. The quantity K is called a contact hamiltonian.

(ii) The Legendre submanifold $\mathcal{L}$ of the thermodynamic space $T^*M \times \mathbb{R}$ defined as the image of the immersion $u \longmapsto (u, \partial\Phi/\partial u(u), \Phi(u))$ is an invariant submanifold. This means that the vector field (4) is tangent to $\mathcal{L}$. Moreover, the vector field (4) restricted to $\mathcal{L}$ becomes identical to the nonlinear Onsager-Casimir vector field (1). We also note that $K|_{\mathcal{L}} = 0$. All statements made in this observation can be readily verified (we verify that eq. (4.1) $|_{\mathcal{L}} \equiv$ eq. (1); eq. (4.2) $|_{\mathcal{L}} \equiv \frac{d}{dt}(\partial\Phi/\partial u)$ and eq. (4.3) $|_{\mathcal{L}} \equiv \frac{d}{dt}\Phi(u,q)$).

The observation (ii) permits us to regard the vector field (4), (5) as a lift of the nonlinear Onsager-Casimir vector field (1). We shall call it a thermodynamical lift. Observation (i) and (ii) imply that an appropriate environment for studying the nonlinear Onsager-Casimir time evolution equation (1) is the contact geometry.

4. THERMODYNAMICS

In this section, we shall explain the connection between the nonlinear Onsager-Casimir time evolution equation (1), or its lift (4), (5), and thermodynamics. First, we recall (following Refs. [13]–[15]) the geometrical formulation of the classical equilibrium thermodynamics.

Example 5.

For the sake of simplicity, we shall consider a one component system confined to the space region $\Omega \subset \mathbb{R}^3$. The volume V of Ω is kept constant. The state variables describing states of macroscopic systems in classical equilibrium thermodynamics are $(e,n) \in \mathbb{R}^2$, e is the energy and n is the number of moles. The individual features of the macroscopic system under consideration are expressed in classical equilibrium thermodynamics by a function $(e,n) \longmapsto s(e,n)$, where $s \in \mathbb{R}$ is the entropy. We shall interpret the function $s(e,n)$ geometrically as a specification of the Legendre submanifold that is the image of $(e,n) \longmapsto \left(e, n, \frac{1}{T}(e,n), -\frac{\mu}{T}(e,n), s(e,n)\right)$, where $\frac{1}{T} = \partial s/\partial e$, $\frac{\mu}{T} = -\partial s/\partial n$, in the five dimensional space with coordinates $\left(e, n, \frac{1}{T}, -\frac{\mu}{T}, s\right)$. By the symbol T we denote temperature, μ denotes the chemical potential. By applying the Legendre involution $\left(e, n, \frac{1}{T}, -\frac{\mu}{T}, s\right) \longmapsto \left(\frac{1}{T}, -\frac{\mu}{T}, e, n, -\frac{PV}{T}\right)$, where $-\frac{PV}{T} = -s + \frac{1}{T}e - \frac{\mu}{T}n$; P is the pressure, we transform the thermodynamic relation $s = s(e,n)$ into its dual (Gibbs-Duhem) form $P = P(\mu, T)$. We introduce the thermodynamic potential $\Phi\left(e, n; \frac{1}{T}, \frac{\mu}{T}\right) = -s + \frac{1}{T}e - \frac{\mu}{T}n$. Solutions of $\partial\Phi/\partial e = 0, \partial\Phi/\partial n = 0$ will be denoted by the symbol (e_{eq}, n_{eq}). We note that $-\frac{PV}{T} = \Phi\left(e_{eq}, n_{eq}, \frac{1}{T}, \frac{\mu}{T}\right)$. As an expample of a macroscopic system we mention an ideal gas. For this system $s/n = R\left(\frac{3}{2}\ln\frac{e}{e_0} + \frac{5}{2}\ln\frac{n}{n_0} + const.\right)$, where e_0, n_0 are parameters and R is a universal constant.

Motivated by the above Example, we propose the following formulation of general thermodynamics. let M be a state space (assumed to be a vector space) and $T^*M \times \mathrm{R}$

the corresponding thermodynamic space. In the previous Example $M \subset \mathbb{R}^2$. The local coordinates in $T^*M \times \mathbb{R}$ that we shall use are: $u \in M, p \in T_u^*M, z \in \mathbb{R}$. In classical equilibrium thermodynamics $u \equiv (e,n), p \equiv \left(\frac{1}{T}, -\frac{\mu}{T}\right), z \equiv s$. Individual features of macroscopic systems are expressed in general thermodynamics through the thermodynamical potential $z = \varphi(u)$, or in the geometrical language, through the specification of a Legendre submanofold $\mathcal{L}$ of $T^*M \times \mathbb{R}$. The Legendre submanifold $\mathcal{L}$ and the thermodynamic potential φ are related by representing the Legendre submanifold as the image of the immersion $u \longmapsto (u, \partial\varphi/\partial u(u), \varphi(u))$. By using the Legendre involution $(u,p,z) \longmapsto (p, u, -z+ < p, u >)$, we can transform the Legendre submanifold given by $z = \varphi(u)$ into another Legendre submanifold giver by $\tilde{z} = \tilde{\varphi}(p); \tilde{z} = -z+ < p, u > |_{u_M}$, u_M is a solution of $\partial\Phi(u,p)/\partial u = 0$, where $\Phi(u,p) = -\varphi(u)+ < p, u >$. The quantity $p = \partial\varphi/\partial u$ is called (thermodynamically) conjugate to u.

Example 6.

Elements of M are pairs of scalar fields $e(\mathbf{r})$ and $n(\mathbf{r}), \mathbf{r} \in \Omega \subset \mathbb{R}^3$ is a position vector, $e(\mathbf{r})$ and $n(\mathbf{r})$ are energy field and number of moles field respectively. This type of state description is used in hydrodynamics. Individual features of macroscopic systems are described by a function $S : M \to \mathbb{R}$. We note that if we choose $S = \frac{1}{V}\int_\Omega d\mathbf{r} s(\mathbf{r})$, where s is the same function of e and n as in Example 5, but now pointwise for all $\mathbf{r} \in \Omega$, then the resulting thermodynamics, called local thermodynamics, is the same as the one introduced in Example 5. An interesting new thermodynamics arises when S depends on $e(\mathbf{r})$ and $n(\mathbf{r})$ also nonlocally (e.g. S depends also on $\partial e/\partial \mathbf{r}$ and $\partial n/\partial \mathbf{r}$).

Example 7.

Elements of the state space M are one particle distribution functions $f : \Omega \times \mathbb{R}^3 \to R_+$; $(\mathbf{r}, \mathbf{v}) \longmapsto f(\mathbf{r}, \mathbf{v}), \mathbf{v}$ denotes velocity of one particle. Thermodynamics is specified for Example by the Boltzmann entropy.

Thermodynamics arising in Examples 6 and 7 is introduced on the levels of description that are more microscopic than the level of description used in classical equilibrium thermodynamics. The question arises as to whether and how such thermodynamics can be reduced to the classical equilibrium thermodynamics. We begin to consider this problem by analysing the following Example.

Example 8.

Let $M \equiv (M, N, \mathcal{M})$, the state space M is the same as in Example 7, the base space N is the equilibrium thermodynamic state space M_{th} introduced in Example 5 and $\mathcal{M} : M \to \mathbb{R}^2; f(\mathbf{r}, \mathbf{v}) \longmapsto (E(f), N(f))$, $E(f) = \int_\Omega d\mathbf{r} \int d\mathbf{v} \frac{1}{2} m v^2 f(\mathbf{r}, \mathbf{v})$ (m is the mass of one particle), $N(f) = \int_\Omega d\mathbf{r} \int d\mathbf{v} f(\mathbf{r}, \mathbf{v})$. Now, we introduce thermodynamic potential $\Phi(f; T, \mu) = -S(f) + \frac{1}{T}E(f) - \frac{\mu}{T}N(f)$ and look for solutions of $\delta\Phi/\delta f(\mathbf{r}, \mathbf{v}) = 0$. We shall denote these solutions (called equilibrium states) by the symbol f_{Meq}. If we insert f_{Meq} into Φ we obtain relation $P = P(\mu, T)$. By applying the Legendre involution in $T^*M_{th} \times \mathbb{R}$ we arrive at $s = s(e,n)$, where $s = S(f_{Meq}), e = E(f_{Meq}), n = N(f_{Meq})$. If the entropy is chosen to be the Boltzmann entropy introduced in Example 7 then we arrive in this way to the thermodynamic relation $s = s(e,n)$ corresponding to the ideal gas (see Example 5).

In the general setting, the problem can be formulated as follows. A Legendre manifold $\mathcal{L}_M$ in the thermodynamic space $T^*M \times \mathbb{R}$ is introduced. We look for its image in another (more macroscopic) thermodynamic space $T^*N \times \mathbb{R}$. The state space M and N are related by representing the state space M as a bundle $M \equiv (M, N, \mathcal{M})$. Elements of $T^*M \times \mathbb{R}$ are denoted in local coordinates by (u, p, z), $u \in M, p \in T_u^*M, z \in \mathbb{R}$. Let the Legendre manifold that we want to reduce be specified by $z = \varphi(u)$ (i.e. the Legendre manifold $\mathcal{L}_M$ is the image of $u \longmapsto (u, \partial\varphi/\partial u(u), \varphi(u))$. Elements of $T^*N \times \mathbb{R}$ are denoted in local coordinates by (v, q, x), $v \in N, q \in T_v^*N, x \in \mathbb{R}$. Since $M \equiv (M, N, \mathcal{M})$, we represent $u \equiv (v, w)$, $v \in N, w \in \mathcal{M}^{-1}(v)$ and correspondingly $p \equiv (q, r)$. We begin the reduction process by the Legendre involution $(v, w, q, r, z) \longmapsto (q, r, v, w, -z + < q, v > + < r, w >)$. Next, we restrict the resulting manifold by requiring $r \equiv 0$. In this way, we obtain a manifold $\mathcal{L}_m^{(N)}$ that is the image of $q \longmapsto (q, u_N(q), \Phi_N(q))$, where $\Phi(q; u) = -\varphi(u) + < q, v >; \Phi_N(q) = \Phi(q; u_N(q))$; $u_N(q)$ are solutions of $\partial\Phi(u; q)/\partial u = 0$. The manifold $\mathcal{L}_M^{(N)}$ is still imbedded in $T^*M \times \mathbb{R}$. Now, we use $\Phi_N(q)$ to construct a Legendre submanifold of $T^*N \times \mathbb{R}$ as the image of $q \longmapsto (q, \partial\Phi_N(q)/\partial q, \Phi_N(q))$. By applying the Legendre involution on this manifold, we obtain a Legendre manifold $\mathcal{L}_N$ given as the image of $v \longmapsto (v, \partial\tilde{\varphi}(v)/\partial v, \tilde{\varphi}(v))$, where $\tilde{\varphi}(v) = -\Phi_N(q_N(v)) + < q_N(v), v >$; $q_N(v)$ are solutions of $\partial(-\Phi_N(q) + < q, v >)/\partial v = 0$. In this way we have arrived from a Legendre manifold $\mathcal{L}_M$ introduced in $T^*M \times \mathbb{R}$ into a Legendre manifold $\mathcal{L}_N$ introduced in $T^*N \times \mathbb{R}$. In other words, we have transformed the thermodynamic potential $z = \varphi(v)$ into the thermodynamic potential $x = \tilde{\varphi}(v)$.

The thermodynamic potential Φ generating the nonlinear Onsager-Casimir dynamics (see Section 2) has been shown in this section to map thermodynamics introduced in $T^*N \times \mathbb{R}$ into a reduced thermodynamics introduced in $T^*N \times \mathbb{R}$. The equilibrium state (solution of $\partial\Phi/\partial u = 0$) that is instrumental in this mapping is the state that is approached as $t \to \infty$.

5. APPLICATIONS

We shall discuss two types of applications of the geometrical formulation of dynamics and thermodynamics presented above. One will be called passive and the other active.

In the passive applications we start with a given time evolution equation (e.g. the Euler, the Navier-Stokes-Fourier hydrodynamic equiations of the Boltzmann and Enskog kinetic equations) and either cast it into the form of the nonlinear Onsager-Casimir equation (1) or look for eq. (1) whose solutions approximate well solutions of the given time evolution equation. If we succeed we can then use all results about solutions of eq. (1). In particular, if $L^0 \equiv 0$, eq. (1) represents a hamiltonian system and we can use all the results known about hamiltonian systems (see e.g. Ref. [5] for an exposition of this type of application). If $L^0 \not\equiv 0$, we expect that the formulation (4), (5) will prove to be as useful for dissipative time evolution equations as eq. (1) with $L^0 \equiv 0$ proved to be useful for nondissipative time evolution equations. At this stage, we have however no result supporting this expectation. The usefulness of the formulation (4), (5) is limited at this stage to the calrification of the role of thermodynamics in dissipative dynamics.

In the active application the starting position is different. Rather than a time evolution equation the starting point is a physical specification of the macroscopic system under consideration. The objective is to formulate the corresponding time evolution equation. We proceed as follows. First, we make a hypothesis that the time evolution equation has the structure of the nonlinear Onsager-Casimir time evolution equation (1). This already guarantees that the time evolution equation, that we want to formulate, is compatible with equilibrium thermodynamics that represents a part of the experimental evidence collected about the macroscopic system under consideration. Second, we use the physical insight in our disposition to suggest the thermodynamical potential Φ

(equilibrium statistical machanics is useful in this context), the Casimir tensor L^c (the Lie group structure of the state space M and the corresponding Lie-Poisson brackets can by used for this purpose – see Refs. [1] and [5]) and the dissipative potential Ψ. Applications of this type have been developed in particular in the context of liquid crystals (see Ref. [9]).

6. REFERENCES

1. P. Libermann and C.-M. Marle, *Symplectic Geometry and Analytical Mechanics*, D. Reidel Pub. Comp. (1987).
2. L. Onsager, Phys, Rev. **37**, 405 (1931), **38**, 2265 (1931).
3. H.G. Casimir, Rev. Mod. Phys. **17**, 343 (1945).
4. J.W. Cahn, Acta Metal. **9**, 795 (1961).
5. D.D. Holm, J.E. Marsden, T. Ratiu and A. Weinstein, Phys. Rep. **123**, 1 (1985).
6. J.J. Moreau, C.R. Acad Sci. Paris **271**, 608 (1970).
7. D.G.B. Edelen, Int. J. Eng. Sci. **10**, 481 (1972).
8. N.G. Van Kampen, Physica A **67**, 1 (1973).
9. M. Grmela, Talk presented at the AMS Workshop, Boulder, July (1983), Contemp. Math. **28**, 125 (1984), Physica **D21**, 179 (1986), Phys. Letters **A130**, 81 (1988), **137**, 342 (1989), J. Phys. A Math. and Gen. **22**, 4375 (1989).
10. A.N. Kaufman, Phys Letters **A100**, 419 (1984).
11. P.J. Morrison, Phys. Letters **A100**, 423 (1984), Physica **D18**, 410 (1986).
12. V.I. Arnold, *Les mthodes mathmatiques de la mcanique classique*, Edition Mir, Moscou (1976).
13. V.I. Arnold, Lecture given at the Gibbs Symposium, Yale University, May (1989)
14. H.B. Callen, *Thermodynamics*, Wiley, New York (1963).
15. R. Hermann, *Geometry, Physics and Systems*, Marcel Dekker, New York (1973).

NORMALIZING CONNECTIONS AND THE ENERGY-MOMENTUM METHOD

D.R. LEWIS[1], J.E. MARSDEN[2*], T.S. RATIU[3**] and J.C. SIMO[4***]

[1]Institute for Mathematics and its Applications, University of Minnesota
Minneapolis, MN 55455, USA

[2]Department of Mathematics, University of California
Berkeley, CA 94720, USA

[3]Department of Mathematics, University of California
Santa Cruz, CA 95064, USA

[4]Division of Applied Mechanics, Stanford University
Stanford, CA 94305, USA

ABSTRACT. The block diagonalization method for determining the stability of relative equilibria is discussed from the point of view of connections. We construct connections whose horizontal and vertical decompositions *simultaneously* put the second variation of the augmented Hamiltonian and the symplectic structure into normal form. The cotangent bundle reduction theorem provides the setting in which the results are obtained.

Introduction In Simo, Posbergh and Marsden [1990a,b], Lewis and Simo [1990] and Simo, Lewis and Marsden [1990], a powerful method for determining the stability of relative equilibria in Hamiltonian systems with symmetry is developed. The technique provides useful tools in the emerging theory of bifurcation of relative equilibria. The main examples treated in these works are rotating systems, such as rotating elastic bodies and rigid bodies with flexible attachments. Coupled rigid bodies are treated in Patrick [1990].

The main feature of these results is a splitting of the space of variations that simultaneously puts the second variation of the augmented Hamiltonian and the symplectic form into a normal form. This splitting is defined by explicit formulas that allow one to readily implement the conditions in complex examples.

---

* Partially supported by NSF grant DMS 87-02502.

** Partially supported by NSF grant DMS 87-01318-01.

*** Partially supported by AFOSR contracts 2DJA-544 and 2DJA-803.

The purpose of this paper is not to pursue the applications of the above ***normal form method*** (or *block-diagonalizaton method*), but rather to explore its intrinsic geometry. The splittings of variables giving the normal form will be shown to be the horizontal-vertical splittings of connections associated with the cotangent bundle reduction theorem. The connections are Ehresmann connections, but as we show, they are affiliated to a large extent with principal bundle connections in the standard sense. One of the connections is the mechanical connection implicitly found in Smale [1970] and used in Guichardet [1984] and Marsden, Montgomery and Ratiu [1990] to study geometric phases (Cartan-Hannay-Berry phases). This latter reference also contains additional relevant references. The mechanical connection occurs in the bundle whose base space is the reduced space itself—in the present work a different and much less obvious connection is needed on a bundle over shape space. (It is only in the abelian case that the two bundles are coincident). These two connections interweave to produce connections on a variety of related bundles and ultimately yield the normal forms.

Another comment is important here. It is clear in advance that splittings block diagonalizing the second variation of the augmented energy must exist—one can use the second variation as a "metric" and take the orthogonal complement of shape space. However, what is remarkable is that *the splittings can be given so explicitly*, in advance of the computation of the second variation and secondly, that *they **simultaneously** bring the symplectic form into normal form*. In particular, this means that the linearized equations of motion are brought into normal form.

There are a number of important questions that warrant future investigation. Among them are the calculations of the curvatures of the connections used here and their role in stability theory. Another is the use of these splittings in optimal control. In Montgomery [1990], it is shown how optimal control strategies are governed by solutions of the equations of motion of a particle in a Yang-Mills field—the field being the mechanical connection. However, this is done in a context (such as zero total angular momenum) where parts of the splitting collapse and the full power of block diagonalization cannot be utilized. In fact, block diagonalization was designed to deal with situations in which the total angular momentum is not zero, such as rotating gravitating bodies, rotating satellites and rotating molecules. We believe that these two lines of investigation can be brought into a closer union.

Acknowledgements We thank A. Bloch, P.S. Krishnaprasad, R. Montgomery, and A. Weinstein for numerous stimulating conversations.

§1 *The Energy-Momentum Method*

This section provides a concise summary of the energy-momentum method following the references cited in the introduction. The energy-momentum method itself first appeared in Marsden, Lewis, Posbergh and Simo [1989] and Simo, Posbergh and Marsden [1990a]. Its development was motivated by the energy-Casimir method (see Holm, Marsden, Ratiu and Weinstein [1985] and references therein), which "fails" due to the lack of an adequate number of Casimir invariants in some key examples in elasticity and fluid dynamics, such as the theory of rods. The energy momentum technique can be traced back to Riemann and Poincaré in specific examples and may be regarded as implicit in Arnold [1966] and Marsden and Weinstein [1974].

Let (P, Ω) be a symplectic manifold, G a Lie group acting by symplectic transformations and let $\mathfrak{g}$ denote its Lie algebra and $\mathfrak{g}^*$ the vector space dual of $\mathfrak{g}$. Let $\mathbf{J} : P \to \mathfrak{g}^*$ be an equivariant momentum map for the action and let $H : P \to R$ be a G-invariant Hamiltonian. In later sections, we specialize to the case of cotangent bundles, but for now we work with general symplectic manifolds. Let X_H be the Hamiltonian vector field of H and let F_t^H be its flow. Recall that by definition of the momentum map, $X_{\langle \mathbf{J},\xi\rangle} = \xi_P$ is the infinitesimal action of $\mathfrak{g}$ on P; equivalently, $F_t^{\langle \mathbf{J},\xi\rangle}(z) = \exp(t\xi)z$ for $z \in P$, where gz denotes the action of $g \in G$ on z. Here and below, $\langle\, ,\rangle$ denotes the natural pairing.

A ***relative equilibrium*** is a point $z_e \in G$ such that $F_t^H(z_e) = \exp(t\xi_e)z_e$ for some $\xi_e \in \mathfrak{g}$; in other words, *a relative equilibrium is a phase space point whose dynamic orbit is also a one-parameter group orbit.* Conservation and equivariance of $\mathbf{J}$ imply that $\xi_e \in \mathfrak{g}_{\mu_e}$ where $\mu_e = \mathbf{J}(z_e)$ is the value of $\mathbf{J}$ at the relative equilibrium z_e and $\mathfrak{g}_{\mu_e}$ is the coadjoint isotropy Lie algebra of μ_e.

There are a number of equivalent characterizations of relative equilibria, but the one most relevant at the moment is as follows.

1.1 Relative Equilibrium Theorem *A point $z_e \in P$ is a relative equilibrium iff there is a $\xi_e \in \mathfrak{g}$ such that z_e is a critical point of the* ***augmented Hamiltonian***

$$H_{\xi_e} := H - \langle \mathbf{J} - \mu_e, \xi_e\rangle \tag{1}$$

This theorem follows from differentiation of the defining condition $F_t^H(z_e) = \exp(t\xi_e)z_e$ with respect to t at $t = 0$.

If μ_e is a regular value of $\mathbf{J}$ (*i.e.,* if the isotropy algebra $\mathfrak{g}_{z_e}$ of z_e is $\{0\}$), then z_e is a relative equilibrium iff $H \mid \mathbf{J}^{-1}(\mu_e)$ has a critical point at z_e. In fact, **1.1** may be viewed as a version of the Lagrange multiplier theorem, where ξ_e is the multiplier.

For theorem **1.2** *and the further developments below, we assume that* μ_e *is a regular value of* $\mathbf{J}$ *and that the action of* G *on* P *is free and proper. We also assume that either* μ_e *is a generic point in* $\mathfrak{g}^*$ *or that the adjoint action of* G_{μ_e} *on* $\mathfrak{g}$ *admits an invariant inner product.*

We refer to Patrick [1990] for more details and examples concerning these assumptions. An example showing that μ_e being a regular value and the action of G on P being free and proper is not enough, is discussed in Krishnaprasad [1989]. Of considerable interest in the theory of bifurcation with symmetry is to investigate what happens when these assumptions fail, especially the assumption that μ_e is not regular; *i.e.*, z_e has nontrivial isotropy (*i.e.*, symmetry) algebra.

The ***energy-momentum method*** proceeds as follows. Choose a linear subspace $\mathcal{S} \subset T_{z_e}\mathbf{J}^{-1}(\mu_e) = \ker T_{z_e}\mathbf{J}(z_e)$ that is transverse to the G_{μ_e}-orbit of z_e, where G_{μ_e} is the coadjoint isotropy subgroup of μ_e. Note that $\mathcal{S}$ is isomorphic to the tangent space to the reduced manifold $P_{\mu_e} = \mathbf{J}^{-1}(\mu_e)/G_{\mu_e}$ (since the G_{μ_e} action is assumed to be free and proper, the quotient is a manifold).

1.2 Theorem *If* $\delta^2 H_{\xi_e}(z_e)$, *the second variation of* H_{ξ_e} *evaluated at* z_e, *is positive or negative definite when restricted to* $\mathcal{S}$, *then* z_e *is* G_{μ_e}*-orbitally stable. That is,* z_e *is dynamically stable mod the* G_{μ_e}*-action on* P.

We note that $\delta^2 H_{\xi_e}(z_e) \,|\, \mathcal{S}$ projects to the second variation of the reduced Hamiltonian on P_{μ_e} at $\pi_{\mu_e}(z_e)$, the projection of z_e; definiteness of this reduced second variation is a well-known sufficient condition for stability.

Now we turn our attention to the case of cotangent bundles. Thus, assume $P = T^*Q$ with its canonical symplectic structure, G acts by cotangent lift and $\mathbf{J}$ is the standard momentum map for this case. Assume that Q carries a Riemannian metric, denoted $\langle\langle\, , \rangle\rangle$ and that G acts on Q by isometries. Finally, assume H is of the form $K + V$ where K is the kinetic energy of the metric $\langle\langle\, , \rangle\rangle$ and $V : Q \to \mathbb{R}$ is a given G-invariant potential. We choose this context to be explicit but we note that much of what is done here generalizes to a large class of G-invariant Lagrangians $L : TQ \to \mathbb{R}$ (see Lewis [1990]).

Define, for each $q \in Q$, the ***locked inertia tensor*** $\mathbb{I}(q) : \mathfrak{g} \to \mathfrak{g}^*$ by

$$\langle \mathbb{I}(q)\,\xi, \eta \rangle = \langle\langle \xi_Q(q), \eta_Q(q) \rangle\rangle , \tag{2}$$

where ξ_Q denotes the infinitesimal generator of $\xi \in \mathfrak{g}$. Note that $\mathbb{I}(q)$ is symmetric relative to the natural pairing $\langle \, , \rangle$. Since we have a free action, $\mathbb{I}(q)$ is invertible. Define

$$\alpha : TQ \to \mathfrak{g} \quad \text{by} \quad \alpha(v_q) = \mathbb{I}(q)^{-1}\mathbf{J}(\mathbb{F}L(v_q)) \tag{3}$$

where $\mathbb{F}L(v_q)$ is the momentum associated to the velocity v_q via the Legendre transform. Thus, α gives the "angular velocity of the locked system". We will use the terms "angular velocity" and "angular momentum" for elements of $\mathfrak{g}$ and $\mathfrak{g}^*$, even though G need not be $\mathbf{SO}(3)$.

1.3 Proposition *The locked angular velocity* α *defines a connection on the principal* G-*bundle* $Q \to S = Q/G$, *if the* G *action is free and proper. We call* α *the* ***mechanical connection.***

Proof The vertical subspace of the G-bundle $Q \to S$ at a point q is the tangent space to the group orbit, $G\cdot q$, so the vertical vectors are the infinitesimal generators $\xi_Q(q)$. The first step in verifying that α is a connection is to check that $\alpha(\xi_Q(q)) = \xi$, which can be seen as follows. Let $p_q \in T_q^*Q$. From $\langle \mathbf{J}(p_q), \xi\rangle = \langle p_q, \xi_Q(q)\rangle$, we obtain

$$\begin{aligned}\langle \nu, \alpha(\xi_Q(q))\rangle &= \langle \nu, \mathbb{I}(q)^{-1}\mathbf{J}(\mathbb{F}L(\xi_Q(q)))\rangle = \langle \mathbb{I}(q)^{-1}\nu, \mathbf{J}(\mathbb{F}L(\xi_Q(q))\rangle \\ &= \langle \mathbb{F}L(\xi_Q(q)), (\mathbb{I}(q)^{-1}\nu)_Q(q)\rangle = \langle \nu, \xi\rangle,\end{aligned}$$

for any $\nu \in \mathfrak{g}^*$, and so

$$\alpha(\xi_Q(q)) = \xi.$$

The second step is to check that α *is equivariant* in the sense that $\alpha(T_q\Phi_h \cdot v_q) = \mathrm{Ad}_h\alpha(v_q)$, where $\Phi_h(q) = h\cdot q$ denotes the left action of h on Q. To show this, we make use of G-invariance of the metric, equivariance of the momentum map $\mathbf{J}$, and the following equivariance property of the locked inertia tensor

$$\mathrm{Ad}_h^*(\mathbb{I}(h\cdot q)\mathrm{Ad}_h\zeta) = \mathbb{I}(q)\zeta. \tag{4}$$

To prove (4), let $h \in G$, $q \in Q$, and $\eta, \zeta \in \mathfrak{g}$. Using G-invariance of the metric,

$$\begin{aligned}\langle \mathrm{Ad}_h^*(\mathbb{I}(h\cdot q)\mathrm{Ad}_h\zeta), \eta\rangle &= \langle \mathbb{I}(h\cdot q)\mathrm{Ad}_h\zeta, \mathrm{Ad}_h\eta\rangle = \langle\langle (\mathrm{Ad}_h\eta)_Q(h\cdot q), (\mathrm{Ad}_h\zeta)_Q(h\cdot q)\rangle\rangle_{h\cdot q} \\ &= \langle\langle T_q\Phi_h\cdot\eta_Q(q), T_q\Phi_h\cdot\zeta_Q(q)\rangle\rangle_{h\cdot q} = \langle\langle \eta_Q(q), \zeta_Q(q)\rangle\rangle_q = \langle \mathbb{I}(q)\zeta, \nu\rangle .\end{aligned}$$

Inverting (4), we obtain

$$\mathrm{Ad}_{h^{-1}} \circ \mathbb{I}(h\cdot q)^{-1} \circ \mathrm{Ad}^*_{h^{-1}} = \mathbb{I}(q)^{-1}.$$

Therefore,

$$\alpha(T_q\Phi_h\cdot v_q) = (\mathbb{I}(h\cdot q)^{-1} \circ \mathbf{J} \circ \mathbb{F}L)(T_q\Phi_h\cdot v_q) = \mathbb{I}(h\cdot q)^{-1}[\mathbf{J}(T_q^*\Phi_{h^{-1}}(\mathbb{F}L(v_q)))]$$

$$= \mathbb{I}(h\cdot q)^{-1}[\mathrm{Ad}^*_{h^{-1}}(\mathbf{J}(\mathbb{F}L(v_q))] = \mathrm{Ad}_h\cdot \mathbb{I}(q)^{-1}[\mathbf{J}(\mathbb{F}L(v_q))] = \mathrm{Ad}_h(\alpha(v_q)).$$

Thus α is G-equivariant and hence is a connection. ■

Another useful observation is that α *is the connection defined by choosing the horizontal spaces to be* $\langle\langle\ ,\ \rangle\rangle$*-orthogonal to the* G*-orbits*; *i.e.*, the spaces of vectors with zero "angular" momentum. Defining $\alpha_\mu : Q \to T^*Q$ by

$$\langle \alpha_\mu(q), v_q\rangle = \langle \mu, \alpha(v_q)\rangle, \tag{5a}$$

we observe that α_μ takes values in $\mathbf{J}^{-1}(\mu)$, a fact valid for any connection; indeed,

$$\langle \mathbf{J}(\alpha_\mu(q)), \eta\rangle = \langle \alpha_\mu(q), \eta_Q(q))\rangle = \langle \mu, \alpha(\eta_Q(q))\rangle = \langle \mu, \eta\rangle,$$

which follows from the proof of Proposition **1.3**.

Another useful observation is that if $\zeta = \mathbb{I}(q)^{-1}\mu$, then

$$\alpha_\mu(q) = \mathbb{F}L(\zeta_Q(q)), \tag{5b}$$

as is readily verified.

The ***amended potential*** V_μ is defined for each $\mu \in \mathfrak{g}^*$ to be

$$V_\mu(q) = H(\alpha_\mu(q)), \tag{6a}$$

as in Smale [1970] and Abraham and Marsden [1978]. One finds from the definitions that

$$V_\mu(q) = V(q) + \tfrac{1}{2}\langle \mu, \mathbb{I}(q)^{-1}\mu\rangle\ . \tag{6b}$$

Here is the verification of (6b); we start with (6a) and compute with the aid of (5b) as follows:

$$H(\alpha_\mu(q)) = \tfrac{1}{2}\langle\langle \alpha_\mu(q), \alpha_\mu(q)\rangle\rangle + V(q) = \tfrac{1}{2}\langle\langle \mathbb{F}L(\zeta_Q(q)), \mathbb{F}L(\zeta_Q(q))\rangle\rangle + V(q)$$

$$= \frac{1}{2}\langle \mathbb{I}(q)\zeta, \zeta\rangle + V(q) = \frac{1}{2}\langle \mu, \mathbb{I}(q)^{-1}\mu\rangle + V(q) = V_\mu(q).$$

For $p_q \in T_q^*Q$, define

$$K_\mu(p_q) = \frac{1}{2}\| p_q - \alpha_\mu(q) \|^2 \tag{6c}$$

and let the functions V_ξ and K_ξ be defined by

$$V_\xi(q) = V(q) - \frac{1}{2}\langle \xi, \mathbb{I}(q)\xi\rangle \tag{7a}$$

and

$$K_\xi(p_q) = \frac{1}{2}\| p_q - \mathbb{F}L\cdot\xi_Q(q) \|^2. \tag{7b}$$

Fix $\xi \in \mathfrak{g}$ and $\mu \in \mathfrak{g}^*$ and let, as in equation (1),

$$H_{\xi,\mu} = H - \langle \mathbf{J} - \mu, \xi\rangle. \tag{8}$$

Then one has the following easily verified identities:

$$H_{\xi,\mu} = K_\xi + V_\xi + \langle \mu, \xi\rangle \tag{9}$$

and

$$K_\mu + V_\mu = H - \langle \mathbf{J} - \mu, \xi\rangle, \tag{10}$$

where

$$\xi(q) = \mathbb{I}(q)^{-1}\mu .$$

To assist in what follows, one can first verify that at a relative equilibrium corresponding to ξ_e (see **1.1**), one has the identities

$$\xi_e = \mathbb{I}(q_e)^{-1}\mu_e \tag{11}$$

and (see (5b) above)

$$p_e = \alpha_{\mu_e}(q_e) = \mathbb{F}L((\xi_e)_Q(q_e)) \tag{12}$$

1.4 Reduced Relative Equilibrium Theorem *The following are equivalent for a point* $z_e \in T^*_{q_e}Q$:

- **i** z_e *is a relative equilibrium*
- **ii** *there is a* $\xi_e \in \mathfrak{g}$ *such that* $z_e = \mathbb{F}L((\xi_e)_Q(q_e))$ *and* q_e *is a critical point of* V_{ξ_e}
- **iii** $z_e = \alpha_{\mu_e}(q_e)$ *and* q_e *is a critical point of* V_{μ_e}, *where* $\mu_e = \mathbf{J}(z_e)$
- **iv** z_e *is a critical point of* $H \mid \mathbf{J}^{-1}(\mu_e)$.

Proof We already noted the equivalence **i** ⇔ **iv** in Theorem **1.1**. The equivalence **i** ⇔ **ii** follows from Theorem **1.1** and equation (9), and the equivalence **iii** ⇔ **iv** follows from equation (10). ■

In condition **iii**, the ξ_e in the definition of relative equilibrium is given by equation (11). We also observe that $K_\mu + V_\mu$ has the correct form of the reduced hamiltonian on P_μ. This is consistent with the cotangent bundle reduction theorem described in the next section, and with the fact that it is $\delta^2 V_\mu$ on the appropriate space that gives the reduced energy-momentum method.

§2 *The Cotangent Bundle Reduction Theorem*

The cotangent bundle reduction theorem of Satzer, Marsden and Kummer gives a realization of the reduced space $P_\mu = J^{-1}(\mu)/G$ in case $P = T^*Q$. The following diagram summarizes the situation:

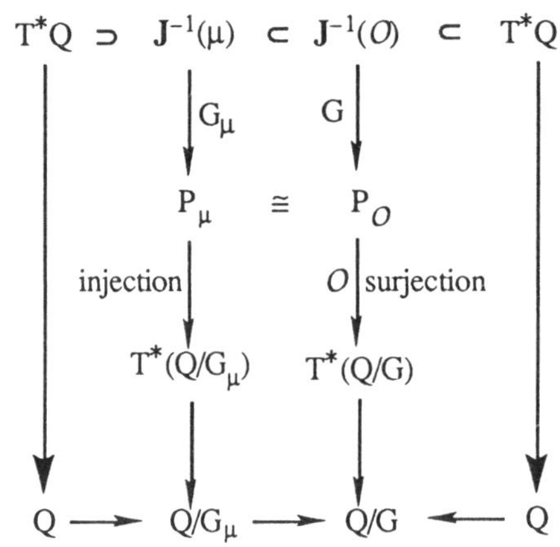

One version of the theorem (see Abraham and Marsden [1978]) says that P_μ *embeds* as a vector subbundle of $T^*(Q/G_\mu)$—this is the injection in the above figure. Another version (see Marle [1976], Kazhdan, Kostant and Sternberg [1978] and Marsden [1981]) says that $P_\mu \cong P_{\mathcal{O}}$ is a *coadjoint bundle* over $T^*(Q/G)$ with fiber the coadjoint orbit $\mathcal{O}$ through μ. Both are proved by a similar technique. We state it as follows:

2.1 Cotangent Bundle Reduction *P_μ is a bundle over $T^*(Q/G)$ with fiber O.*

Step 1 Reduction at zero

$$(T^*Q)_0 \cong T^*(Q/G). \tag{1}$$

Here $(T^*Q)_0 = \mathbf{J}^{-1}(0)/G$, since $G_0 = G$. We can identify this quotient with $T^*(Q/G)$ by observing that $\beta_q \in \mathbf{J}^{-1}(0)$ if and only if $\beta_q \in [\mathfrak{g}\cdot q]^A$, *i.e.*, $\langle \beta_q, \eta_Q(q)\rangle = 0$ for all $\eta \in \mathfrak{g}$; hence the pairing of β_q with vectors modulo generators $\eta_Q(q)$ is well-defined and we can regard β_q as an element of $T^*_{[q]}(Q/G)$. The symplectic form on $T^*(Q/G)$ is the canonical one. (See Abraham and Marsden [1978].)

Step 2 Orbit reduction *The reduced space $(T^*Q)_\mu = \mathbf{J}^{-1}(\mu)/G_\mu$ can be identified with the quotient $\mathbf{J}^{-1}(O)/G$, where O is the coadjoint orbit through μ.* (See Marsden [1981], Marsden, Weinstein, Ratiu, Schmidt, and Spencer [1983] and references therein.)

Step 3 Shifting *Use the shift map Σ defined by $\Sigma(z) = z - \alpha_{J(z)}(q)$ (where $z \in T^*_qQ$) to map $\mathbf{J}^{-1}(O)$ to $\mathbf{J}^{-1}(0)$.*

Letting $\alpha_z \in T^*_qQ$ be defined by $\alpha_z = \alpha_{J(z)}(q)$, we can write $\Sigma(z) = z - \alpha_z$. We claim Σ is equivariant with respect to the G-action. To see this, let $h \in G$ and note that

$$\begin{aligned}\langle T^*_q\Phi_h \cdot \alpha_{h\cdot z}, v_q\rangle &= \langle \alpha_{h\cdot z}, T_q\Phi_h \cdot v_q\rangle = \langle \mathbf{J}(h\cdot z), \alpha(T_q\Phi_h \cdot v_q)\rangle \\ &= \langle \mathrm{Ad}^*_{h^{-1}}\mathbf{J}(z), \mathrm{Ad}_h\alpha(v_q)\rangle = \langle \alpha_z, v_q\rangle .\end{aligned}$$

Thus, $T^*_q\Phi_h \cdot \alpha_{h\cdot z} = \alpha_z$ and so Σ is equivariant and hence drops to the quotient, producing the desired map

$$(T^*Q)_\mu = \mathbf{J}^{-1}(O)/G \to \mathbf{J}^{-1}(0)/G = T^*(Q/G). \tag{2}$$

This map has fiber O, *i.e.*, $\Sigma \circ \alpha_\mu = 0$ for all $\mu \in O$, so our assertion is proved. ■

Below, we show how the canonical symplectic structure restricted to the subspace $\mathcal{S}$ can be identified with the Poisson structure on the bundle $P_{\mu_e} \to Q/G$. The calculation of the Poisson structure for general reductions of principal bundles is given in Marsden, Montgomery, and Ratiu

[1984]. (See also Lewis, Marsden, and Ratiu [1987] for an application to the dynamics of systems with free boundaries.)

§3 *The Rigid-Internal Splitting*

We now describe the splitting of the space of variations in the energy-momentum method. First, fix a point $z = p_q \in T_q^*Q$ and fix $\mu = \mathbf{J}(p_q)$. Define

$$\mathcal{V} = \{\delta q \in T_qQ \mid \langle\langle \delta q, \zeta_Q(q)\rangle\rangle = 0 \text{ for all } \zeta \in \mathfrak{g}_\mu\} \tag{1a}$$

and let

$$\mathcal{S} = \{\delta z = (\delta q, \delta p) \in T_z(T^*Q) \mid T_z\mathbf{J}(z)\cdot\delta z = 0 \text{ and } \delta q \in \mathcal{V}\} \tag{1b}$$

This is the choice of $\mathcal{S}$ taken in the energy-momentum method—other choices are also possible, but we choose (1b) in this paper for definiteness.

Next, we split $\mathcal{V} = \mathcal{V}_{\mathrm{RIG}} \oplus \mathcal{V}_{\mathrm{INT}}$ in the following way. Define

$$\mathcal{V}_{\mathrm{RIG}} = \{\eta_Q(q) \in T_qQ \mid \eta \in \mathfrak{g}_\mu^\perp\} \tag{2}$$

where $\mathfrak{g}_\mu^\perp$ is the orthogonal complement to $\mathfrak{g}_\mu$ in $\mathfrak{g}$ with the locked inertia metric. (This choice of orthogonal complement depends on q, but we do not include this in the notation). From (1a) it is clear that $\mathcal{V}_{\mathrm{RIG}} \subset \mathcal{V}$ and that $\mathcal{V}_{\mathrm{RIG}}$ has the dimension of the coadjoint orbit through μ. Next, define

$$\mathcal{V}_{\mathrm{INT}} = \{\delta q \in \mathcal{V} \mid \langle \eta, [\mathbf{D}\mathbb{I}(q)\cdot\delta q]\cdot\xi\rangle = 0 \text{ for all } \eta \in \mathfrak{g}_\mu^\perp\} \tag{3a}$$

where $\xi = \mathbb{I}(q)^{-1}\mu$. An equivalent definition is

$$\mathcal{V}_{\mathrm{INT}} = \{\delta q \in \mathcal{V} \mid [\mathbf{D}\mathbb{I}(q)^{-1}\cdot\delta q]\cdot\mu \in \mathfrak{g}_\mu\}, \tag{3b}$$

which is clear from (3a). The definition of $\mathcal{V}_{\mathrm{INT}}$ has an important mechanical interpretation in terms of the objectivity of the centrifugal force in case $G = \mathbf{SO}(3)$; see Simo, Lewis and Marsden [1990] and Simo, Posbergh and Marsden [1990a,b].

Define the ***Arnold form*** $\mathcal{A}_\mu : \mathfrak{g}_\mu^\perp \times \mathfrak{g}_\mu^\perp \to \mathbb{R}$ by

$$\mathcal{A}_\mu(\eta, \zeta) = \langle \mathrm{ad}_\eta^*\mu, \chi_{(q,\mu)}(\zeta)\rangle = \langle \mu, \mathrm{ad}_\eta\chi_{(q,\mu)}(\zeta)\rangle, \tag{4a}$$

where $\chi_{(q,\mu)} : \mathfrak{g}_\mu^\perp \to \mathfrak{g}$ is defined by

$$\chi_{(q,\mu)}(\zeta) = \mathbb{I}(q)^{-1}\mathrm{ad}^*_\zeta\mu + \mathrm{ad}_\zeta\mathbb{I}(q)^{-1}\mu \tag{4b}$$

The Arnold form appears in Arnold's [1966] stability analysis of relative equilibria in the special case $Q = G$. At a relative equilibrium, the form $\mathcal{A}_\mu$ is symmetric, as is verified either directly or by recognizing it as the second variation of V_ξ on $\mathcal{V}_{\mathrm{RIG}} \times \mathcal{V}_{\mathrm{RIG}}$ (see Simo, Lewis and Marsden [1990] for details). At a relative equilibrium, the form $\mathcal{A}_\mu$ is degenerate as a symmetric bilinear form on $\mathfrak{g}_\mu^\perp$ when there is a non-zero $\zeta \in \mathfrak{g}_\mu^\perp$ such that

$$\mathbb{I}(q)^{-1}\mathrm{ad}^*_\zeta\mu + \mathrm{ad}_\zeta\mathbb{I}(q)^{-1}\mu \in \mathfrak{g}_\mu\,; \tag{4c}$$

in other words, when $\mathbb{I}(q)^{-1} : \mathfrak{g}^* \to \mathfrak{g}$ has a *nontrivial symmetry* relative to the coadjoint-adjoint action of $\mathfrak{g}$ (restricted to $\mathfrak{g}_\mu^\perp$) on the space of linear maps from $\mathfrak{g}^*$ to $\mathfrak{g}$. (When one is not at a relative equilibrium, we say the Arnold form is *non-degenerate* when $\mathcal{A}_\mu(\eta, \zeta) = 0$ for all $\eta \in \mathfrak{g}_\mu^\perp$ implies $\zeta = 0$.) This means, for $G = \mathbf{SO}(3)$ that $\mathcal{A}_\mu$ is non-degenerate if μ is not in a multidimensional eigenspace of $\mathbb{I}^{-1}$. Thus, *if the locked body is not symmetric (i.e., a Lagrange top), then the Arnold form is non-degenerate.*

3.1 Proposition *If the Arnold form is non-degenerate, then*

$$\mathcal{V} = \mathcal{V}_{\mathrm{RIG}} \oplus \mathcal{V}_{\mathrm{INT}} \tag{5}$$

Indeed, non-degeneracy of the Arnold form implies $\mathcal{V}_{\mathrm{RIG}} \cap \mathcal{V}_{\mathrm{INT}} = \{0\}$ and, at least in the finite dimensional case, a dimension count gives **3.1**. In the infinite dimensional case, the relevant ellipticity conditions are neded.

The split (5) can now be used to induce a split of the phase space

$$\mathcal{S} = \mathcal{S}_{\mathrm{RIG}} \oplus \mathcal{S}_{\mathrm{INT}}\,. \tag{6}$$

Using a more mechanical viewpoint, Simo, Lewis and Marsden [1990] show how $\mathcal{S}_{\mathrm{RIG}}$ can be defined by extending $\mathcal{V}_{\mathrm{RIG}}$ from positions to momenta using *superposed rigid motions*. For our purposes, the important characterization of $\mathcal{S}_{\mathrm{RIG}}$ is

$$\mathcal{S}_{\mathrm{RIG}} = T_q\alpha_\mu \cdot \mathcal{V}_{\mathrm{RIG}} \tag{7}$$

so $\mathcal{S}_{\mathrm{RIG}}$ is isomorphic to $\mathcal{V}_{\mathrm{RIG}}$. Since α_μ maps Q to $\mathbf{J}^{-1}(\mu)$ and $\mathcal{V}_{\mathrm{RIG}} \subset \mathcal{V}$, we get $\mathcal{S}_{\mathrm{RIG}} \subset \mathcal{S}$. Define

$$\mathcal{S}_{\rm INT} = \{\delta z \in \mathcal{S} \mid \delta q \in \mathcal{V}_{\rm INT}\}; \tag{8}$$

(6) holds if the Arnold form is non-degenerate. Next, we write

$$\mathcal{S}_{\rm INT} = \mathcal{W}_{\rm INT} \oplus \mathcal{W}^*_{\rm INT}, \tag{9}$$

where $\mathcal{W}_{\rm INT}$ and $\mathcal{W}^*_{\rm INT}$ are defined as follows:

$$\mathcal{W}_{\rm INT} = T_q\alpha_\mu \cdot \mathcal{V}_{\rm INT} \tag{10a}$$

and

$$\mathcal{W}^*_{\rm INT} = \{\upsilon(\gamma) \mid \gamma \in [\mathfrak{g}\cdot q]^A\} \tag{10b}$$

where $\mathfrak{g}\cdot q = \{\zeta_Q(q) \mid \zeta \in \mathfrak{g}\}$, $[\mathfrak{g}\cdot q]^A \subset T^*_qQ$ is its annihilator, and $\upsilon(\gamma) \in T_z(T^*Q)$ is the vertical lift of $\gamma \in T^*_qQ$; in coordinates, $\upsilon(q^i, \gamma_j) = (q^i, p_j, 0, \gamma_j)$. The vertical lift is given intrinsically by taking the tangent to the curve $\sigma(s) = z + s\gamma$ at $s = 0$.

§4 *The Normal Form*

4.1 Block Diagonalization Theorem *In the splittings introduced in* §3 *and at a relative equilibrium,* $\delta^2H_\xi(z_e)$ *and the symplectic form* Ω_{z_e} *have the form:*

$$\begin{array}{cc} & \begin{array}{ccc} \mathcal{S}_{\rm RIG} & \mathcal{W}_{\rm INT} & \mathcal{W}^*_{\rm INT} \end{array} \\ \delta^2H_\xi(z_e) = & \begin{bmatrix} \left[\begin{array}{c}\text{Arnold}\\ \text{form}\end{array}\right] & 0 & 0 \\ 0 & \delta^2V_\mu & 0 \\ 0 & 0 & \delta^2K_\mu \end{bmatrix} \end{array}$$

and

$$
\begin{array}{c} \\ \Omega_{z_e} = \end{array}
\begin{array}{c}
\quad \mathcal{S}_{\mathrm{RIG}} \qquad\qquad \mathcal{W}_{\mathrm{INT}} \qquad\qquad \mathcal{W}^*_{\mathrm{INT}} \\
\left[\begin{array}{ccc}
\begin{bmatrix}\text{Coadjoint orbit}\\ \text{symplectic form}\end{bmatrix} & \begin{bmatrix}\text{Internal-Rigid}\\ \text{coupling}\end{bmatrix} & 0 \\
-\begin{bmatrix}\text{Internal-Rigid}\\ \text{coupling}\end{bmatrix} & & \\
 & \begin{bmatrix}\text{Canonical symplectic}\\ \text{form plus a}\\ \text{magnetic term}\end{bmatrix} & \\
0 & &
\end{array}\right]
\end{array}
$$

Here and below, we write $\xi = \xi_e$ and $\mu = \mu_e$ for simplicity. The terms appearing in the formula for Ω_{z_e} will be explained below. We illustrate some steps in the proof of the theorem as follows.

4.2 Lemma $\delta^2 H_\xi(z_e)\cdot(\Delta z, \delta z) \;=\; 0$ *for all* $\Delta z \in \mathcal{S}_{\mathrm{RIG}}$ *and* $\delta z \in \mathcal{S}_{\mathrm{INT}}$.

To prove this, recall that H_ξ can be written (see equation (10) of §**1**):

$$H_\xi(z) \;=\; V(q) + \frac{1}{2}\mathbf{J}(z)\cdot\mathbb{I}^{-1}(q)\mathbf{J}(z) - [\mathbf{J}(z) - \mu]\cdot\xi + K(\Sigma(z)). \tag{1}$$

Given $\mu \in \mathfrak{g}^*$, define the functions $\rho : T^*Q \to \mathfrak{g}$ and $\mathcal{Q}: T^*Q \to \mathbb{R}$ by

$$\rho(z) = \frac{1}{2}\mathbb{I}^{-1}(q)(\mathbf{J}(z) + \mu) - \xi \tag{2}$$

and

$$\mathcal{Q}(z) = \langle \mathbf{J}(z) - \mu, \rho(z)\rangle\, . \tag{3}$$

We can now rewrite H_ξ in the form

$$H_\xi(z) \;=\; K(\Sigma(z)) + \mathcal{Q}(z) + V_\mu(q). \tag{4}$$

This regrouping of H_ξ is convenient because, as we saw before, at a critical point $z_e \in \mathbf{J}^{-1}(\mu)$, each of the three terms has zero first variation. We shall show that the second variations of these terms restricted to variations of the form $(\Delta z, \delta z) \in \mathcal{S}_{\mathrm{RIG}} \times \mathcal{S}_{\mathrm{INT}}$ equal zero. The first variation of the term $K(\Sigma(z)) = \frac{1}{2}\| p - \alpha_z \|^2$ equals zero at a relative equilibrium, since $p_e = \alpha_{z_e}$. The first variation of the term $\mathcal{Q}(z)$ is

$$\mathbf{D}\mathcal{Q}(z_e)\cdot\delta z = \langle \mathbf{DJ}(z_e)\cdot\delta z, \rho(z_e)\rangle + \langle \mathbf{J}(z_e) - \mu, \mathbf{D}\rho(z_e)\cdot\delta z\rangle = 0,$$

since $\mathbf{J}(z_e) = \mu = \mathbb{I}(q_e)\xi$ implies $\rho(z_e) = 0$. Thus, $\mathbf{D}H_\xi(z_e) = 0$ if and only if $\mathbf{D}V_\mu(q_e) = 0$.

By construction, $T_{z_e}\Sigma\cdot\Delta z = (\Delta q, 0)$ for $\Delta z \in \mathcal{S}_{RIG}$; hence

$$\mathbf{D}^2(K\circ\Sigma)(z_e)(\Delta z, \delta z) = \langle T_{z_e}\Sigma\cdot\Delta z, T_{z_e}\Sigma\cdot\delta z\rangle = 0$$

for $\Delta z \in \mathcal{S}_{RIG}$ and arbitrary δz. Next, we show that $\mathbf{D}^2\mathcal{Q}(z_e)\,|_{\mathcal{S}\times\mathcal{S}} = 0$. Let $\delta z, \widetilde{\delta z} \in \mathcal{S}$; then

$$\begin{aligned}\mathbf{D}^2\mathcal{Q}(z_e)(\delta z, \widetilde{\delta z}) &= \langle \mathbf{J}(z_e) - \mu, \mathbf{D}^2\rho(z_e)(\delta z, \widetilde{\delta z})\rangle + \langle \mathbf{DJ}(z_e)\cdot\delta z, \mathbf{D}\rho(z_e)\cdot\widetilde{\delta z}\rangle \\ &\quad + \langle \mathbf{DJ}(z_e)\cdot\widetilde{\delta z}, \mathbf{D}\rho(z_e)\cdot\delta z)\rangle + \langle \mathbf{D}^2\mathbf{J}(z_e)(\delta z, \widetilde{\delta z}), \rho(z_e)\rangle = 0,\end{aligned}$$

since $\mathbf{J}(z_e) = \mu$, $\mathbf{DJ}(z_e)\,|_{\mathcal{S}} = 0$, and $\rho(z_e) = 0$.

Finally, we show that $\mathbf{D}^2V_\mu(q_e)(\Delta q, \delta q) = 0$ for $\Delta q \in \mathcal{V}_{RIG}$ and $\delta q \in \mathcal{V}_{INT}$. Let $\Delta q = \eta_Q(q_e)$ and $\zeta = (\mathbf{D}\mathbb{I}(q_e)^{-1}\cdot\delta q)\mu$. Since $\mathbf{D}\mathcal{V}_\mu(q_e) = 0$, we have

$$\begin{aligned}\mathbf{D}^2\mathcal{V}_\mu(q_e)(\Delta q, \delta q) &= \mathbf{D}(\mathbf{D}\mathcal{V}_\mu(q_e)\cdot\Delta q)\cdot\delta q = \mathbf{D}(\mathrm{ad}^*_\eta\mu\cdot \mathbb{I}(q_e)^{-1}\mu)\cdot\delta q \\ &= (\mathrm{ad}^*_\eta\mu\cdot(\mathbf{D}\mathbb{I}(q_e)^{-1}\cdot\delta q)\mu = \mathrm{ad}^*_\eta\mu\cdot\zeta = 0,\end{aligned}$$

since $\mathrm{ad}^*_\eta\mu \in (\mathfrak{g}_{\mu_e})^A$ and $\delta q \in \mathcal{V}_{INT}$ implies $\zeta \in \mathfrak{g}_\mu$. ■

We refer the reader to Simo, Lewis and Marsden [1990] for the calculation of $\delta^2 H_\xi$ on the remaining blocks.

Now we turn to the symplectic form $\Omega_{z_e} = \Omega(z_e)$. In Simo, Lewis, and Marsden [1990] it is shown that for $\Delta q \in \mathcal{V}_{RIG}$, $\Delta p = T_{q_e}\alpha_\mu\cdot\Delta q$ is given by $\Delta p = v(\mathbb{F}L\cdot\zeta_Q(q_e)) - T^*_{z_e}\eta_Q\cdot p_e$, where

$$\Delta q = \eta_Q(q_e) \text{ and } \zeta = \mathbb{I}(q_e)^{-1}\mathrm{ad}^*_\eta\mu\,.$$

Using this notation, we have

4.3 Lemma

i *For* $\delta z \in T_{z_e}P$,

$$\Omega(z_e)(\Delta z, \delta z) = (\mathbf{DJ}(z_e)\cdot \delta z)\cdot \eta - \langle\langle \zeta_Q(q_e), \delta q\rangle\rangle\,.$$

ii *For* $\delta z \in \mathcal{S}_{\mathrm{INT}}$, *the* ***rigid internal coupling*** *terms are*

$$\Omega(z_e)(\Delta z, \delta z) = -\langle\langle \zeta_Q(q_e), \delta q\rangle\rangle = -\langle \mu_e, [\eta, \alpha(\delta q)]\rangle .$$

iii *For* $\Delta z = T_{z_e}Z_\mu \cdot \eta_Q(q_e)$ and $\widetilde{\Delta z} = T_{z_e}Z_\mu \cdot \tilde{\eta}_Q(q_e) \in \mathcal{S}_{\mathrm{RIG}}$, *the* ***coadjoint orbit, or Lie-Poisson*** *terms are*

$$\Omega(z_e)(\Delta z, \widetilde{\Delta z}) = -\mu_e \cdot [\eta, \tilde{\eta}] .$$

iv *For given variations* $\delta z = T_{q_e}\alpha_\mu \cdot \delta q$ and $\widetilde{\delta z} = T_{q_e}\alpha_\mu \cdot \widetilde{\delta z} \in \mathcal{W}_{\mathrm{INT}}$, *the* ***magnetic terms*** *are*

$$\Omega(z_e)(\delta z, \widetilde{\delta z}) = \mathbf{d}\alpha_\mu(\widetilde{\delta q}, \delta q) = \mathbf{d}\alpha_\xi(\widetilde{\delta q}, \delta q),$$

where $\alpha_\xi : Q \to T^*Q$ *is the one form given by* $\alpha_\xi(q) = \mathbb{F}L(\xi_Q(q))$.

v *For* $\delta z = T_{q_e}\alpha_\mu \cdot \delta q \in \mathcal{W}_{\mathrm{INT}}$, *and* $\widetilde{\delta z} = (0, \widetilde{\delta z}) \in \mathcal{W}^*_{\mathrm{INT}}$, *the* ***canonical terms*** *are*

$$\Omega(z_e)(\delta z, \widetilde{\delta z}) = \langle \widetilde{\delta z}, \delta q\rangle .$$

vi *For given variations* $\delta z, \widetilde{\delta z} \in \mathcal{W}^*_{\mathrm{INT}}$, *we have* $\Omega(z_e)(\delta z, \widetilde{\delta z}) = 0$.

Proof

$$\begin{aligned}\Omega(z_e)(\Delta z, \delta z) &= \langle \delta p, \Delta q\rangle - \langle \Delta p, \delta q\rangle \\ &= \langle \delta p, \eta_Q(q_e)\rangle + \langle p_e, T_{q_e}\eta_Q \cdot \delta q\rangle - \langle\langle \zeta_Q(q_e), \delta q\rangle\rangle \\ &= (\mathbf{D}\mathbf{J}(z_e)\cdot\delta z)\cdot\eta - \langle\langle \zeta_Q(q_e), \delta q\rangle\rangle .\end{aligned}$$

Hence **i** holds. If $\delta z \in T_{q_e}\mathbf{J}^{-1}(\mu_e)$, then $\mathbf{D}\mathbf{J}(z_e)\cdot\delta z = 0$, so the first equality in **ii** holds as well. Also,

$$\langle\langle \zeta_Q(q_e), \delta q\rangle\rangle = \langle \alpha_{\mathrm{ad}^*_\eta \mu}(q_e), \delta q\rangle = \langle \mathrm{ad}^*_\eta \mu, \alpha(\delta q)\rangle = \langle \mu, [\eta, \alpha(\delta q)]\rangle .$$

Thus we have proved **ii**. Statement **iii** follows from **ii** and the fact that $\alpha(\eta_Q(q_e)) = \eta$. To show **iv**, we check that

$$\begin{aligned}\Omega(z_e)(\delta z, \widetilde{\delta q}) &= \langle \mathbf{D}\alpha_\mu(q_e)\cdot\widetilde{\delta q}, \delta q\rangle - \langle \mathbf{D}\alpha_\mu(q_e)\cdot\delta q, \widetilde{\delta q}\rangle \\ &= \mathbf{D}(\langle \alpha_\mu(q_e), \delta q\rangle)\cdot\widetilde{\delta q}, - \mathbf{D}(\langle \alpha_\mu(q_e), \widetilde{\delta q}\rangle)\cdot\delta q - \alpha_\mu(q_e)\cdot \pounds_{\delta q}\widetilde{\delta q}\end{aligned}$$

$$= \mathbf{d}\alpha_{\mu_e}(\widetilde{\delta q}, \delta q),$$

where $\pounds$ is the Lie derivative—in this calculation, δq and $\widetilde{\delta q}$ are temporarily extended to vector fields. On the other hand,

$$\begin{aligned}\langle \mathbf{D}\alpha_{\mu_e}(q_e)\cdot\widetilde{\delta q}, \delta q\rangle &= \langle \mathbf{D}\alpha_\xi(q_e)\cdot\widetilde{\delta q}, \delta q\rangle + \langle\langle [\mathbf{D}\mathbb{I}^{-1}(q_e)\cdot\widetilde{\delta q})\cdot\mu_e]_Q(q_e), \delta q\rangle\rangle \\ &= \langle \mathbf{D}\alpha_\xi(q_e)\cdot\widetilde{\delta q}, \delta q\rangle,\end{aligned}$$

since $\widetilde{\delta q} \in \mathcal{V}_{\mathrm{INT}}$ implies $v = (\mathbf{D}\mathbb{I}^{-1}(q_e)\cdot\widetilde{\delta q})\mu_e \in \mathfrak{g}_{\mu_e}$ and so $\langle\langle v_Q(q_e), \delta q\rangle\rangle = 0$ by defintion of $\mathcal{V}$. Thus, we can replace α_{μ_e} by α_ξ to obtain $\Omega(z_e)(\delta z, \widetilde{\delta z}) = \mathbf{d}\alpha_\xi(\widetilde{\delta q}, \delta q)$, which completes the proof of **iv**. Parts **v** and **vi** follow from the definition of the canonical symplectic form. ■

Notice that if $\delta^2 H_\xi$ is to be definite, the sign must be positive since the term $\delta^2 K_\mu$ is always positive unless it is absent altogether. These results lead to:

4.4 Reduced Energy-Momentum method *Assume that the* $\delta^2 K_\mu$ *block is nontrivial (in the finite dimensional case, this means that* $\dim Q \neq \dim G$).

Necessary and sufficient conditions for formal stability are that the following two conditions hold:

i *The Arnold form* $\mathcal{A}_\mu : \mathfrak{g}_\mu^\perp \times \mathfrak{g}_\mu^\perp \to \mathbb{R}$ *be positive definite*

and

ii *The second variation* $\delta^2 V_\mu(q_e)$ *restricted to* $\mathcal{V}_{\mathrm{INT}}$ *be positive definite.*

In the case that the block $\delta^2 K_\mu$ is trivial, then necessary and sufficient conditions for formal stability are that the Arnold form be definite, either positive or negative, a case that was treated already by Arnold [1966]. If the group G is abelian, then it follows from work of Smale [1970] (see also Abraham and Marsden [1978]) that the condition for formal stability is just positive definiteness of $\delta^2 V_\mu$ since then the Arnold block is trivial and the system reduces to a simple mechanical one (with a magnetic term) on $T^*(Q/G)$. Thus, *the reduced energy-momentum method can be thought of as a synthesis of the work of Arnold and Smale.*

§5 *The Normalizing Connection*

We begin by describing connections on the level of configuration spaces. These are then lifted via the mechancal connection to phase space as in §3. We consider the following sequence:

$$Q \xrightarrow{\varphi_\mu} Q/G_\mu \xrightarrow{\sigma_\mu} Q/G = S$$

where φ_μ and σ_μ are the natural projections. Let γ_μ be the connection on $Q \to Q/G_\mu$ whose horizontal space is metric orthogonal to the G_μ orbits. This is the mechanical connection regarded on the G_μ bundle $Q \to Q/G_\mu$

At a particular $q \in Q$ (fixed in the discussion), the corresponding space $\mathcal{V}$ is

$$\mathcal{V} = \mathrm{hor}(\gamma_\mu) \tag{1}$$

Thus, $\mathcal{V}$ defines γ_μ and vice versa.

5.1 Proposition $\mathrm{hor}_{\gamma_\mu}(\mathrm{vert}\, \sigma_\mu) = \mathcal{V}_{\mathrm{RIG}}$, *where* $\mathrm{hor}_{\gamma_\mu}$ *is the horizontal lift with respect to* γ_μ *and* $\mathrm{vert}\, \sigma_\mu$ *is the vertical space of* σ_μ :

$$\mathrm{vert}\, \sigma_\mu = \{ w \in T_{\varphi_\mu(q)}(Q/G_\mu) \mid T\sigma_\mu \cdot w = 0 \} \tag{2}$$

Proof Let $\eta_Q(q) \in \mathcal{V}_{\mathrm{RIG}}$, so $\eta \in \mathfrak{g}_\mu^\perp$. Thus $T\sigma_\mu \cdot T\varphi_\mu \cdot \eta_Q(q) = T(\sigma_\mu \circ \varphi_\mu) \cdot \eta_Q(q)$. This is zero because $\sigma_\mu \circ \varphi_\mu$ projects $\zeta_Q(q)$ to zero for all $\zeta \in \mathfrak{g}$. Since $\mathcal{V}_{\mathrm{RIG}} \subset \mathcal{V}$, $\mathcal{V}_{\mathrm{RIG}}$ is horizontal, so we have proved

$$\mathcal{V}_{\mathrm{RIG}} \subset \mathrm{hor}_{\gamma_\mu}(\mathrm{vert}\, \sigma_\mu)$$

We prove the inclusion $\supset$ as follows. If $w_q \in \mathrm{hor}_{\gamma_\mu}(\mathrm{vert}\, \sigma_\mu)$, then $T\sigma_\mu \cdot T\varphi_\mu \cdot w_q = 0$, so $w_q = \eta_Q(q)$ for some $\eta \in \mathfrak{g}$. But if $\eta_Q(q)$ is γ_μ-horizontal, then $\eta \in \mathfrak{g}_\mu^\perp$, so $\mathrm{hor}_{\gamma_\mu}(\mathrm{vert}\, \sigma_\mu) \subset \mathcal{V}_{\mathrm{RIG}}$. ■

Now consider the split

$$\mathcal{V} = \mathcal{V}_{\mathrm{RIG}} \oplus \mathcal{V}_{\mathrm{INT}} \subset T_q Q$$

In view of the propositon, we can identify $\mathcal{V}$ with $T_{[q]}(Q/G_\mu)$ and $\mathcal{V}_{\mathrm{RIG}}$ with the vertical space. Thus, $\mathcal{V}_{\mathrm{INT}}$ *defines a connection, say* δ_μ on the bundle

$$Q/G_\mu \xrightarrow{\sigma_\mu} Q/G = S$$

The connections γ_μ and δ_μ induce connections on the other bundles in the cotangent bundle reduction theorem. (See the figure in §2). For example, the split $\mathcal{S} = \mathcal{S}_{RIG} \oplus \mathcal{S}_{INT}$ may be viewed as the vertical-horizontal split of a connection on the bundle

$$P_{\mathcal{O}} \to T^*S$$

with fiber the coadjoint orbit $\mathcal{O}$. Similarly, the split $\mathcal{S}_{INT} = \mathcal{W}_{INT} \oplus \mathcal{W}^*_{INT}$ is a connection on the bundle $T^*S \to S$ pulled up to $P_{\mathcal{O}}$ by the projection $P_O \to T^*S$, as in formulas (7-10) of §3.

Another question is interesting here. We saw that σ_μ is a principal bundle connection on $Q \to Q/G_\mu$. In what sense can δ_μ be viewed as a principal connection? There are two ways to answer this question. The ***first method*** allows μ to vary while the ***second method*** fixes μ. Turning to the first method, define the manifold

$$\tilde{Q} = \{(q, \nu) \in Q \times \mathfrak{g}^* \mid \mathbb{I}(q)^{-1}\nu \in \mathfrak{g}_\nu\} \tag{3}$$

Under our assumptions, $\mathbb{I}(q)$ is invertible, and this is enough to show that $\tilde{Q}$ is a manifold, as a transversality argument shows. In fact, the map $(q, \nu) \mapsto \mathbb{I}(q)^{-1}\nu$ is transversal to $\mathfrak{g}_\nu$ and so $\tilde{Q}$ is a manifold of dimension that of $Q \times \mathfrak{g}_\nu$.

Remark Enlarging the configuration space from Q to $Q \times G$ where we imagine the system together with an orthonormal frame, is a natural procedure, as in Krishnaprasad and Marsden [1987]. The conditions $\mathbb{I}(q)^{-1}\mu \in \mathfrak{g}_\mu$ on $Q \times \mathfrak{g}^*$ and $\mathbf{J}(z) - \mu = 0$, on $T^*(Q \times G)$, may be viewed as "locking devices" that lock the orthonormal frame to the structure. Reduction by G produces $T^*Q \times \mathfrak{g}^*$ and μ is eliminated by the condition $\mathbf{J}(z) = \mu$ (*i.e.*, the momentum map for the total action of G on $T^*(Q \times G)$ is zero.)

Let G act on $\tilde{Q}$ by the product of the action on Q with the coadjoint action on $\mathfrak{g}^*$. By equivariance properties of $\mathbb{I}$ noted earlier, this action preserves $\tilde{Q}$. Consider the principal bundle

$$\tilde{Q} \to \tilde{Q}/G \tag{4}$$

and note that the quotient space $\tilde{Q}/G$ has the dimension of $S \times \mathfrak{g}_\mu$.

5.2 Proposition *The choice of horizontal space* $\mathcal{V}_{\mathrm{INT}} \oplus T_\mu \mathcal{O} \subset T_{(q,\mu)}\tilde{Q}$ *defines a principal G-connection on the bundle* (4).

The connection form ε associated with this connection can be described as follows. Define the map $\beta_{(q,\nu)} : T_{(q,\nu)}\tilde{Q} \to \mathfrak{g}_\nu$ by $(\delta q, \delta\nu) \mapsto \mathbb{P}_\nu(\alpha(\delta q))$, where $\mathbb{P}_\nu$ is the orthogonal projection onto $\mathfrak{g}_\nu$. Also, define the map $\beta^\perp_{(q,\nu)} : T_{(q,\nu)}\tilde{Q} \to \mathfrak{g}^\perp_\nu$ by

$$\beta^\perp_{(q,\nu)}(\delta q, \delta\nu) = \chi^{-1}_{(q,\nu)}(\mathbb{P}^\perp_{(q,\nu)}\{[\mathbf{D}\mathbb{I}(q)^{-1}\cdot \delta q]\cdot \nu\}) = \chi^{-1}_{(q,\nu)}(\mathbb{P}^\perp_{(q,\nu)}\{\mathbb{I}(q)^{-1}\delta\nu\})$$

where $\mathbb{P}^\perp_{(q,\nu)}$ is the projection onto $\mathfrak{g}^\perp_\nu$. Now set

$$\varepsilon(\delta q, \delta\nu) = \beta_{(q,\nu)}(\delta q, \delta\nu) + \beta^\perp_{(q,\nu)}(\delta q, \delta\nu) \tag{5}$$

It is straight forward to check that the form ε is the connection one form for the connection in **5.2** and that it is a G-connection.

This connection can be lifted up to $\tilde{P} = \{z \in T^*Q \mid \mathbb{I}(q)^{-1}\mathbf{J}(z) \in \mathfrak{g}_{\mathbf{J}(z)}\}$ as before. Note that the inverse image of a point $(q, \mu) \in \tilde{Q}$ under the projection of $\tilde{P}$ to $\tilde{Q}$ is $\mathbf{J}^{-1}(\mu) \cap T^*_qQ$.

The ***second method*** proceeds as follows. Here we keep μ fixed and define the manifold

$$Q_\mu = \{q \in Q \mid \mathbb{I}(q)^{-1}\mu \in \mathfrak{g}_\mu\}\,. \tag{6}$$

Note that Q_μ is a fiber of $\tilde{Q}$ obtained by fixing ν at the value μ. The tangent space to Q_μ is given by linearizing the defining condition; one finds that

$$T_qQ_\mu = \{\delta q \in T_qQ \mid \eta\cdot(\mathbf{D}\mathbb{I}(q)\cdot\delta q)\xi = 0 \text{ for all } \eta \in \mathfrak{g}^\perp_\mu\} \cong \mathcal{V}. \tag{7}$$

From this and the fact that G_μ acts on Q_μ, we see that *the tangent to the quotient space* Q_μ/G_μ *is isomorphic to* $\mathcal{V}_{\mathrm{INT}}$. Thus, if we take $\mathcal{V}$ to be given by (7) and regard it as being isomorphic to the original definition ((1a) of §3), then the split $\mathcal{V} = \mathcal{V}_{\mathrm{RIG}} \oplus \mathcal{V}_{\mathrm{INT}}$ is just the horizontal-vertical split of the principal connection on the bundle $Q_\mu \to Q_\mu/G_\mu$ defined by declaring the horizontal space to be the metric orthogonal to the G_μ-orbits. In this viewpoint, one replaces the sequence $Q \to Q/G_\mu \to Q/G$ and its corresponding connections with the sequence $Q \supset Q_\mu \to Q_\mu/G_\mu$. Then one lifts the connections to phase spaces as before and blends them with the connection corresponding to the splitting $\mathcal{S}_{\mathrm{INT}} = \mathcal{W}_{\mathrm{INT}} \oplus \mathcal{W}^*_{\mathrm{INT}}$.

References

R. Abraham and J. Marsden [1978] *Foundations of Mechanics*. Second Edition, Addison-Wesley Publishing Co., Reading, Mass.

V.I. Arnold [1966] Sur la géometrie differentielle des groupes de Lie de dimenson infinie et ses applications a l'hydrodynamique des fluids parfaits. *Ann. Inst. Fourier. Grenoble* **16**, 319-361.

V.I. Arnold [1968] On an a priori estimate in the theory of hydrodynamical stability. *Izv. Vyssh. Uchebn. Zaved. Mat. Nauk.* **54**, 225-226.

V.I. Arnold [1978] *Mathematical Methods of Classical Mechanics*. Graduate Texts in Math. **60**, Springer-Verlag.

V.I. Arnold [1988] *Encyclopedia of Mathematics: Dynamical Systems III*. Springer-Verlag, N.Y.

A. Guichardet [1984] On rotation and vibration motions of molecules. *Ann. Inst. H. Poincaré* **40**, 329-342.

D.D. Holm, J.E. Marsden, T. Ratiu and A. Weinstein [1985] Nonlinear stability of fluid and plasma equilibria. *Physics Reports* **123**, 1-116.

D. Kazhdan, B. Kostant and S. Sternberg [1978] Hamiltonian group actions and dynamical systems of Calogero type. *Comm. Pure Appl. Math.* **31**, 481-508.

P.S. Krishnaprasad [1989] Eulerian many-body problems. *Cont. Math. AMS* **97**, 187-208.

P.S. Krishnaprasad and J.E. Marsden [1987] Hamiltonian structure and stability for rigid bodies with flexible attachments, *Arch. Rat. Mech. An.* **98**, 137-158.

M. Kummer [1981] On the construction of the reduced phase space of a Hamiltonian system with symmetry. *Indiana Univ. Math. J.* **30**, 281-291.

D. Lewis [1989] Nonlinear stability of a rotating planar liquid drop. *Arch. Rat. Mech. Anal.* **106**, 287-333.

D. Lewis [1990] Lagrangian block diagonalization (*IMA preprint*).

D. Lewis and J.C. Simo [1989] Nonlinear stability of rotating pseudo-rigid bodies. *Proc. Royal Society London* (preprint).

C.M. Marle [1976] Symplectic manifolds, dynamical groups and Hamiltonian mechanics. *Differential Geometry and Relativity*. M. Cahen and M. Flato, eds., D. Reidel, Boston.

J.E. Marsden [1981] *Lectures on Geometric Methods in Mathematical Physics. SIAM, CBMS Conf. Series* **37**.

J.E. Mardsen, R. Montgomery and T. Ratiu [1984] Gauged Lie-Poisson structures. *Cont. Math. AMS*, **28**, 101-114.

J.E. Mardsen, R. Montgomery and T. Ratiu [1990] Reduction, symmetry, and geometric phases in mechanics. *Memoirs of the AMS* (to appear).

J.E. Marsden, J.C. Simo, D. Lewis, and T.A. Posbergh [1989] A block diagonalization theorem in the energy momentum method. *Cont. Math. AMS* **97**, 297-313.

J.E. Marsden and A. Weinstein [1974] Reduction of symplectic manifolds with symmetry. *Rep. Math. Phys.* **5**, 121-130

R. Montgomery [1990] Optimal control of deformable bodies, isoholonomic problems, and sub-Riemannian geometry. *Comm. Math Phys.*(to appear).

G. Patrick [1990] *Nonlinear stability of coupled rigid bodies.* Ph.D. Dissertation, Department of Mathematics, University of California, Berkeley.

J.C. Simo, D. Lewis, and J.E. Marsden [1990] Stability of relative equilibria I: The reduced energy momentum method. *Arch. Rat. Mech. Anal.* (to appear)

J.C. Simo, J.E. Marsden and P.S. Krishnaprasad [1988] The Hamiltonian structure of nonlinear elasticity: The material, spatial, and convective representations of solids, rods, and plates. *Arch. Rat. Mech. An.* **104**, 125-183.

J.C. Simo, T.A. Posbergh and J.E. Marsden [1990a] Stability of coupled rigid bodies and geometrically exact rods: block diagonalization and the energy-momentum method, *Physics Reports,* (to appear)

J.C. Simo, T.A. Posbergh and J.E. Marsden [1990b] Stability of relative equilibria II: Three dimensional elasticity. *Arch. Rat. Mech. Anal,* (to appear)

S. Smale [1970] Topology and Mechanics. *Inv. Math.* **10**, 305-331, **11**, 45-64.

ON REDUCTION OF ALMOST REGULAR MOMENTUM MAPS

JEDRZEJ ŚNIATYCKI

Department of Mathematics and Statistics, The University of Calgary
Calgary, Alberta T2N 1N4, Canada

ABSTRACT. Zero levels of momentum maps have regular strata which admit Marsden-Weinstein reduction. The contribution of the singular strata can be analyzed in terms of algebraic reduction. The relationship between two algebraic reduction schemes and the Marsden-Weinstein reduction of the regular stratum is discussed. Possible applications to Yang-Mills theory are outlined.

1. CONSTRAINTS IN YANG-MILLS THEORY

Yang-Mills fields are connections in a principal fibre bundle over the space-time manifold X. The field equations in temporal gauge can be split into the evolution equations and the constraint equations for the Cauchy data which consist of the induced connections on the Cauchy surface and the canonically conjugate momenta. The space of Cauchy data P has a weak symplectic form ω, and the evolution equations can be put in the usual Hamiltonian form. The group G of time independent gauge transformations has a Hamiltonian action in P with the momentum map $J : P \to g^*$. The constraint equations are

$$J = 0 \,. \tag{1}$$

In physical applications points in P are supposed to describe physical states of the system. However, only the points satisfying the constraint equations are dynamically admissible. Moreover, Cauchy data related by a gauge transformation describe the same physical state. Thus, physical states are given by the points in the reduced phase space

$$R = J^{-1}(0)/G \tag{2}$$

of the G orbits in the constraint set $J^{-1}(0)$.

2. MARSDEN-WEINSTEIN REDUCTION

If $0 \in g^*$ is a regular value of J, then the constraint set is a submanifold of P. If, moreover, the action of G in $J^{-1}(0)$ is free, then R is a quotient manifold of $J^{-1}(0)$, and there is a unique symplectic form ω_R on R such that

$$\rho^* \omega_R = \iota^* \omega \,, \tag{3}$$

where $\iota : J^{-1}(0) \to P$ is the inclusion map and $\rho : J^{-1}(0) \to R$ is the canonical projection. In this case the symplectic manifold (R, ω_R) is called the Marsden-Weinstein reduced phase space, [1].

3. DIRAC ALGEBRA OF CLASSICAL OBSERVABLES

If 0 is not a regular value of J, then $J^{-1}(0)$ need not be a submanifold of P and we may encounter singularities. The question arises how to describe the reduction for singular momentum maps. Since the aim is of this exercise is to determine the "true degrees of freedom" of a physical system with constraints we have to go back to the physical interpretation.

Elements of the Poisson algebra of (P,ω), denoted $\mathcal{P}(P,\omega)$, are interpreted as dynamical variables. For each $f \in \mathcal{P}(P,\omega)$ and each $p \in P$, the evaluation $f(p)$ corresponds to the measurement of the dynamical variable f in the state classical state p. Since dynamical variables are observed in terms of their measurements on physical states, we cannot distinguish by measurements between dynamical variables with difference vanishing on $J^{-1}(0)$. Moreover, observable dynamical variables should have the same values on states related by gauge transformations. Hence, observable dynamical variables form the space $\mathcal{O}$ of G-invariant elements in the quotient of $\mathcal{P}(P,\omega)$ by the ideal $\mathcal{I}$ consisting of functions vanishing on $J^{-1}(0)$,

$$\mathcal{O} = (\mathcal{P}(P,\omega)/\mathcal{I})^G \,. \tag{4}$$

If $\mathcal{I}$ is a Poisson subalgebra of $\mathcal{P}(P,\omega)$, then $\mathcal{O}$ inherits the structure of a Poisson algebra. Following Dirac, [2], we say in this case that the constraints are first class, and we shall refer to $\mathcal{O}$ as the Dirac algebra of observables.

4. HOMOMORPHISM THEOREM

Suppose that $J^{-1}(0)$ has an open dense regular stratum S_0 to which one can apply Marsden-Weinstein reduction obtaining the reduced phase space (R_0,ω_{R_0}). Then, the constraints are first class and there is a Poisson algebra monomorphism

$$\varphi : \mathcal{O} \to \mathcal{P}(R_0,\omega_{R_0}) \tag{5}$$

defined by the restrictions of functions to S_0.

Under assumptions above singularities of $J^{-1}(0)$ can be taken into account by restricting $\mathcal{P}(R_0,\omega_{R_0})$ to functions satisfying appropriate boundary conditions at the singularity. For an appropriate choice of a polarization the quantization of (R_0,ω_{R_0}) would be insensitive to the boundary conditions, and the obtained quantum system would be independent of the structure of the singularities.

Many physically interesting solutions of Yang-Mills equations, namely those invariant under non-trivial subgroups of the gauge group, correspond to points in the singular strata of the constraint set. Hence, it is somewhat surprising that their contribution to the quantum theory could be ignored.

5. REDUCED POISSON ALGEBRA

The problem arises if there is an algebraic structure associated with constraints which would give more information than the Dirac algebra $\mathcal{O}$ about the singularities of the constraints. Such a structure can be obtained if instead of the ideal $\mathcal{I}$ one considers the ideal $\mathcal{J}$ generated by the momenta J_ξ, for $\xi \in g$. The space $\mathcal{A}$ of G-invariant elements in the quotient of $\mathcal{P}(P,\omega)$ by $\mathcal{J}$,

$$\mathcal{A} = (\mathcal{P}(P,\omega)/\mathcal{J})^G \,, \tag{6}$$

has the structure of a Poisson algebra. It is called the reduced Poisson algebra of the system. In the regular finite dimensional case $\mathcal{A}$ and $\mathcal{O}$ are naturally isomorphic to the Poisson algebra of the Marsden-Weinstein reduced phase space. For singular constraints the reduced Poisson algebra gives somewhat different information about the constraints than the Dirac algebra $\mathcal{O}$; for example $\mathcal{A}$ is always a Poisson algebra while $\mathcal{O}$ is a Poisson algebra only for first class constraints.

6. EXAMPLE

Consider a symplectic space (P, ω) with $P = \mathbf{R}^4$ and

$$\omega = dp_1 \wedge dx_1 + dp_2 \wedge dx_2 .$$

Let $G = \mathbf{R}$ act on P by the Hamiltonian flow of

$$J = x_1^2 + x_2^2 .$$

In this case 0 is not a regular value of J and the Marsden-Weinstein reduction does not apply. The space R of G orbits in $J^{-1}(0) \cong \mathbf{R}^2$ coincides with $J^{-1}(0)$ and the algebra $\mathcal{O}$ is isomorphic to the algebra $C^\infty(\mathbf{R}^2)$ in which the Poisson bracket vanishes identically. On the other hand side the reduced Poisson algebra $\mathcal{A}$ is non-abelian. It is isomorphic to the Poisson algebra of formal power series

$$f(z, w) = \sum_{n=0}^{\infty} a_n(w) \frac{z^n}{n!} ,$$

where the coefficients a_n are entire analytic functions of a complex variable w and the coefficient a_0 is real, with the Poisson bracket

$$\{f, h\} = \frac{\partial f}{\partial z}\frac{\partial h}{\partial w} - \frac{\partial f}{\partial w}\frac{\partial h}{\partial z} .$$

It seems that in this case the reduced Poisson algebra $\mathcal{A}$ contains the required information about the constraints. A generalization of geometric quantization to a quantization of $\mathcal{A}$ gives results equivalent to the geometric quantization of (P, ω) and a subsequent imposition of the quantum constraint condition requiring that the admissible quantum states should be G-invariant [3].

7. ALMOST REGULAR SYSTEMS

For regular systems with constraints the Dirac algebra $\mathcal{O}$ is isomorphic to the reduced Poisson algebra $\mathcal{A}$. For singular constraints these two algebras may be quite distinct as shown in the example above. We say that a system with constraints is almost regular if the Dirac algebra $\mathcal{O}$ and the reduced Poisson algebra $\mathcal{A}$ are isomorphic.

The question arises which constraint systems are almost regular. In particular is Yang-Mills theory almost regular? If Yang-Mills theory were almost regular we could expect that the quantization of the Marsden-Weinstein reduction of the regular stratum in an appropriate polarization would be sufficient.

Examples of constraints with singularities have been studied by several authors, refs. [4] through [9]. Some of these examples will be reviewed below.

8. SIMPLE CONICAL SINGULARITY

Consider $P = \mathbf{R}^{2n+2}$ with the canonical symplectic form $\omega = \sum_{i=0}^{n} dp_i \wedge dx_i$, and let $G = \mathbf{R}$ act on P by the Hamiltonian flow of

$$J = (p_0)^2 - \sum_{i=1}^{n} (p_i)^2 .$$

Almost regularity of this system follows from its analysis in ref. [4]. Here we give an alternative proof of its almost regularity based on the Mather Division Theorem, cf. ref. [10].

In order to prove almost regularity of a constraint system it clearly suffices to establish the equality of the ideals $\mathcal{I}$ and $\mathcal{J}$. Since $\mathcal{J} \subseteq \mathcal{I}$ it suffices to show that every function $f \in C^\infty(\mathbf{R}^{2n+2})$ which vanishes on $J^{-1}(0)$ is divisible by J. By the Mather Division Theorem, for each $x = (x_0, x_1, ..., x_n)$ in $\mathbf{R}^{n+1}$, there exist a neighbourhood U of $(x, 0) \in \mathbf{R}^{2n+2}$, a smooth function q, and smooth functions r_0 and r_1 independent of p_0 such that

$$f|_U = qJ + p_0 r_1 + r_0 \ .$$

Since $f = 0$ whenever $J = 0$, it follows that

$$r_0 = -\sqrt{\sum_{i=1}^{n} (p_i)^2 r_1} \ ,$$

and

$$r_0 = +\sqrt{\sum_{i=1}^{n} (p_i)^2 r_1} \ .$$

Hence, $r_0 = 0$, and $r_1 = 0$, which implies that $f|_U = qJ$. Using partition of unity we obtain that f is divisible by J. Therefore, $f \in \mathcal{J}$, which proves almost regularity of our example.

9. ANGULAR MOMENTUM CONSTRAINT

Consider the phase space $P = \mathbf{R}^{6n}$ of n particles in $\mathbf{R}^3$. Its symplectic form is

$$\omega = \sum_{a=1}^{n} \sum_{i=1}^{3} dp_{ai} \wedge dx_{ai} \ .$$

The action of the rotation group $G = SO(3)$ in $\mathbf{R}^3$ induces a Hamiltonian action in (P, ω) with the momentum map given by the usual angular momentum J. It was shown in ref. [6] that the constraint given by $J = 0$ is almost regular.

The dynamics of homogeneous Yang-Mills fields reduces to the dynamics of n particles with angular momentum constraints, [11]. Hence, one can use the results of almost regularity of angular momentum constraints to show almost regularity of homogeneous Yang-Mills fields, [7].

10. STRUCTURE OF SINGULARITIES OF MOMENTUM MAPS

The structure of the zero levels of momentum maps appearing in physics has been analyzed by Arms, Marsden and Moncrief, [12]. They have shown that, for each point $p \in J^{-1}(0)$, there exists a neighbourhood U of p and a splitting of $J|_U$ into components J_1 and J_2 such that 0 is a regular value of J_1 and J_2 restricted to $J_1^{-1}(0)$ corresponds to the momentum map j of the stability group of p. Since this action has a fixed point at p, the Darboux Theorem can be used to show that there exist symplectic coordinates in which this action is linear and hence in which j is quadratic.

11. SPECULATIONS ON ALMOST REGULARITY OF YANG-MILLS THEORY

In the physics papers on quantized Yang-Mills fields the problem of singularities of constraints is usually ignored. This would be justified if Yang-Mills theory were almost regular. Thus it is of interest to try to extend the notions discussed above to field theories.

In field theory the gauge group G is infinite dimensional and therefore one has to be careful about the definition of the ideal $\mathcal{J}$. The usual algebraic definition of the ideal generated by a set of functions is probably not suitable since it involves only finite sums. It seems that in field theory one should consider an ideal $\overline{\mathcal{J}}$ consisting of functions of the form $< J, F >$, where F is a map from the Lie algebra of G to P *and* $<,>$ *denotes* evaluation.

Since in field theory not all functions admit Hamiltonian vector fields, the Poisson algebra does not consist of all smooth functions on P. For algebraic reduction one should consider the intersection

$$\mathcal{J} = \overline{\mathcal{J}} \cap \mathcal{P}(P, \omega) \ .$$

It is an ideal in the associative algebra structure of $\mathcal{P}(P,\omega)$. However, it is not clear if $\mathcal{J}$ is closed under Poisson bracket.

Many results in finite dimensions are based on the partition of unity. For infinite dimensional manifolds we need partitions of unity contained in $\mathcal{P}(P,\omega)$. It is not known when such partitions of unity exist.

If the difficulties listed above were overcome one would have a reasonable chance to check if Yang-Mills theory is almost regular. Actually, the results on the angular momentum constraints together with the structure of the singularities seem to indicate that one might prove an appropriate version of almost regularity, which would imply that all the physical information about the theory is contained in the Poisson algebra of the reduction of the regular stratum and the boundary conditions at the singularity.

REFERENCES

[1] J.E. Marsden and A. Weinstein, Rep. Math. Phys. **5** (1974), 121-130.

[2] P.A.M. Dirac, Can. J. Math. **2** (1950), 129-148.

[3] J. Śniatycki and A. Weinstein, Lett. Math. Phys.**7** (1983), 155-161.

[4] M.J. Gotay, J. Math. Phys. **25** (1984), 2154-2159.

[5] M.J. Gotay and L. Bos, J. Diff. Geom. **24** (1986), 181-203

[6] J.M. Arms, M.J. Gotay and G. Jennings, "Geometric and Algebraic Reduction for Singular Momentum Maps", to appear in Adv. in Math.

[7] M.J. Gotay, "Reduction of Homogeneous Yang-Mills Fields", to appear in J. Geom. Phys.

[8] J. M. Arms, R. Cushman and M.J. Gotay, "A Universal Reduction Procedure for Hamiltonian Group Actions", to appear in the Proceedings of the Workshop on the Geometry of Hamiltonian Systems, M.S.R.I. Berkeley, 1989.

[9] J.M. Arms, M.J. Gotay and Wilbour, Nucl. Phys. **B** (Proc. Suppl.) **6** (1989), 384-389.

[10] M. Golubitsky and V. Guillemin, **Stable Mappings and Their Singularities**, Springer Verlag, New York, 1973, *p*. 95.

[11] J. Śniatycki, "On Quantization of Yang-Mills Fields", in **Differential Topology, Geometry and Related Fields, and Their Applications to the Physical Science and Engineering**, J.M. Rassias (ed.), Taubner Verlag, Leipzig 1985.

[12] J.M. Arms, J.E. Marsden and V. Moncrief, Comm. Math. Phys. **78** (1981), 455-478.

ORIGINAL TITLES AND ABSTRACTS

INVITED LECTURES

ABLOWITZ, MARK J.

Program in Applied Mathematics
University of Colorado at Boulder
Boulder, CO 80309, USA

ASPECTS OF SOLITONS AND COMPUTATION: Recent studies have shown that standard discretizations of certain integrable nonlinear evolution equations can lead to spurious (chaotic) results at intermediate values of the mesh size. Examples include many of the well known equations. Schemes which show no signs of difficulty will be presented as well. If time permits, a discussion of soliton cellular automata will also be given.

ADAMS, M.R.

Department of Mathematics
University of Georgia
Athens, GA 30602, USA

ISOSPECTRAL FLOW IN LOOP ALGEBRAS AND MOMENT MAP EMBEDDINGS: The theory of rank-r isospectral Hamiltonian deformations of a fixed $N \times N$ matrix, as developed by Adams, Harnad, and Previato [Commun. Math Phys. **117**, 451–500 (1988)] will be presented. A moment map from the Poisson manifold of rank-r matrices into the dual of the positive part of a loop algebra, $(\widetilde{gl}(r)^+)^*$, is constructed so that the Adler–Kostant–Symes flows on $(\widetilde{gl}(r)^+)^*$ pull back to isospectral deformations. This theory gives a unified treatment of Moser's representation of classical integrable Hamiltonian systems in terms of isospectral deformations as well as giving explicit models for invariant finite dimensional subspaces of solutions to integrable partial differential equations arising from commutativity conditions of AKS flows (i.e. generalized AKNS systems). The theory will be demonstrated through examples and if time permits a duality theorem relating AKS flows in $(\widetilde{gl}(r)^+)^*$ to those in $(\widetilde{gl}(N)^+)^*$ will be described. The material in this lecture is background for the lecture of J. Hurtubise and J. Harnad.

ANDERSON, R.L.

Department of Physics and Astronomy
University of Georgia
Athens, GA 30602, USA

REMARKS ON INTEGRABLE HIERARCHIES IN FINITE DIMENSIONS: In this paper we consider the hierarchy of vector fields generated by recursively applying an operator to a given Hamiltonian vector field. We give a simple set of conditions which guarantee that the hierarchy is integrable. The result presented on the generation of an integrable hierarchy by recursion, is similar to the theorem of Magri on biHamiltonian systems, but only one sympletic structure is used so there are fewer hypotheses to check. We illustrate the use of a recursion operator for certain oscillator systems and for the Korteweg-de

Vries 2–soliton sector. One of the examples of 2 harmonic oscillators is presented as an analogue for the KdV 2–solitons. Our results on the 2–soliton sector are obtained by direct calculation with Hirota's soliton formulas. [Joint work with M.R. Adams and R. Varley]

BATES, L.M.

Department of Mathematics
University of Calgary, Calgary
Alberta, T2N 1N4, CANADA

REDUCTION AND EXOTIC SYMPLETIC STRUCTURES: We give an explicit example of an exotic sympletic structure on $\mathbf{R}^4$.

BLOCH, A.M.

Department of Mathematics
The Ohio State University
OH 43210, USA

GRADIENT AND HAMILTONIAN FLOWS ASSOCIATED WITH THE TODA LATTICE: In this talk we discuss some recent results on the nonperiodic Toda Lattice obtained with R. Brockett, H. Flaschka and T. Ratiu. We show how the generalized Toda lattice equations in Lax pair form may be exhibited as a gradient flow on an adjoint orbit of a compact Lie group. We also discuss a Hermitian formulation of the Toda equations, and indicate how Moser's gradient flows fit into our picture as gradient flows on projective space with respect to the Kähler metric. We also prove some convexity results regarding the Toda flows. In particular we show how an isospectral manifold of Jacobi matrices may be mapped via a moment map onto a convex polytope. Some connections of this work with linear programming and least squares optimization are pointed out.

BROCKETT, R.W.

Division of Applied Sciences
Harvard University, Pierce Hall
Cambridge, MA 02138, USA

SORTING WITH FINITE AND INFINITE DIMENSIONAL HAMILTONIAN SYSTEMS: In this talk I will describe a Hamiltonian System which, in its finite dimensional incarnation, sorts a list of numbers via a continuous analog of the successive compare and exchange method used in standard computer algorithms. The infinite dimensional version of these equations flows in such a way as to rearrange a smooth function of one variable defined on a finite interval so as to obtain a monotone decreasing function which is equimeasurable with the given function.

DAVID, D.
Theoretical Division and Center for Nonlinear Studies
Los Alamos National Laboratory, MS B284
Los Alamos, NM 87545, USA

INTEGRABILITY AND CHAOS IN COUPLED NONLINEAR OPTICAL BEAMS: We present a study, using dynamical systems methods, of a system composed of two nonlinearly coupled optical beams counterpropagating in a nonlinear medium. In the approximation of travelling-wave motion, this setup is described as a Hamiltonian system, with canonical Poisson bracket, on the manifold $\mathbf{C}^2 \times \mathbf{C}^2$ coordinatized by two complex 2-component electric field amplitudes. The Hamiltonian function is invariant under several actions of the rotation group; this allows a lowering of the number of degrees of freedom in the following manner. A first S^1 invariance reduces the phase space to a product of three-spheres, $S^3 \times S^3$. A second S^1 symmetry then induces a reduction to a product of two-spheres, $S^2 \times S^2$; this is accomplished via a Hopf mapping. Finally, in the case of an isotropic propagation medium, the system possesses a further conserved quantity identifiable as the sum of the angular momenta of the beams, allowing for a reduction of the phase space to a single sphere S^2. We present a geometrical description of the dynamics on the reduced space by determining its qualitative features. Specifically, we determine the fixed points of the dynamics and their stability types, and the various homoclinic connections present in the phase portrait; these special orbits correspond to solutions behaving like kinks and solitons. The dynamical equations contains additional physical parameters that may cause bifurcations to occur when varied. We present these, in particular degenerate ones. Last, we investigate the existence of induced chaos under certain classes of periodic Hamiltonian perturbations of the reduced dynamics. These may possibly break the symmetry associated with the conservation of the total angular momentum, thereby lifting back the dynamics to $S^2 \times S^2$. Specifically, we use the Melnikov method to demonstrate analytically that the perturbations considered either imply horsehoe chaos, or Arnold diffusion when the dimensionality permits it.

DICKEY, L.
Department of Mathematics
University of Oklahoma
Waltham, MA 02154, USA

ANOTHER EXAMPLE OF A τ-FUNCTION: The τ-function was suggested by Japanese mathematicians for the Kadomtsev–Petviashvili hierarchy of integrable systems and for its reductions. It is important to extend the definition of the τ-function to all integrable systems. Here we do this for hierarchies of equations generated by linear first-order differential operators (AKNS or Dubrovin systems). The group-theoretical meaning of this τ-function still remains unclear.

GOTAY, MARK
Department of Mathematics
U.S. Naval Academy
Annapolis, MD 21402, USA

THE SYMPLETIC GEOMETRY OF BOSONIC BRST THEORY: I show that BRST symmetry has a natural bosonic analogue in sympletic geometry. In fact, bosonic BRST theory arises as a purely sympletic construction, which can be naturally viewed as a specific instance of sympletic induction. In this context, both the BRST charge and the Maurer–Cartan form appear as the "components" of a momentum map on an extended symplectic phase space. An application to the quantization of constrained classical systems with nonunimodular symmetries is discussed and it is shown, under certain circumstances, that both the processes of induction and reduction commute with quantization. The importance of these considerations is emphasized in the particular case of pseudo-rigid bodies.

GRMELA, MIROSLAV

Département de génie chimique
École Polytechnique de Montréal
Montréal, QC, H3C 3J7, Canada

THERMODYNAMICS IN GRADIENT HAMILTONIAN FLOWS: Every gradient Hamiltonian dynamical system is naturally associated with thermodynamics. The thermodynamics are essentially a specification of geometrical structure of the space of slow dynamical variables. My aim is to introduce this general interpretation of thermodynamics, show that it naturally extends the well known geometrical interpretation of classical equilibrium thermodynamics, and illustrate its usefulness on several well known and new examples.

HARNAD, J.

Department of Mathematics, Concordia University and
Centre de Recherches Mathématiques, Université de Montréal
C.P. 6128 – A, Montréal, QC, H3C 3J7, CANADA

LIOUVILLE GENERATING FUNCTION FOR ISOSPECTRAL FLOW IN LOOP ALGEBRAS: The symplectic structure on rational coadjoint orbits of the loop algebra $\widetilde{gl}(r)^+$ is shown to reduce under the Hamiltonian $GL(r)$ action to a natural sympletic structure on the (singular) bundle of Jacobi varieties over the space of admissable spectral curves. The latter follows from the algebraic geometry of the embedding of the spectral curves in an ambient ruled surface. The divisors of the corresponding line bundles over spectral curves provide natural Darboux coordinates in which the Liouville generating function for isospectral flow is expressed in separated form. Certain non-reduced invariant sympletic submanifolds of the coadjoint orbits are also given separating Darboux coordinates with natural algebro-geometric interpretation in terms of extended Jacobi varieties. These results provide an entirely Hamiltonian approach to the linearization of flow on Jacobi varieties, equally valid for the case of singular spectral curves arising, e.g. in the finite gap solutions of the coupled cubically nonlinear Schrödinger equation. [Joint work with M. Adams and J. Hurtubise.]

HOLM, D.D.

Theoretical Division and Center for Nonlinear Studies
Los Alamos National Laboratory
MS B284 Los Alamos, NM 87545, USA

HAMILTON CHAOS IN NONLINEAR OSCILLATORS: Newly developed nonlinear methods are used to investigate the problem of chaotic dynamics in two identical anharmonic oscillators interacting as a resonant classical Hamiltonian system. This problem has two apparently disparate applications: It describes either (1) a single polarized laser pulse propagating as a traveling wave in an anisotropic cubically nonlinear lossless medium, or (2) the dynamics of high excitation states of ABA triatomics whose energy states are parameterized using the Darling–Denisson Hamiltonian. Invariance of the Hamiltonian function for the coupled oscillator system under phase shifts of the two-component complex oscillator modes allows reduction of the phase space to the two-sphere, S^2, on which the problem becomes completely integrable. The fixed points (corresponding to equilibrium polarization states for the polarized laser pulse; and to local modes, normal modes, and their generalizations for triatomics molecules) are identified, and the complete bifurcation diagram is presented describing how the phase portraits on S^2 change as the orbits deform and reconnect when parameters in the Hamiltonian vary. (These bifurcations describe polarization instabilities for the laser pulse and local–normal mode transitions for the triatomics). Homoclinic and heteroclinic orbits on S^2 are identified in each parameter domain. By using the Melnikov method, homoclinic chaos is analytically shown to develop when the parameters in the Hamiltonian are periodically perturbed. The perturbed orbits then show sensitivity to intial conditions, even in the absence of resonance island overlap. The mechanism for this sensitivity to initial conditions is shown to be a Smale horsehoe in the Poincaré map for the perturbed oscillator system.

HURTUBISE, J.

Department of Mathematics
McGill University
Montreal, QC, H3A 1G5, CANADA

THE ALGEBRAIC GEOMETRY OF ISOSPECTRAL FLOWS: Many Hamiltonian systems (e.g. the "Moser systems") can be embedded into loop algebras, where the equations governing the flows are expressed in Lax form. We present a version of the method for the integration of these flows on the Jacobian of their spectral curve which emphasizes the geometry of the situation and which allows one to handle, for example, the case of singular curves. We will then discuss the symplectic structure of the adjoint orbits in the loop algebra in terms of the geometry of these curves. [Joint work with M. Adams and J. Harnad]

KAUFFMAN, L.H.

Department of Mathematics, Statistics and Computer Science,
The University of Illinois at Chicago
Chicago, Illinois 60680, USA

ON THE QUANTUM GROUP $SL(2)q$ AND THE UNIVERSAL R-MATRIX: We show that the quantum group $SL(2)q$ arises naturally from topological considerations and from abstract tensor ideas related to the original Jones polynomial and to the Penrose theory of spin networks. In this context there are two forms of invariance. The knot diagrams map to abstract tensor diagrams that are (for appropriate choice of matrices) internally invariant under the action of $SL(2)q$ and externally invariant under the Reidemeister moves. The R-matrix provides the connection between the internal and external invariance. The universal R-matrix is discussed from this point of view.

KUPERSHMIDT, B.A.
The University of Tennessee Space Institute
Tullahoma, TN 37388, USA

POISSON BRACKETS ON MATRIX SPACES : A system of quadratic Poisson brackets makes $\mathrm{Mat}_{n \times n}$ into a Lie–Poisson semigroup with respect to multiplication. The determinant is a Casimir; thus, $GL(n), SL(n)$, and $\{x \varepsilon \mathrm{Mat}_{n \times n} \mid \det(x) = 0\}$ are Lie–Poisson subsemigroups in $\mathrm{Mat}_{n \times n}$. This Poisson structure can be extracted in the quasiclassical limit from Drinfel'd's fomulae for the generators and relations of the quantum group $(SL_h(n))^*$. This Poisson structure has the additional property of not being invariant with respect to reorderings of the basis, but being invariant with repect to the dilatations of the basis. More generally, it is an example of a conformal–Hamiltonian space with the conformal factor 1; other examples will be given, when the conformal factor is 2.

Using matrix-valued water wave equations as an example, obstructions to the Hamiltonian formalism in associative rings will be discussed; these boil down to variational derivatives of a Hamiltonian becoming operators, in contrast to the commutative case where they are just functions.

LAGARIAS, J.C.
AT & T Bell Laboratories
Murray Hill, NJ 07974
USA

MONOTONICITY AND INTEGRABILITY PROPERTIES OF PROJECTED TODA FLOWS AND LINEAR PROGRAMMING FLOWS : This talk describes flows on finite dimensional spaces that exhibit both "gradient-like" and "completely integrable" behavior. These include: Flows associated to interior-point linear programming algorithms including Karmarkar's algorithm, and projections of finite nonperiodic Toda lattice flows. The latter are spectrum-increasing flows. Spectrum-increasing flows are characterized using a certain Lie bracket decomposition.

MARSDEN, J.E.
Department of Mathematics
University of California
Berkeley, CA 94720, USA

MECHANICS AND SYMMETRY – DYNAMICS AND BIFURCATION OF ROTATING STRUCTURES [Aisenstadt Lectures]: This will be a series of three lectures on the applications of geometric mechanics to the problem of the dynamics, stability, and bifurcations of rotating structures. The structures we have in mind will be mostly elastic systems, such as coupled rigid bodies, or a rigid body with a geometrically exact beam. However, the methods also apply to the problem of stability of rotating gravitational fluid masses in the spirit of Riemann and Poincaré. The method is based on recent developments in the separation and block diagonalization of rotational and internal variations that make the computations tractible. This structure is also important for the study of bifurcations and chaotic solutions, such as has been shown in the problem of three coupled planar rigid bodies. Connections with the phases of Hannay and Berry will also be demonstrated.

MATHIEU, PIERRE
Département de physique
Université Laval
Québec, QC, G1K 7P4, CANADA

$N = 1,2$ SUPERSYMMETRIC KORTEWEG-DE VRIES EQUATIONS: There is an unique integrable space supersymmetric extension of the Kortweg-de Vries equation. Some results related to this equation are reviewed: (second) Hamiltonian structure, Miura transformation, integrable deformation, Lax formulation and relation with the supersymmetric sine-Gordon equation. It is also pointed out that, as in the bosonic case, the supersymmetric Korteweg-de Vries equation is a member of a quasi-hamiltonian pair of super evolution equations, the transformation relating the two equations being the super Schwarzian derivative. On the other hand there are three integrable $N = 2$ space supersymmetric extensions of the Kortweg-de Vries equation. We report on all the known results concerning these systems and indicate open problems.

McKEAN, H.
Courant Institute of Mathematical Sciences
New York University
New York, NY 10012, USA

BREATHERS: Abstract withdrawn.

MOSER, J.
Forschungsinstitut für Mathematik
Eidegenössische Technische Hochschule
CH-8092,Zürich, SWITZERLAND

DISCRETE VERSIONS OF SOME CLASSICAL INTEGRABLE SYSTEMS AND FACTORIZATION OF MATRIX POLYNOMINALS: An integrable discrete system is discussed which corresponds to the discretization of the motion of the spinning top. Another example is an integrable discretization of the Neumann problem. The integrability of these systems and the construction of Lax pairs is based on a factorization of matrix polynomials. This represents joint work with A. Veselov.

NAKAMURA, Y.
Department of Mathematics
Gifu University
Gifu 501-11, JAPAN

CYCLIC-TODA HIERARCHY AND APPLICATIONS: Let $A(\mathbf{t}) \in F^{k\times k}$, $B(t) \in F^{k\times 1}$, $C(t) \in F^{1\times k}$ be functions of $\mathbf{t} = (t_0,\ t_1,\ \ldots..,\ t_{k-1}) \in F^k$ where $F = \mathbf{R}$ or $\mathbf{C}$. The cyclic-Toda hierarchy over F is the system $(0 \leq j \leq k-1)$ of integrable nonlinear and linear PDE's

$$\frac{\partial A}{\partial t_j} = [F_j(a),\ A], \tag{1}$$

$$\frac{\partial B}{\partial t_j} = (A^j + F_j(A))\, B,\ \frac{\partial C}{\partial t_j} = -CF_j(A), F_j(A) - (A^j)L^* = (A^j)L \tag{2}$$

with the inital value $\{A(0), B(0), C(0)\} = \{A_0, B_0, C_0\}$ such that $(zI - A_0B_0)$ is surjective and $\begin{pmatrix} zI & - & A_0 \\ & C_0 & \end{pmatrix}$ is injective for all $z \in \mathbf{C}$.

If A_0 is similar to a Jacobi matrix over $\mathbf{R}$, then (i) is the system of the usual finite nonperiodic Toda equations introduced by J. Moser. In this case, we shall call the sub-hierarchy of (1) and (2) the Jacobi–Toda hierarchy. It is to be noted that the parametric rational function $f(z;t)$ defined by $f(z;t) = C(t)(zI - A(t))^{-1}B(t)$ lies in simply connected components of the space of rational functions of degree k. Namely the Cauchy index of such a rational function always equals $t \pm k$. This fact was observed by R. Brockett and P.S. Krishnaprasad.

Consider the general case where $f(z;t)$ is given by every solution of the cyclic–Toda hierarchy. We shall prove the hierarchy completely parametrizes the space $R^F(k)$ of rational functions over F of the form

$$f(z) = \frac{p(z)}{q(z)} = \frac{p_{k-1}z^{k-1} + \dots + p_0}{z^k + q_{k-1}z^{k-1} + \dots + q_0},$$

where $p(z)$ and $q(z)$ are coprime polynomials and $p_j \in F$, $q_j \in F$.

The space $R^R(k)$ can be viewed as the parameter space of controllable and observable linear dynamical systems of McMillan degree k. The cyclic–Toda hierarchy describes a shift of the time axis for the linear systems. The invariant-tori theorem in classical mechanics asserts that each cyclic–Toda flow is diffeomorphic to a certain cylinder.

It is known that $R^C(k)$ is homeomorphic to a moduli space $M(k)$ of $SU(2)$ Yang–Mills–Higgs k–monopoles in gauge theory (Nahm–Hitchin–Donaldson theorem). We see the cyclic–Toda hierarchy causes a phase shift of k–monopoles. Furthermore, we shall find a generalized cyclic–Toda hierarchy induces an integrable flow on the moduli space $N(SU(m), k)$ of $SU(m)$ k–instantons.

Finally we remark that the cyclic–Toda hierarchy can be regarded as a nice finite dimensional sector of the infinite Toda hierarchy which is a discrete version of M. Sato's KP hierarchy.

PREVIATO, E.

Department of Mathematics
Boston University
Boston, MA 02215, USA

INTEGRABLE SYSTEMS ON SPACES OF HIGHER–RANK VECTOR BUNDLES OVER RIEMANN SURFACES: This work is intended to answer specific questions on Krichever's version of the Calogero–Moser system (CMK for short) and raise others that may lead to a generalization.

The CMK is a completely integrable system in, say, $2n$ degrees of freedom. A point of the phase space can be interpreted as a rank–n bundle over a fixed elliptic curve X together with a twisted endomorphism, or a line bundle over a moveable curve that covers X . Thus, combining results of N. Hitchin and A. Treibich – J.L. Verdier, the phase space can be viewed as the cotangent bundle to the complement of a hyperplane in a linear system over a ruled surface S. On the generic leaves, the system can be identified with the elliptic KP flows; the question (suggested by H. Airault – H.P. McKean – J. Moser) then arises: describe the equilibrium loci for (essentially) the Nth flow, where the system reduces to the Nth KdV. Case $N = 2$ (A. Treibich – J.L. Verdier): there exists a finite (nonzero) number of connected KdV equilibrium loci

for all n; each has dimension $\leq \gamma(n) = \max\{g \in \boldsymbol{N} \mid g(g+1) \leq n\}$ and its closure is stratified by generalized Jacobians. Case $N = 3$: there exists a finite (nonzero) number of Galois–Boussinesq equilibrium loci for all n for exactly one X. This result was obtained in collaboration with the late J.-L. Verdier.
Questions/Work in progress. (a) There is a $\boldsymbol{C}$ action that should be got rid of by reduction; it was not apparent in the CMK model, while it corresponds to the automorphism group of the surface S. (b) The CMK system should be generalized to the case when C is not elliptic by replacing as above Hitchin's cotangent bundle by the one whose fibre is $H^1(E^* \otimes E \otimes O(-1))$. (c) A formula for Boussinesq–elliptic operators can be found; to seek an analog of Ince's theorem. (d) The equilibrium loci of the CMK systems are a clue to the number-theoretic significance of KP, as shown by recent work of A.R. Its and V.Z. Enol'skii.

RATIU, T.

Department of Mathematics
University of California
Santa Cruz, CA 95064, USA

MECHANICS AND DYNAMICS OF THE HEAVY TOP: We examine the stability of relative equilibria of the heavy top and the Hamiltonian Hopf bifurcation in the fast slow transition. The heavy top also has other bifurcations and features that will be discussed, such as Berry's phase, chaos via Melnikov's method, and monodromy.

REYMAN, A.G.

Steklov Mathematical Institute
Leningrad Branch
Fontanka 27, Leningrad 191011, USSR

MULTIDIMENSIONAL INTEGRABLE TOPS: We review recent results on the multidimensional extension of the classical integrable cases of motion of a rigid-body. These new integrable systems are an outcome of the general r–matrix construction applied to loop algebras, which also provides Lax representations and the possibility to solve the equations of motion in terms of suitable theta functions. One of the most interesting examples here is the generalized Kowalewski top where improvement is achieved even in the classical 3-dimensional case. (References: The Kowalewski top 99 years later: a Lax pair, generalizations and explicit solutions. *Commun. Math. Phys.* **122**, 321–354 (1989) (with A.I. Bobenko and M. A. Semenov–Tian Shansky).)

SAINT-AUBIN, Y.

Centre de recherches mathématiques and
Département de mathématiques et statistiques
Université de Montréal
CP 6128 – A, Montréal, QC, H3A 3J7, CANADA

THE HIDDEN SYMMETRY OF $U(n)$ PRINCIPAL σ MODELS: Uhlenbeck unraveled recently the structure of the finite action solutions of the Euclidean 2–dimensional unitary σ models (i.e. harmonic maps in the unitary group). Following her study, Wood showed that any solution can be factorized in a unique way in terms of basic unitons. Using their results, we study the action on this nonlinear σ model of the infinite-dimensional algebra known as "hidden" or "nonlocal" symmetry discovered in the early eighties by several people (Dolan, Ge and Wu, Devchand and Fairlie, Ueno and Nakamura). On the finite action solutions, the algebra turns out to be finite-dimensional and its commutation rules can be given explicitly. The most striking result here however is that these symmetries cannot be linked to usual symmetries (as described, for example, in Olver's book). Indeed, on a given solution, their action depends on the number of unitons in Wood's factorization and this number can be constructed only through a nonlinear process.

SNIATYCKI, JEDRZEJ

Department of Mathematics and Statistics
University of Calgary
Calgary, Alberta , CANADA

REDUCTION OF ALMOST REGULAR MOMENTUM MAPS: Zero levels of almost regular momentum maps have open dense regular strata which admit Marsden–Weinstein reduction. The contribution of the singular strata can be analyzed in terms of algebraic reduction. The relationship between various algebraic reduction schemes and the Marsden–Weinstein reduction of the regular stratum is studied for almost regular momentum maps. Implications for quantization are discussed.

van MOERBEKE, P.

Institut de Mathématiques Pures et Appliquées
Université Catholique de Louvain
2 Chemin du Cyclotron
Louvain-La-Neuve, BELGIQUE

THE GEOMETRY OF INTEGRABLE SYSTEMS: Integrable system possess a number of striking features, like the role played by the Fay trisecant identity, the effect of the Bruhat decomposition for groups, the Grassmannian point of view and the Painlevé analysis. These features turn out to be all intimately related.

We shall mainly discuss the Toda lattices. The Toda lattices, with non-nearest neighbour interaction, are integrable only when the interaction is governed by the Dynkin diagrams for the Kac–Moody Lie algebras. In this lecture, I intend to give a thorough geometrical description of the invariant tori for the periodic Toda lattices, in terms of Lie–theoretical notions. The same techniques can also be used in other situations.

VINET, LUC

Laboratoire de Physique Nucléaire, Département de Physique
Université de Montréal
Montréal, QC, H3T 1C5, CANADA

BERRY POTENTIALS AND APPLICATIONS: We shall review how Berry potentials arise in the context of the adiabatic and the Born–Oppenheimer approximations. A situation borrowed from nuclear physics, namely the dipole quadrupole dynamics, will be used for illustration. Symmetry aspects will also be discussed.

WEINSTEIN, A.

Department of Mathematics
University of California
Berkeley, CA 94720, USA

AFFINE POISSON STRUCTURES: A Poisson structure on a Lie group G is called multiplicative if the map from $G \times G$ to G is a Poisson map. A more general class of interesting Poisson structures on groups are those for which the map $(x, y, z) \rightarrow yx^{-1}z$ is a Poisson map from $G \times G \times G$ to G. Such a structure on G is called affine; the natural setting for this kind of structure is not a group but rather an "affine space", by which we mean a space X on which a ternary operation $(x, y, z) \rightarrow yx^{-1}z$ is defined and satisfies the appropriate axioms.

To further encompass the geometric structures which occur in the theory of hamiltonian systems with symmetry, such as the flat affine structure on the tori of a completely integrable system, we introduce the notion of Poisson structure on an affinoid space. On an affinoid space (a notion introduced previously by J. Pradines and A. Kock under other names), the operation $(x, y, z) \rightarrow yx^{-1}z$ is only partially defined, its domain depending on a pair af simple foliations of X. The appropriate compatibility condition for a Poisson structure is now that the set of "parallelograms" $(x, y, z, yx^{-1}z)$ forms a coisotropic submanifold of $X \times X \times X \times X$.

The talk will begin with basic properties of affine Poisson structures and will continue with examples from the theory of hamiltonian symmetry groups and dual pairs.
(TALK CANCELLED DUE TO THE CALIFORNIA EARTHQUAKE.)

WILSON, G.

Department of Mathematics
Imperial College
London, ENGLAND

ANTIPLETIC PAIRS IN THE HAMILTONIAN FORMALISM OF EVOLUTION EQUATIONS: Let λ be a closed 2–form on a manifold. If λ is non-degenerate, it can be inverted to form a skew tensor l of the opposite type: it is this tensor that enters the definition of the Poisson bracket. In the case of infinite dimensional systems of KdV type, the tensor l is non degenerate, hence is not usually considered to arise in this way. Nevertheless, the typical situation is that the degenerate tensor l is indeed obtained by "inverting" a degenerate 2–form λ. The degeneracy of λ is remedied by its invariance under some symmetry group. The most interesting example is when l defines the Adler–Gel'fand–Dickii bracket.

WINTERNITZ, P.

Centre de recherches mathématiques
Université de Montréal
C.P. 6128 – A, Montréal,
Québec, H3C 3J7, CANADA

KAC–MOODY–VIRASORO SYMMETRIES OF INTEGRABLE NONLINEAR EVOLUTIONS EQUATIONS: The Lie algebras of the symmetry groups of multidimensional integrable evolution equations, such as the Kadomtsev–Petviashvili equation, the Davey–Stewartson equation, and the three wave interaction equations are infinite dimensional and of the

Kac–Moody–Virasoro type. "Conditionally integrable equations" such as those occuring in the Jimbo–Miwa hierarchies also have such Kac–Moody–Virasoro symmetries, but the corresponding local point transformations can only be applied to solutions of the considered equation that simultaneously solve all preceeding equations in the hierarchy. Subgroup classification methods developed for finite dimensional Lie algebras are extended to Kac–Moody–Virasoro algebras and applied to obtain new solutions of the three wave interaction equations.

POSTER SESSIONS

M. S. ALBER
Department of Mathematics
University of Pennsylvania
Philadelphia, PA 19104–6395, USA

GEOMETRIC ASYMPTOTICS FOR INTEGRABLE SYSTEMS : Recently complex geometric asymptotics have been studied in connection with the algebraic-geometric methods for nonlinear problems and new pointwise Hamiltonians. In [1] a general method is suggested for constructing semiclassical solutions (modes) as functions of several complex variables on the Jacobian varieties of compact multisheeted Riemann surfaces. Corresponding quantum conditions are defined as conditions of finiteness on the number of sheets in the Riemann surface. This approach enables one to investigate geometric asymptotics for different problems from one uniform point of view. Important real cases can be obtained from the general complex solutions. A transport theorem proven for the case of integrable systems facilitates the construction of semiclassical solutions (whispering gallery modes) for the Jacobi problem of geodesics on n–dimonsional quadrics and in domains bounded by quadrics. As a result of a special collapsing construction reflected and diffracted modes in both elliptic and hyperbolic cases are also found for the investigation of the geometric theory of diffraction. The general method is applied to the sine-Gordon and nonlinear Schrdinger equations as well as to other integrable equations introduced in [2].

M. BERGVELT
Department of Mathematics
University of Georgia
Athens, GA 30602, USA

HEISENBERG ALGEBRAS, GRASSMANNIANS AND ISOSPECTRAL CURVES : This work is concerned with relating the various geometric and algebraic settings used for studying Liouville integrable nonlinear partial differential evolution equations. The first framework for the case of the KdV equation uses the representation theory of the affine

[1] M.S. Alber, On integrable systems and semiclassical solutions of the stationary Schrödinger equations, *Inverse Problems* **5** (1989), 131–148.

[2] S.J. Alber and M.S. Alber, Hamiltonian formalism for finite-zone solutions of integrable equations, *C.R. Acad. Sc. Paris* **301** (1985), 777–780.

Kac–Moody algebra of $A_1^{(1)}$: the Plücker equations describing the group orbit through the highest weight vector of the basic representation in the principal construction of $L(\Lambda_0)$ give the Hirota bilinear equations for the KdV hierarchy. This construction uses the flows of the principal Heisenberg subalgebra of $A_1^{(1)}$ which correspond to the time variables of the KdV hierarchy. The next framework is more geometric: the principal Heisenberg flows are realized as loop group flows on the infinite Grassmannian $Gr^{(2)}$. Finally, there is an algebro-geometric construction using linear flows of line bundles over Riemann surfaces X which are branched covers $f : X \to CP^1$. Other hierarchies of equations involve other Kac–Moody algebras and/or other constructions of the basic representation using other Heisenberg algebras. For example, the AKNS–Toda system may be studied using the homogeneous Heisenberg algebra of $A_1^{(1)}$. The choice of a conjugacy class of maximal Heisenberg subalgebras yields a construction of the basic module and a corresponding hierarchy of "integrable" equations in Hirota bilinear form.

Here we show that in the algebro-geometric setting the choice of Heisenberg algebra corresponds to the choice of Riemann surfaces together with a specific branching of the mapping $f : X \to CP^1$ over the point at infinity. These surfaces are invariant under the corresponding Heisenberg flow in the infinite Grassmannian, which give rise to a hierarchy of soliton equations in Lax-form. (Joint work with M. Adams)

J.P. PARÉ

Département de mathématiques appliquées
École Polytechnique
Montréal, QC, H3C 3A7, CANADA

KALUZA-KLEIN APPROACH TO MOTION OF PARTICLES AND STRINGS WITH SPIN AND NONABELIAN CHARGE: Both the Hamiltonian and the Lagrangian formulation of charged spinning particles and strings in a background Yang-Mills field is developed within the nonabelian Kaluza–Klein framework extended to include Grassmann variables to describe spin. In addition to the general relativistic coupling of spin and Riemannian curvature, obtained more usually by considering an extended matter distribution, and the coupling of the nonabelian charges (for particles) and currrents (for strings) to the background gauge field, we obtain a further coupling between spin and gauge field strength. (Joint work with J. Harnad)

R.J. SCHILLING

Department of Mathematics
Louisiana State University
Baton Rouge, LA 70803–4913, USA

THE AUGMENTED DIVISOR AND ISOSPECTRAL PAIRS: The Neumann system of an algebraic second order scalar operator was shown to be isospectral with respect to an $l \times l$ matrix L and a 2×2 matrix Z. This isospectral pair is given in the coordinates of the Neumann problem by these formulas

$$Z + e_{2,1} + \sum_{r=1}^{l} x^r \otimes u^r = e_{2,1} + x^T(\lambda - \mathbf{a})^{-1}u$$

and

$$L = \mathbf{a}z^2 + (x_1 \otimes u_1 + x_2 \otimes u_2)z + x_1 \otimes u_2 = \mathbf{a}z^2 + x(l_2 z - e_{2,1})u^T.$$

Our notational conventions are these: $x^r = (x_1^r, x_2^r)^T$ belong to $\mathbf{C}^2$ for $r = (1, ..., l)$, $x_\alpha = (x_\alpha^1, ..., x_\alpha^l)$ belongs to $\mathbf{C}^l$ for $(\alpha = 1, 2)$ and x is the $l \times 2$ matrix with columns x_1 and x_2. The matrix $\boldsymbol{a}$ is diagonal and it contains the frequencies of the Neumann oscillators. The same conventions are used for the u–coordinates. J. Moser considered a reduced form of our version of the Neumann problem in which one has $u_1 = x_2$ and $u_2 = x_1$. He showed that the Neumann system is an isospectral deformation of the following rank two perturbation of a:

$$P = a + x_1 \otimes x_2 - x_2 \otimes x_1 - x_1 \otimes x_1.$$

Constants of motion may be obtained from the characteristic polynomials of these matrices. Adams, Harnad and Hurtubise (AHH) showed that the coefficients of the characteristic polynomials of the isospectral matrices Z and L generate the same algebra of functions. The term duality was used in (AHH) to describe the relationship between the isospectral matrices. We suggest the phrase isospectral pair in place of dual pair.

These matrices were interpreted by Adams, Harnad, Previato in a beautiful setting involving Hamiltonian group actions and the Poisson geometry of momentum mappings. With their set up in mind, one may show the matrix

$$Z - 2e_{2,1} = -e_{2,1} + \sum_{r=1}^{l} x^r \otimes u^r$$

is the momentum mapping corresponding to a certain infinitesimal action on $\mathbf{C}^{2l}$ by the subalgebra of the loop algebra in the parameter λ over $gl(2, \mathbf{C})$ consisting of all polynomial loops. This result requires a rather unusual representation of the dual of the subalgebra of polynomial loops. By the Adler/Kostant/Symes theorem, the Uhlenbeck polynomials, which are the coefficients of the characteristic polynomial of Z, are involutive. By the results of (AHH), the algebra generated by the coefficients of L is involutive.

In this report we shall derive isospectral matrices and construct their representations in Neumann coordinates using Baker function techniques. The primary spectral problem is that of a scalar differential operator of order n. We shall derive one member; namely the order n analogue Z, of the isospectral pair using a rational extension of the Krichever homomorphism. The other isospectral matrix L is derived from a translation in the Jacobian of the underlying spectral curve. The key to our approach is a fundamental relationship between the Baker functions of the primary and secondary spectral problems. This relationship is originally due to H. Flaschka. This report amounts to a further development of that relationship.

M.-A. WISSE
Département de mathématiques et statistiques
Université de Montréal
C.P. 6128 – A, Montréal, QC, H3C 3J7, CANADA

FINITE DIMENSIONAL ISOSPECTRAL FLOW AND SOLUTIONS OF COUPLED SYMMETRIC SPACE NONLINEAR SCHRÖDINGER EQUATIONS: Coupled nonlinear Schrödinger equations corresponding to the classical hermitian symmetric spaces are considered as commutativity conditions for isospectral hamiltonian flows on finite dimensional coadjoint orbits of loop algebras. Their pull-back by means of an injective moment map gives rise to an isospectral flow of matrices. The Abel map induces a linear flow on the Jacobi variety of an algebraic curve and thus the systems can be integrated to yield solutions in terms of quotients of theta functions. [Joint work with J. Harnad]

LIST OF PARTICIPANTS

ABLOWITZ, M.J.
Program in Applied Sciences
University of Colorado
Boulder, CO 80369-0426
USA

ADAMS, M.R.
Department of Mathematics
University of Georgia
Athens, GA 30602
USA

ALBER, Mark S.
Department of Mathematics
University of Pennsylvania
Philadelphia, PA 19104-6395
USA

ANDERSON, R.L.
Department of Physics and Astronomy
University of Georgia
Athens, GA 30602
USA

BAIDER, Alberto
Department of Mathematical Sciences
Hunter College of the City Univ. of N.Y.
New York, NY 10021
USA

BATES, L.M.
Department of Mathematics
University of Calgary
Calgary, Alberta T2N 1N4
CANADA

BERGVELT, M.
Department of Mathematics
University of Georgia
Athens, GA 30602
USA

BLOCH, A.M.
Department of Mathematics
The Ohio State University
Columbus, OH 43210
USA

BOR, Gil
Department of Mathematics
University of Berkeley
Berkeley, CA 94720
USA

BROCKETT, R.W.
Department of Mathematics
Harvard University, Pierce Hall
Cambridge, MA 02138
USA

BUGAJSKA, K.
Department of Physics
University of Western Ontario
London, Ontario N6A 5B9
CANADA

DAVID, D.
Center for Nonlinear Studies, MS-B258
Los Alamos National Laboratory
Los Alamos, NM 87545
USA

DICKEY, L.
Department of Mathematics
University of Oklahoma
Norman, OK 73019
USA

GOTAY, M.J.
Department of Mathematics
U.S. Naval Academy
Annapolis, MD 21402-5000
USA

GRMELA, M.
Département de génie chimique
École Polytechnique
CP 6079 – A, Montréal, Québec
H3C 3A7, CANADA

GRUNDLAND, M.
Département de mathématiques
Université du Québec à Montréal
Trois-Rivières, Québec G9A 5H7, CANADA
and
CRM
Université de Montréal
CP 6128 – A, Montréal, Québec H3C 3J7
CANADA

HARNAD, J.P.
Department of Mathematics and Statistics
Concordia University
Montreal, Quebec, CANADA
and
Centre de Recherches Mathématiques
Université de Montréal
CP 6128 – A,
Montreal, Quebec H3C 3J7
CANADA

HOLM, D.D.
Theoretical Division
Los Alamos National Laboratory
Los Alamos, NM 87545
USA

HURTUBISE, J.C.
Department of Mathematics
McGill University
Montreal, Quebec H3A 2K6
CANADA

HUSSIN, Véronique
CRM
CP 6128 – A, Université de Montréal
Montréal, Québec H3C 3J7
CANADA

KAUFFMAN, L.
Department of Mathematics
University of Illinois at Chicago
Chicago, IL 60680
USA

KUMMER, M.
Department of Mathematics
University of Toledo
Toledo, OH 43606
USA

KUPERSHMIDT, B.A.
Space Institute
University of Tennessee at Tullahoma
Tullahoma, TN 37388
USA

LAGARIAS, J.C.
AT & T Bell Laboratories
Room 2C–373
Murray Hill, NJ 07974
USA

LEVI, D.
Department of Physics
Rome University 1
La Sapienza
ITALY

LEWIS, A.
104-44 Caltech
Pasadena, CA 91125
USA

LI, W.
Department of Mathematics
Clarkson University
Potsdam, NY 13676
USA

MARCHILDON, L.
Département de Physique
Université du Québec à Trois-Rivières
Trois-Rivières, Québec G9A 5H7
CANADA

MARSDEN, J.E.
Department of Mathematics
University of California
Berkeley, CA 94720
USA

MATHIEU, P.
Département de physique
Université Laval
Québec, Québec G1K 7P4
CANADA

McKEAN, H.P.
Courant Institute of Mathematical Sciences
New York, NY 10012
USA

MOSER, J.K.
Department of Mathematics
Eidgenössische Technische Hochschule
CH-8092 Zurich
SWITZERLAND

NAKAMURA, Y.
Department of Mathematics
GIFU University
Gifu 501 - 11
JAPAN

NIJHOFF, Frank
Department of Mathematics
Clarkson University
Potsdam, NY 13676
USA

PAQUIN, G.
Département de mathématiques appl.
École Polytechnique de Montréal
CP 6079 - A, Montréal, Québec H3C 3A7
CANADA

PARE, J.P.
Département de mathématiques appl.
École Polytechnique de Montréal
CP 6079 - A, Montréal, Québec H3C 3A7
CANADA

PREVIATO, E.
Department of Mathematics
Boston University
Boston, MA 02215
USA

RATIU, Tudor
Department of Mathematics
University of California
Santa Cruz, CA 95064
USA

REYMAN, A.G.
Steklov Mathematical Institute
Leningrad Branch (LOMI)
Fontanka 27, Leningrad 191011
USSR

ROD, D.L.
Department of Mathematics
University of Calgary
Calgary, Alberta T2N 1N4
CANADA

ROM-KEDAR, V.
Department of Mathematics
University of Chicago
Chicago, IL 60628
USA

SAINT-AUBIN, Y.
CRM
Université de Montréal
CP 6128 - A, Montréal, Québec H3C 3J7
CANADA

SAID, Sakhi
École Polytechnique de Montréal
CP 6079 - A, Montréal, Québec H3C 3A7
CANADA

SCHILLING, R.
Department of Mathematics
Louisiana State University
Baton Rouge, LS 70808
USA

SCHLOMIUK, D.
Département de mathématiques
Université de Montréal
CP 6128 - A, Montréal, Québec H3C 3J7
CANADA

SNIATYCKI, J.
Department of Mathematics
University of Calgary
Calgary, Alberta T2N 1N4
CANADA

TRATNIK, M.
Los Alamos National Laboratory
Center for Nonlinear Studies
Los Alamos, NM 87545
USA

van MOERBEKE, P.
Institut de mathématiques pures et appl.
Université Catholique de Louvain
B-1348 Louvain-la-Neuve
BELGIQUE

VASSILIOU, P.
School of Information Science - Engineering
University of Canberra
Belconnen, A.C.T. 2616
AUSTRALIA 2616

VINET, L.
Laboratoire de physique nucléaire
Université de Montréal
CP 6128 - A, Montréal, Québec H3C 3J7
CANADA

WILSON, G.
Department of Mathematics
Imperial College
London SW7 2AZ
UK

WINTERNITZ, P.
CRM,
Université de Montréal
CP 6128 - A, Montréal, Québec H3C 3J7
CANADA

WISSE, M.A.
Département de mathématiques
Université de Montréal
CP 6128 - A, Montréal, Québec H3C 3J7
CANADA

AUTHOR INDEX

Printed by
Marquis, Montmagny, Qc